Urban Sustainability

The Urban Sustainability Book Series is a valuable resource for sustainability and urban-related education and research. It offers an inter-disciplinary platform covering all four areas of practice, policy, education, research, and their nexus. The publications in this series are related to critical areas of sustainability, urban studies, planning, and urban geography.

This book series aims to put together cutting-edge research findings linked to the overarching field of urban sustainability. The scope and nature of the topic are broad and interdisciplinary and bring together various associated disciplines from sustainable development, environmental sciences, urbanism, etc. With many advanced research findings in the field, there is a need to put together various discussions and contributions on specific sustainability fields, covering a good range of topics on sustainable development, sustainable urbanism, and urban sustainability. Despite the broad range of issues, we note the importance of practical and policy-oriented directions, extending the literature and directions and pathways towards achieving urban sustainability.

The series will appeal to urbanists, geographers, planners, engineers, architects, governmental authorities, policymakers, researchers of all levels, and to all of those interested in a wide-ranging overview of urban sustainability and its associated fields. The series includes monographs and edited volumes, covering a range of topics under the urban sustainability topic, which can also be used for teaching materials.

Ali Cheshmehzangi · Sue Roaf
Editors

Persian Vernacular Architecture

Lessons from Master Builders of Iran on Climate Resilient Design

Springer

Editors
Ali Cheshmehzangi
School of Architecture, Design
and Planning
University of Queensland
Brisbane, QLD, Australia

Sue Roaf
School of Energy, Geoscience,
Infrastructure and Society
Heriot-Watt University
Edinburgh, Scotland, UK

ISSN 2731-6483 ISSN 2731-6491 (electronic)
Urban Sustainability
ISBN 978-981-96-1115-7 ISBN 978-981-96-1116-4 (eBook)
https://doi.org/10.1007/978-981-96-1116-4

This Springer imprint is published by the registered company Springer Nature Singapore Pte Ltd.
The registered company address is: 152 Beach Road, #21-01/04 Gateway East, Singapore 189721, Singapore

We describe this as "not your everyday space" and an inspirational representation of art, architecture, space, scale, and ambiance. Photo is captured by Pedram Raz, openly available to use under the Unsplash License

For Ostads, for Architects, for Vernacular Lovers.
For all those dedicated educators and researchers who do not simply give up.

In other words, a dedication to the dedicated.

Acknowledgements

We remain committed to including and appreciating everyone's contributions to this exciting and important project. We extend our gratitude to all authors, co-authors, chapter contributors, and collaborators who have shared their expertise, passion, and creativity. In addition, we thank Dr. Ali Rad Yousefnia of the University of Queensland, for providing a spectacular image of Persepolis (or 'Takht-e Jamshid' in Persian), located near the City of Shiraz in Iran. This image will be used for all books under this particular theme of 'Architecture, Construction, and Heritage' as part of the *Urban Sustainability* Book Series.

This book covers an important topic of research, where we hope knowledge is shared, exchanged, and extended—if not immediately re-invented. There is more to do to cover the topic of Persian or Iranian Vernacular Architecture, and we hope we are part of a much larger initiative to highlight the realities, knowledge, and lessons from Iran; and most important of all, the fast-disappearing ones that should tickle our minds when we think of a good architecture. This appreciation extends to our attempt and ongoing work, recognising the dedication and vision of all contributors and master builders or Ostads in Iran, who have directly and indirectly helped shape this book. The knowledge of dedicated Iranian Ostads is something that we cannot simply neglect, it is something that needs more than just recognition and retention, and it is something that needs deep and thorough thinking to advance our new knowledge of better architecture for generations to come. Our contribution is now, and hopefully a tangible part of an evolution of a rather slow vernacular architecture. Such collaborative effort fosters new dialogues, networking opportunities, and a valuable platform for research exchange among scholars, master builders, and practitioners or vernacular experts worldwide. Most importantly, it helps bring back the *real* resilience thinking and *genuine* sustainability practices that are embedded in the extended—but yet gradually fading—knowledge of vernacular architecture.

We always strive for innovation and improvement, particularly in the realms of sustainability and vernacularity. There are still many lessons to be learnt, and many that could help us achieve even better architectural design thinking and paradigms. We value and celebrate the knowledge from the real knowledge holders that enriches work like this. The movement towards better architecture should continue with a greater force to ensure this never-ending learning cycle is sustained vigorously.

About This Book

Persian Vernacular Architecture aims to address a significant challenge in contemporary design and architecture, i.e., **"to learn from the vernacular"**. Vernacular architecture refers to traditional and indigenous building styles that have evolved in response to local climate, materials, and cultural practices. These styles often embody 'extreme' design principles, as they support life without relying on modern mechanical systems for heating and cooling. The core premise of the book is that these time-tested design solutions offer valuable lessons for *real* sustainable and climate-resilient architecture today.

To validate this premise, the book explores various methods and directions to substantiate the argument that vernacular styles are inherently sustainable and resilient. It critiques the reliance on purely modelled studies in modern architectural research, emphasising that such studies must be grounded in empirical evidence from real buildings. Modelled studies often make assumptions that can only be validated through the examination of existing vernacular structures, which have endured the test of time and environmental challenges. Hence, the case study examples explore lessons from master builders of Iran on climate resilient design.

The book advocates for a return to the vernacular as a source of inspiration and knowledge. Hence, a genuine dedication to dedicated Ostads or master builders of Iran, who have contributed extensively to making and co-creating climate-resilient architecture and climate-sensitive design. This is more than just a box-ticking sustainability exercise and delves more deeply into the building performance and its impact on our everyday life, our health and wellbeing, and us as human beings. The book posits that contemporary architects and designers can learn valuable lessons by studying the adaptive strategies and sustainable practices embedded in the traditional vernacular architecture of Iran. These lessons are crucial for addressing modern challenges such as climate change, resource scarcity, and the need for energy-efficient buildings. By revisiting and analysing vernacular architecture, the book aims to bridge the gap between traditional knowledge and contemporary design practices. It provides a comprehensive exploration of how vernacular principles can be integrated into modern architectural education and practice. The goal is not only to preserve traditional wisdom but also to innovate upon it, creating buildings that are both culturally relevant and environmentally sustainable.

The contributing case study chapters of the book include an excellent range of practical recommendations for architects, researchers, and policymakers. By fostering a deeper understanding of vernacular architecture, the book seeks to inspire a new generation of architects to create buildings that are harmonious with their environment and resilient to future challenges. Through this approach, we believe lessons from vernacular buildings ought to offer the promise of reintegrating real *Resilience* into the design of the next generation of buildings.

Prologue

> *Three Principles on which all buildings should be based are: (1) Design for a climate, (2) design for both the physical and social environment, (3) design for time, be it day or night, a season or the whole lifetime of a building, so designing a building that will adapt over time.*
>
> *—Sue Roaf*

The extraordinary thing about the *Vernacular Architecture* of Iran is that has been evolving for millennia. It is *Slow Architecture* that for several centuries might change little, but then it rapidly blossoms in a short period of time into a flowering of craftsmanship from which unique and extraordinary buildings emerge, powered by new money, new populations, and their itinerant craftsmen. Vernacular buildings can be divided into great *Statement Buildings*, emblems of wealth and ambition, and then the *Ordinary Vernacular* that follow in their wake absorbing some of the physical and technical advances made by the great Ostads, the master craftsmen of their Age.

Nowhere in the world is this more apparent than in Iran where examples of humanity's finest architecture are to be found. Some of the most beautiful buildings and gardens in the world are there, along with technically extraordinary achievements from the pre-mechanical world: buildings that can create Paradisian coolth in the hottest desert summers; great structures in which ice from the coldest days of winter were preserved to provide ice for summer drinks; water cisterns that supply cool water in the heat of the day, and air conditioning systems that use cold mountain air to cool summer living rooms on the plains. The following pages include examples of buildings that are perfectly adapted to their local climates, and urban spaces that serve and inspire their communities with pleasing forms interwoven in many different patterns into the cultures and climates of the diverse landscapes of this ancient country.

This book is very timely in attempting to capture some of the fast-disappearing lessons of Iranian Vernacular buildings. The hope is to record them, analyse and learn from them, and share and trumpet their lessons in an attempt to slow down, or in some cases halt the onslaught of the worst aspects of *Fast Architecture*. Modern international style of buildings that emerged only half a century ago now dominate development in many Iranian cities, employing forms that are fundamentally alien to the local climates and landscapes of this hot and dry country. How many industrialised, fully mechanised glass and steel buildings have tragically been built over vital examples of how to live in deep harmony with nature?

The march of time is inevitable, just as is the increasing need for mechanical cooling in a heating world. The way forward is to mix the best of the old with the best of the new, braided together to create truly affordable and sustainable buildings and lifestyles for the hotter climates of the future. The urgent challenge now is to find a better way forward, not least because the stability of a society will increasingly depend on *Comfort Justice* for all, where it is not just the rich who can stay cool, but everyone.

The systems we rely on to survive in the built environment are increasingly brittle. Even now the soaring costs, availability, and reliability of energy supplies, for instance, are causing many of our current life support systems to fail, not least during ever more extreme weather events. Lessons from Vernacular buildings offer the promise of reintegrating real *Resilience* into the design of the next generation of buildings. After all, they have kept populations thermally safe and comfortable for millennia without the use of mechanical heating and cooling, and by looking for the lessons of how they did that future generations can re-employ some of the wisdom of the ancients to protect future generations from increasingly less secure environmental conditions.

But the glue that holds societies together is not just about the bricks and mortar of settlements. It is made up of the ways in which buildings are used over time and the social and cultural order and niceties of how people interact with each other across their daily lives within them. Many of the chapters here touch on the relationships between buildings and their occupants, and the beliefs and structures of the societies they co-evolved to exist in.

Each chapter in this book contains gems of understanding about how the extraordinary architecture of Iran looked, worked, evolved, and mattered. Unfortunately, there are so few written records in Iran about life in and around the buildings, how and why they were built, and their brilliant technical developments, for instance in the way they moved and directed air in windcatchers, stored heat and coolth buildings water and ice, and how the Ostads shaped and sized each element of a building to optimise its performance. The only records we often have left are the buildings themselves which is why this book is so important.

Each person who has contributed to a chapter here has embarked on a journey themselves of discovery, motivated by a deep respect and often love for the buildings they have chosen to study. Researching tirelessly to delve into the ways and whys of how the buildings work they have each salvaged unique understandings of their buildings in each of their physical and historical settings. An additional important contribution to these recordings of the Vernacular of Iran has been the range of effective research methodologies employed to draw out the essence of their study subjects. These methodologies can in the future be replicated and developed by the next generation of teachers and scholars to explore and record different subjects to add to the corpus of existing knowledge on one of the finest assemblages of Vernacular buildings of any country in the world. Great international scholars of the past concentrated on recording the most magnificent of Iran's buildings, but lessons to be gleaned from the humble Vernacular may have more important impacts on the buildings of the future.

Science, art, and philosophy have lost their once-valued places in many limited contemporary conversations around ideas of sustainable architecture and design. Readers will find that rich, profound, and resonant stories of science, art, and philosophy are hard-baked into the walls of many of the buildings described here. We believed, when joining hands with our fellow authors and standing with them on the shoulders of the great Ostads of the past that in this book, and within these buildings lie some seeds of the new design paradigms, new thinking, and re-learnt knowledge that will be so important for the next generation of buildings that are fit for purpose in the unpredictable and very different decades ahead. Here lie some seeds of a Resilient New Vernacular. We hope readers can enjoy gathering them up and planting them in their own buildings for the future.

Edinburgh, UK Sue Roaf
Brisbane, Australia Ali Cheshmehzangi

Contents

Editors and Contributors

About the Editors

Ali Cheshmehzangi is Professor and Head of the School of Architecture, Design and Planning (ADP) at The University of Queensland, Brisbane, Australia. He is among the top 30 global scholars in the urban sustainability research area. With a career spanning over two decades, he has made significant contributions to the academic and professional communities, with a focus on sustainable and environmentally conscious design. He also remains affiliated with the Network for Education and Research on Peace and Sustainability (NERPS) at Hiroshima University, Japan. So far, Ali has published over 500 journal papers, articles, conference papers, book chapters, and reports. He also has 34 other academic books, three of which have received awards at the international, national, provincial, and municipal levels. He also has received international awards and recognition for his research on urban resilience studies and sustainability research, as well as Vice-Chancellor's award for his impactful contribution to higher education.

Sue Roaf is an emeritus professor of Architectural Engineering at Heriot-Watt University, Edinburgh, and the winner of the 11th Farabi International Award for Iranian Studies in 2020. She is also an Honorary Professor at the School of Architecture, Design and Planning (ADP), The University of Queensland, Brisbane, Australia. An award-winning architect, teacher, and author/editor of 24 books her Ph.D. was on the Windcatchers of Yazd, Iran. She is best known for her pioneering work on domestic solar energy and eco-design. In 1995, she built a house in Oxford designed to minimise emissions from her lifestyle with the first photovoltaic roof in Britain. Using low-energy construction techniques, high thermal mass, and wood-burning, many of its climatic features were inspired by Persian vernacular architecture as explained in her internationally best-selling book on Ecohouse Design. Her public services have included seven years as an Oxford City Councilor, membership of the UK Architect's Registration Board, and five years leading research on Climate Adaptation for Scotland. She has organised and chaired many international conferences including those on comfort at the extremes.

Contributors

Mohammad Hamed Abdi Department of Urban Planning and Design, University of Bojnord, Bojnord, Iran;
Research Group for Urban Development, University of Antwerp, Antwerpen, Belgium

Sadaf Alikhani Huckabee College of Architecture, Texas Tech University, Lubbock, TX, USA

Maryam Alsadat Ziaie Mazinan Department of Architecture, Hakim Sabzevari University, Sabzevar, Iran

Mojgan Amiribolvardi Shiraz, Iran

Mohammad Anvar Adibhesami Department of Architecture, School of Architecture and Environmental Design, Iran University of Science and Technology, Tehran, Iran

Peyman Bahrami Doost Department of Architecture and Urban Planning, Shahid Rajaee Teacher Training University, Tehran, Iran

Hamidreza Bani Ansari Department of Architecture and Urban Planning, Isfahan (Khorasgan) Branch, Islamic Azad University, Tehran, Iran

Kimia Chenary Department of Urban Planning and Design, University of Bojnord, Bojnord, Iran

Ali Cheshmehzangi School of Architecture, Design and Planning, The University of Queensland, Brisbane, Australia

Parisa Derakhshesh Department of Urban Design, Faculty of Engineering, North Tehran Branch, Islamic Azad University, Tehran, Iran

Nasim Ebneali Department of Architecture, School of Architecture and Environmental Design, Iran University of Science and Technology, Tehran, Iran

Maryam Ekhtiari Department of Architecture, School of Art & Architecture, Shiraz University, Shiraz, Iran

Hassan Estaji Department of Architecture and Urbanism, Hakim Sabzevari University, Sabzevar, Iran

Maryam Ghasemi Department of Architecture, Faculty of Architecture, Eastern Mediterranean University, Famagusta, Turkey

Elham K. Hassani Department of Architecture and Design, Faculty of Architecture, University of Sapienza, Rome, Italy

Aminreza Iranmanesh Faculty of Architecture and Fine Arts, Final International University, Girne, Türkiye

Atefeh Karbasi Faculty of Architecture and Urban Planning, Shahid Beheshti University, Tehran, Iran

Hirou Karimi Department of Architecture, Faculty of Architecture, Eastern Mediterranean University, Famagusta, Turkey

Azadeh Khaki Ghasr Department of Architecture, Faculty of Architecture and Urban Planning, Shahid Beheshti University, Tehran, Iran

Asma Mehan Architectural Humanities and Urbanism Lab (AHU_Lab), Huckabee College of Architecture, Texas Tech University, Lubbock, TX, USA

Kourosh Momeni Architecture and Planning Department, Jundi-Shapur University of Technology, Dezful, Iran

Fatemeh Pajani Department of Urban Planning and Design, Faculty of Arts and Architecture, Tarbiat Modares University, Tehran, Iran

Hossein Pourmehdi Ghaemmaghami Department of Architecture, Faculty of Architecture and Urban Planning, Shahid Beheshti University, Tehran, Iran

Ali Rad Yousefnia UQ College (UQC), The University of Queensland, Brisbane, Australia

Rouhollah Rahimi Department of Architecture, University of Mazandaran, Babolsar, Iran

Mohsen Riahi Department of Architecture, School of Art and Architecture, University of Kurdistan, Sanandaj, Iran

Sue Roaf Heriot Watt University, Edinburgh, Scotland, UK;
School of Architecture, Design and Planning, The University of Queensland, Brisbane, Australia

Hossein Soltan Rahmati Shiraz, Iran

Mahdiyeh Tabatabaei Department of Architecture, Field of Architecture and Design Cultures, University of Bologna, Bologna, Italy

Amir Tayyebi Architecture and Planning Department, Jundi-Shapur University of Technology, Dezful, Iran

Hadi Torkshavand Yazd, Iran

Persian Vernacular Buildings—Lessons on Resilient Design

Sue Roaf and Ali Cheshmehzangi

Abstract This chapter highlights the significance of Iran's rich architectural heritage in a changing world and climate. Iran's vernacular architecture resulted from the gradual co-evolution of climates, buildings and many cultures over millennia that created some of the most sophisticated passive structures in the world, capable of keeping people not only thermally safe by comfortable but at best sensually delighted within their walls and gardens. As climates become more extreme around the world lessons from the traditional Iranian houses and technologies can provide important insights for today's architects into how to create secure and durable shelters for the future, using their practices involving thermal mass, water, advanced natural ventilation systems and spatial arrangements. At the crossroads of routes from major civilisations to the East and West Persian buildings were enriched by exposure to the knowledge and craftsmen of passing dynasties who left in their wake a dazzling architectural legacy. First recorded in books by awed European travellers, the chapter touches on the antiquity of the settlements in Iran and the various viewpoints from which the buildings were originally recorded, as ancient monuments, works of art, backdrops to history, technologies—and more importantly—as sources for study on human practices and behaviours. Since the 1920 the recording of Persian vernacular has been undertaken through the efforts of visiting scholars and Iranian government agencies. Since the 1950s its study has moved more into the realm of academics and universities in Iran and abroad using new methods and tools to open new avenues of enquiry. From the depth of wisdom being accumulated by this and previous generations of scholars come new principles of resilient design that can teach the world much about thermally safer design in a heating world.

Keywords Persian vernacular · Monuments · Technology · Passive design · Extreme climate · Master craftsmen · Architecture

S. Roaf (✉)
Heriot Watt University, Edinburgh, Scotland, UK
e-mail: S.Roaf@hw.ac.uk

S. Roaf · A. Cheshmehzangi
School of Architecture, Design and Planning, The University of Queensland, Brisbane, Australia

A. Cheshmehzangi and S. Roaf (eds.), *Persian Vernacular Architecture*, Urban Sustainability, https://doi.org/10.1007/978-981-96-1116-4_1

1 Introduction

Iran is a very hot country with large arid deserts at its heart set amid diverse landscapes with maritime and montane to sub-tropical climates. It was also home to one of the first settled and civilised areas of the world, peopled over millennia by many different cultures and societies, each bringing their own unique contributions to, and legacy from, the structures, settlements and cities they occupied.

Perhaps it is that very antiquity, combined with its location at the heart of such major crossroads for goods, populations, ideas and material cultures that has resulted in huge variety and extreme sophistication of many of the vernacular buildings described in this book. From China and India in the east to Europe, Egypt and Africa to the West, Iran was a melting pot for architecture. Many of the technologies and skills garnered from adjacent regions were absorbed into Persian buildings and improved by their master craftsmen, then re-exported abroad. Like their carpets, few buildings were more exquisite than the finest made in Iran.

1.1 A Short Lens: Vital Lessons from the Vernacular in a Heating World

Persian vernacular buildings are not only treasured because of their beauty but because they can be of the utmost importance in a heating world. It is vital that the climatic design lessons of Persian vernacular are not lost. They hold many important insights into how buildings can passively withstand higher and lower ambient temperatures in extreme climates. Traditional Persian buildings, using advanced designs, construction techniques and technologies, provide indoor climates during climate extremes that range from being survivable or at least acceptable, to comfortable and at best sensually delightful conditions. But to do so required co-evolution of three forces: the local climate over days and seasons, the buildings themselves and the lifestyles of their occupants. All are covered in the chapters that follow.

By contrast, many *modern* buildings both in Iran and globally perform poorly in local climates, conforming to a twentieth century international pattern book being over-glazed, lightweight and poorly oriented making them overheat, often all year. They overcome this fault by importing grid energy to keep occupants thermally safe indoors, relying on power from grids that increasingly fail during extreme weather events. The main problem is that the global climate is heating, leading to ever more extreme weather conditions and climate trends.

The rate at which this is happening is escalating with weather records around the world being broken almost daily. We are at a point of climate emergency. Armstrong McKay et al. [1] analysed nine 'core' and seven 'impact' *Climate Tipping Points* affecting the functioning of Earth's systems and human welfare. They estimated the crossing of ten Tipping Points becomes likely within the Paris Agreement range of 1.5 °C. For mean Global temperatures, and the rest above 1.5 °C. In October 2024 the Chair of the IPCC warned that 1.5 °C is officially out of reach, 2.0 °C demands heroic action and 3.0 °C is now the likely outcome [2]. The time for urgent and heroic action is now, in part referencing the pre-electric design prototypes of local vernaculars that operated without grid energy.

The next generation of buildings must provide a high level of passive protection against heat and cold, but the resilience of buildings is not a simple issue that can be plugged into. Resilience is not just a function of just a building form or fabric but reflects the physical needs, aspirations, behaviours and customs of constantly evolving societies, economies, cultures, beliefs and climates. Resilient design requires complex solutions of the types embodied in most historic buildings where people, buildings and climate have co-evolved. The chapters in this book importantly shed new light not only on physical buildings themselves but how they were lived in, understood and adapted to by their occupants. Unfortunately, many of those in the past who recorded historic buildings left little or no record of their occupants and how they were inhabited. Perhaps we will never know the full story of how and why many of them evolved over time, over a very long time.

1.2 A Very Long Lens: Ancient Foundations of Persian Vernacular Traditions

Persian vernacular buildings are so advanced in their performance not least because many have been evolving for millennia. Iran was one of the earliest settled regions in the history of modern man. During the early Palaeolithic period from 800,000 to 200,000 ago prehistoric hominins already lived on the Iranian plateau. Between 200,000 and 40,000 ago early Stone Age Neanderthals were present in both the Zagros mountains and central Iran during this period and by 13,000–11,000 Before Common Era (BCE) flint blades using hunters, gatherers and fishers appeared along the Zagros mountains in South-West Iran [3].

Evidence of a very early man-made Neolithic settlement was excavated at Ganj Dareh in the Zagros mountains, radiocarbon dated to around 10,000 years ago. A cluster of two-story structures with many small cubicles built using several methods of brick-making and wall construction included unusual mortice-and-tenon devices, portholes, plugs and storage bins made of prefabricated clay slabs. These well-made buildings were occupied by goat herders who lived apparently without growing their own food [4].

The agricultural transition was already underway not far away in the Fertile Crescent some 12,000 years ago where in Pre-Pottery Neolithic villages cereals and legumes were domesticated. A DNA genome taken from an early Neolithic woman from Ganj Dareh showed it was inhabited by a population genetically most similar to hunter-gatherers from the Caucasus, but distinct from the adjacent Neolithic Anatolian people who later brought food production into Europe. People from Ganj Dareh apparently lived to the east of the 'Cradle of Civilisation' in current Iraq between the Tigris and Euphrates rivers, but were genetically isolated from them [5].

From earliest times across the diverse Persian landscapes, very different populations co-existed, each of whom may have contributed more or less to the extraordinary cohort of vernacular buildings found in Iran today. It is conceivable that many brilliant technologies touched on in the chapters here, like the houses, ice-houses or water cisterns and water supply systems that enabled desert cities to flourish, were ten thousand years in the making.

2 Viewpoints on the Vernacular

The chapters in this book are written by a generation who largely have never lived in the buildings they write about, although some may have visited elders in them. Some older scholars were lucky enough to spend time in their study buildings, benefitting from first-hand knowledge of how the buildings were used. Modern researchers have to rely on the extant physical structures and at best, written or spoken first or second hand accounts of behaviours and attitudes of former inhabitants of the buildings.

Every generation tends to have ingrained views on how historic buildings should be recorded and rooted in the societies of the time. The twentieth century scholars, largely armed only with a tape measure, alidade and camera set out to largely record buildings, as inert monuments and works of art.

2.1 Persian Monuments as Works of Art

Buildings in the twentieth century were viewed and described from the point of view of their structures and aesthetics as individual works of art. Perhaps their greatest twentieth century historian was Arthur Upham Pope, an American scholar who in 1938 and 1939 published his six-volume *Survey of Persian Art* [6]. He described the architecture of pre- and post-Islamic periods as individual entities reflecting the distinctive styles and features such as domes, arches and tile and plasterwork of particular of traditions over time in such detail that established Persian architecture as unrivalled in the East. His tireless work and publications laid the foundations for a popular understanding of a range of great Persian monuments including the world-famous tomb and palaces of Persepolis, Golestan, Ali Qapu and Pasargadae [7, 8]. They were described in terms of the people who built and paid for them, but seldom those who lived in, or tended them.

Despite the sheer number of historic Iranian monuments, efforts to record and preserve them are unparalleled in the Middle East. In 1921 a Department of Antiquities with a modest museum was created by the Ministry of Education in Tehran. In 1922 the National Monument Council was established and in 1928 the Department of National Antiquities was established with the French architect André Godard as

Director. He remained in place until 1967. In 1930 the Parliament enacted an antiquities law, setting out precise regulations for the classification and conservation of monuments and historical sites. The register of protected sites grew rapidly from 1930 to 1955 to include 419 monuments and historical sites. By late 1975 that number had increased to 1224. The law then forbade the restoration or alteration of any registered monument without permission and supervision by the antiquities service. A section for the protection of historic buildings was established within the department, in order to oversee the conservation of monuments. Persian architects and builders who were particularly experienced in traditional techniques were employed by the service, helping to preserve traditional construction skills and techniques. The small archaeological museum in Tehran grew into a major institution housing a collection of monuments and antiquities and a centre for scientific research with its own official journal *Athār-é Īrān.*

Education was seen as key to the survival of the monuments and in 1935 a department training archaeologists was established. In 1940, the School of Fine Arts opened and later became the Faculty of Fine Arts at the University of Tehran in 1949, where Iranian/Persian students were scientifically trained in both traditional and modern architecture with many later employed in the conservation of monuments. In 1965 the National Organization for Preservation of the Historic Monuments of Iran was established by royal decree in the Ministry of Culture and Arts, subsequently opening technical bureaus in the provincial capitals [9]. Each regional department was headed by an archaeologist seconded by a civil engineer and experienced local architects, many of whom played important roles in preserving what buildings they could in the face of the modern developments rolled out across the face of many Iranian cities in the last half-century. Importantly regional centres trialled advanced conservation techniques for traditional constructions and materials including wood, stone, plaster ceramics, etc. ensuring the continuation and enhancement of the skills of those involved in the preservation of architectural heritage.

In the decade 1978–88 after the Islamic revolution and the Iraq war many Persian monuments fell into repair or were damaged by groups who saw them as emblems of monarchies, iniquity and oppression, regardless of building type or period. In 1983, concerned about the level of destruction being wrought Ayatollah Ruhollah Khomeini issued an edict declaring that the artistic and cultural wealth of Persia belongs to the people and must be respected: "*Above everything is culture*" because it "*edifies man, and for this aim the prophets were sent*". The importance of experts in historic conservation is again being recognised and encouraged, not least by the growing financial importance of the burgeoning tourism industry in Iran which is heavily dependent on its beautiful buildings and the need to remedy the war-damaged areas of the country. Currently, restoration and conservation activities are coordinated by the Ministry of Science and Higher Education which also publishes a quarterly journal *Majalla-ye mīrāṯ-e farhangī.*

The proliferation of excellent universities around Iran has also been important for those interested in recording and developing new understandings of both the structures and also how they worked climatically and culturally. The faculty of Fine Arts and Architecture led the way with a cohort of academics at the University of

Tehran, many of whom had gained experience in working with historic buildings in regional centres. The diaspora of academics abroad also contributes to studies of the vernacular. Scholarship on the subject is consolidating around centres of excellence in the field and through publications such as the Journal of Architecture in Hot Dry Climate, published by the Faculty of Art and Architecture at Yazd University [10].

2.2 Monuments as Stage Sets for History

Because of a general paucity of historic texts from the regions the corpus of well-preserved historic buildings is therefore doubly important, not least as stage sets for the continuous flow of dynasties back and forth across the country. Each period had its own architectural styles and ambitions. From the Achaemenids and the Sassanids to the Seljuqs, Ilkhanids, Timurids and Qajars, successive rulers build extraordinary monuments to reflect their own glory in the forms and facades of their buildings. Historians list stylistic features of each age as indicators of the dates and reach of different cultures. Tile and plaster work, domes, arches and decorative features are minutely detailed and explained as time capsules of their historic contexts [11]. Many are preserved either because of their structural solidity or the fact they were built in hot dry climates.

3 Technological Vernacular

The skills of Iranian master craftsmen and builders, i.e., the Ostads, who created the finest buildings were famous, be it for their wood, glass, tile or brickwork and the builders they trained and exported to regions around them in the Persian Gulf, Iraq, Afghanistan, Pakistan, India and the north. But they were not just artists, they were great engineers and technologists, too. One important records of these vernacular technologies was published by Hans Wulff in 1966 in his book on The *Traditional Crafts of Persia*, detailing their spread and influence on countries in the East and West [12].

Whole buildings themselves were the technologies utilising energy from the sun, wind and water. Some of the earliest examples of water and windmills, water cisterns and ice houses were built there. Many, including the caravanserai around which they could cluster in the deserts, were fortunately recorded by scholars including Maxime Siroux who in the 1930s and 1940s published his classic book on the buildings along the caravan routes through Iran [13]. A later useful book on desert technologies was by Beazley and Harverson in 1982 [14]. Already by then the importance of understanding how desert buildings worked was being realised, so that lessons could be learnt and reapplied by modern architects in similar climates. A seminal work published on the subject in the 1970s was published by Mahmood Tavassoli on hot and dry climate architecture, and he has continued to update the ideas behind it up to the present in his current book on the design of cities in the desert [15]. It is this aspect of Persian vernacular buildings that could prove increasingly valuable in a heating world, seeing buildings as machines for cooling, heating and in living for people in extreme climates.

3.1 The Anthropology of Architecture

Anthropology is the scientific study of humanity, concerned with human behaviour, human biology, cultures, societies, and linguistics. Buildings function in a three-way relationship between climate and the environment, the building and the people in it. Early anthropological studies in Iran were often of remote, or unusual populations with much work done on nomadic tribes and their lifestyles and material cultures [16]. As new technologies and fields developed so too did the ability to link people to the structures they occupied but also to building attributes they had never previously measured. Thus issues of thermal comfort in buildings came to the fore and Iranian scholars developed proficiency in relating buildings to climate and behaviours through such studies.

Shahin Heydari, in his Ph.D. thesis on Thermal Comfort in Iranian Courtyard Housing, began the burgeoning interest in thermal comfort practices within traditional building designs [17]. He has been followed by many researchers linking building performance to climate and behaviours, lifestyles and comfort lending more insight into the relationship between lifestyles and time in buildings and different climates [18].

Many of the chapters in this book that investigate the workings of vernacular buildings incorporate methodologies from art, architectural and technology historians as well as from the field of anthropology and thermal comfort researchers. Many of these approaches have been developed in Western universities and transported back to Iran. How much is lost in the gap between Euro-centric mindsets, methods and research approaches, and the locally grounded, profound understanding of how the ancient buildings and technologies actually work in the climates and cultures they operate in? There are some problems bridging that gap.

3.2 Post-colonial Thinking

Readers will have noted that many of the papers quoted were by European scholars. This is not least because before the 1970s most widely available publications on Iranian architecture were written and published by them. Inevitably, Western scholars often got it wrong. For example, the renowned French architect, Andre Godard [19] made many assumptions in his publications on Persian buildings that were demonstrably wrong [20]. Mohammad-Karim Pirnia [21], perhaps the most renowned Iranian scholar and a contemporary of Godard, challenged Godard on a range of facts. For instance, he demonstrated that the origins of pointed arches in Iran can be traced back to ancient buildings such as the Elamid and Meadian temples from three thousand years ago. Such was Godard's arrogance that he demeaned Persian scholars' views while spending relatively little time in Iran himself and incorporating the research of Persian scholars into his own work. Pirnia strongly criticised Godard's book, pointing out numerous inaccuracies including the misidentification of several buildings [22]. What is evident in this book is that the scholars writing it are in the main Iranian, and most have a close connection with the buildings they study.

From that profound local exposure to the buildings comes a deeper understanding not only of how they look or work but of what they mean. Pirnia wrote a number of books on the subject, particularly on the principles [23] and styles of Iranian Architecture [24]. Importantly he set out in them five principles and later six historical styles for Iranian architecture. In the 1990s these included: (1) human scale, (2) inward-looking, (3) self-sufficiency, (4) avoiding non-essentials and (5) structural rigidity and proportion. Until the time of Pirnia's death in 1998, he kept the order of the principles and constantly struggled to prove their accuracy and enrich them with explanations and examples [25].

While one cannot dismiss the enormous contribution made to the recording of Persian buildings by Europeans, it is really important that such studies are now carried out not by visiting scholars, but by people who live and breathe the subject and can learn from the buildings enough to understand how they can inform the more resilient buildings now needed to survive in the more extreme climates of the future. Pirnia embodied the genius of a Persian thinker and architect and is still revered by many with reason. It is a revelation to find that his profound understanding of the meaning of Persian vernacular could be reframed as a global definition of a resilient building: self-sufficient, efficient in providing space and comfort, structurally sound and culturally appropriate. By moving beyond the shackles of twentieth century Western thinking Pirnia provided a blue-print for future buildings, learning from the old buildings he understood so well.

4 Persian Vernacular: Pastiche or Respect?

Often one hears people denigrating traditional forms that are re-used as cultural symbols on new buildings. But many features of Persian buildings are so simple and soul-touchingly beautiful; so why not use them to enhance the often bleak and soulless faces of modern buildings in any part of the world? Two wonderful examples of the benefits of doing so were designed by the great Persian Nader Ardelan in Iran and Kuwait. The windcatchers on the roof of the Tehran Museum of Modern Art completed in 1977 were not used as ventilators but as light catchers. Inspired by the informal, vernacular of desert villages of Iran they were a potent symbol of the old reused for new purposes. Above all, they were a symbol of respect for the beauty of historic Iranian architectural forms and modern architectural inventiveness [26].

Ardalan's use of attractive windcatchers to hide the ventilation systems on the very popular Souq Sharq on Kuwait's waterfront is an example of how beauty is profitable. Ardalan created that visual selling point to his scheme from a profound knowledge of the psychological and philosophical understandings that underpin the proportions of the towers. This he demonstrated in his book 'The Sense of Unity: The Sufi Tradition in Persian Architecture' [27]. As in Pirnia's written work, Ardalan showcases the importance of 'meaning' in architectural forms, whether technical or aesthetic, that can be hugely important in informing better buildings for a different future.

4.1 The Traditional Technology Trap

When recording the buildings of the past, regardless of nationality there are a number of 'Traditional Technology Traps' that researchers can fall into. The first rule must be to 'beware of Assumptions'. Real problems can arise from applying assumptions pertinent to one location and culture to buildings in another. Equally assumptions can lose their meaning easily as they are passed from one person to another, rather as happens in Chinese Whispers. For instance, a common mistake is to assume that if a windcatcher works like this in Yazd, it must also work like that in Cairo or Jeddah [28]. Wrong answer, and potentially an expensive/uncomfortable mistake if applied in real designs. Even important architects such as Hassan Fathy fell into this trap in his Saudi Palaces to which he imported Egyptian windcatchers [29].

Another Technology Trap is encountered by assuming that the forces that led to the development of a particular building feature in the past are the same as those encountered today. For instance, many assume the largest windcatchers were built because they work better than smaller windcatchers in the climate of Yazd, the famous 'City of Windcatchers'. Possibly, a not all right answer. Many great windcatchers of the mid to late nineteenth century were paid for from the profits of the Opium Wars between England and China at that period when overnight Yazd became one of the richest opium-producing regions of the world. A very visible way to demonstrate wealth was to build a showy windcatcher rising high above the blank mud walls of the streets and alleys along which their houses were built [30].

Many thermal conditioning systems built in old Persian houses and palaces were of such complexity and sophistication that even with our limited modern understanding of the forces and sources of heat and coolth moving through traditional buildings and the limiting tools we use to analyse them, they can either take us beyond our former understanding of how the buildings work, or leave us with the wrong answers. For instance, the simulation tools we use to understand and adapt the workings of traditional windcatchers are really not designed to easily replicate the complicated interactions of external and internal airflow linked to the thermal mass of homes or shed light on how the spaces they serve actually provided free cooling at different times of day or year for their original occupants [31]. Simulation models are certainly not able to routinely link the thermal performance of a basement in summer to both the through-flow of air from a tower above to the water from a distant mountain aquifer, connected via an underground water channel (Qanat) [32, 33].

Another lens through which the scope of a study can be seen is the focal length of its author. Researchers need to have a clear idea of the focal length of their own view of a study. If one sets out to understand how Hammams (bath houses) work two different people might approach the subject from different viewpoints at different focal lengths. One author will write on the evolution and decline of public baths across Iran while another might focus on the temperatures that were actually occupied in the ablutive pathways from street to clean and back again in the buildings [34]. Thus, we ask, is such a study historical or technical or a description of a work of art and where does the study fit into other studies in the field?

5 Conclusion

Persian vernacular architecture is profoundly important, not just to architectural or art historians but to architects around the world searching to create more resilient buildings in a heating world. It is fitting that the greatest of the Persian vernacular scholars really put their finger on how to create buildings we can all survive in comfortably going forward. Mohammad-Karim Pirnia was right. The resilient buildings we need now will have to be human scale, self-sufficient, avoiding non-essentials and having structural rigidity and proportion and safely inward looking. Where better to learn how to design such buildings than in the vernacular buildings of Iran? The studies contained in this book are important as historical records but also as repositories for opportunities for new thinking about the design of resilient buildings of a different and more extreme future.

References

1. Armstrong McKay DI, Staal A, Abrams JF (2022) Exceeding 1.5°C global warming could trigger multiple climate tipping points. Science 377(6611). https://doi.org/10.1126/science.abn7950
2. Skea J (2024) It's too late to save Britain from overheating, says UN climate chief. The Telegraph, 6th Oct. At: https://www.telegraph.co.uk/business/2024/10/06/too-late-save-britain-overheating-climate-chief-jim-skea/
3. https://en.wikipedia.org/wiki/Palaeolithic_Era_in_Iran
4. Smith PEL (1990) Architectural innovation and experimentation at Ganj Dareh, Iran. World Archaeol 21(3):323–335. JSTOR: 124833. https://doi.org/10.1080/00438243.1990.9980111
5. Gallego-Llorente M, Connell S, Jones E, Merrett DC, Jeon Y, Eriksson A, Siska V, Gamba C, Meiklejohn C, Beyer R, Jeon S, Cho YS, Hofreiter M, Bhak J, Manica A, Pinhasi R (2016) The genetics of an early Neolithic pastoralist from the Zagros, Iran. Sci Rep 6:31326. https://doi.org/10.1038/srep31326
6. Pope AU (1938–1939) A survey of persian art from prehistoric times to the present. In Six volumes published under the auspices of the American Institute for Iranian Art and Archaeology, with Phyllis Ackerman. Oxford University Press, London & New York

7. Pope AU (1965) Persian architecture. Thames & Hudson, London
8. Wilber DN (1955) The architecture of Islamic Iran: the Il Khānid period. Princeton University Press, Princeton
9. Mousavi A (2004) Ernst Herzfeld, politics, and antiquities legislation in Iran. In: Gunter A, Hauser S (eds) Ernst Herzfeld and the development of near eastern studies 1900–1950. Brill
10. Yazd University's Journal of Architecture in Hot and Dry Climate. Available from: https://smb.yazd.ac.ir/?lang=en
11. O'Kane B (2021) Studies in Persian architecture. Edinburgh University Press. ISBN: 9781474474849
12. Wulff H (1966) The traditional crafts of Persia: their development, technology, and influence on eastern and western civilizations. The MIT press, Cambridge, Mass
13. Siroux M (1949) Caravanserails d'Iran et petites constructions routieres. Institut Francais D'archeologie Orientale, Cairo
14. Beazley E, Harverson M (1982) Living with the Desert, Wilts (England). Aris and Philips Ltd.
15. Tavassoli M (2002) Urban structure and architecture in the hot arid zone of Iran. Payam & Peivand, Tehran, Iran
16. Feilberg CG (1944) La Tente Noire: contribution ethnographique à l'histoire culturelle des nomads. Gyldendal, Copenhagen
17. Heidari S (2000) Thermal comfort in Iranian courtyard housing. PhD thesis, University of Sheffield. Accessed from: https://etheses.whiterose.ac.uk/10239/1/327670.
18. Foruzanmehr A (2017) Thermal comfort in hot dry climates: traditional dwellings in Iran. Routledge, London
19. Godard A (1965) Art of Iran. George Allen & Unwin, Sydney
20. Salari Sardari M (2024) Andre Godard and Maxime Siroux: disentangling the narrative of French colonialism and modern architecture in Iran. Iran Stud 57:265–293. https://doi.org/10.1017/irn.2024.10
21. https://en.wikipedia.org/wiki/Mohammad_Karim_Pirnia
22. Pirnia MK (2004) Iranian architecture of methodologies. In: Compiling and gathering: memarian Gholamhossein, 3rd edn. Memar publishing, pp 30–34
23. Pirnia MK (2002) Memorabilia of the professor Karim Pirnia. Edited by Akbar Qalam Siyah. Pirnia institute, Yazd
24. Qayyoomi Bidhendi M, Abdollahzadeh MM (2015) Architecture, land, and man rereading, reviewing and criticism of Pirnia's suggested principles for Iranian architecture. Kimiya-ye-Honar 3:72–99
25. Nadimi H, Abouie R, Moradi Z (2020) Explaining the concept of architecture stylistics by introducing a new approach in Iran's architecture stylistics. J Arch Hot Dry Clim 7(10):1–29
26. Ardalan Associates LLC (n.d.) From: https://www.ardalanassociates.com/nader-ardalan.html
27. Ardalan N, Bakhtiar L (1973) The sense of unity: the Sufi tradition in Persian architecture. University of Chicago Press, Chicago
28. Roaf S (1990) The traditional technology trap: stereotypes of middle eastern traditional building types and technologies. Trialog 25:26–33
29. Fathy H (1986) Natural energy and vernacular architecture: principles and examples with reference to hot and arid climates. The University of Chicago Press Ltd., London
30. Roaf S (2008) The traditional technology trap (2): more lessons from the windcatchers of Yazd. In: Proceedings of the PLEA conference, Dublin, Ireland, Oct 2008. Available from: http://web5.arch.cuhk.edu.hk/server1/staff1/edward/www/plea2018/plea/2008/content/papers/oral/PLEA_FinalPaper_ref_739.pdf
31. McCabe C, Roaf S (2013) The wind towers of Bastakiya: assessing the role of the towers in the whole house ventilation system using dynamic thermal modelling. Archit Sci Rev 56(2):183–194. https://doi.org/10.1080/00038628.2012.723398
32. Roaf S (2008) Resource quality cascades in traditional low exergy technologies: the Qanats and Badgirs of Yazd. In: Proceedings of the smart and sustainable built environments conference, Delft

33. Roaf S (1989) Settlement, form and qanat routes in the Yazd province. In: Beaumont P, Bonine ME, McLachlan KS, McLachlan A (eds) Qanat, kariz, and khattara: traditional water systems in the Middle East and North Africa. Middle East & North African Studies Press, London, Wisbech, Cambs, England, pp 59–61
34. Roaf S (2016) Sustainable building design: learning from nineteenth-century innovations. J Archit Conserv 22(3):233–236. https://doi.org/10.1080/13556207.2017.1290961

Learning from the Past

From Ostads to Architects: Evolution of Iranian Architectural Practices in Residential Buildings

Sadaf Alikhani and Asma Mehan

Abstract Given the increasing frequency of severe weather conditions, it is crucial to reassess our design strategies to establish architectural principles that protect individuals' emotional and physical health and general welfare. Iranian master builders, known as Ostads, have historically devised effective methods to tackle climate change challenges and improve human comfort. Vernacular architecture in Iran showcases a continuity between its components, local construction processes, climatic adaptation, and cultural integration. Iran's architectural solutions vary in response to different climatic zones. Vernacular and indigenous Iranian architecture, with a history spanning at least two millennia, demonstrates its adaptability through seamless integration within very different surrounding environments and climates and changing societies over time. The extraordinary buildings created by the Ostads of Iran have effectively enhanced the comfort and well-being of occupants while protecting populations from often already extreme climates. Contemporary architects can utilise Iranian traditional architecture to create cost-effective, comfortable, socially embraced, and environmentally aware communities prepared to withstand the even harsher weather conditions of the future. This research compares traditional and modern architecture of Iran's residential buildings, designed and constructed from the Qajar era to the present day. The primary emphasis is on analysing specific examples within the Köppen regional division, which is one of the best-known and most efficient approaches for climatic zoning. The Köppen climate classification scheme which uses monthly and annual threshold values of precipitation and temperature (Rahimi et al. in Theoret Appl Climatol 141(1–2):183–199, 2020 [35]).

Keywords Vernacular architecture · Housing · Traditional Iranian building technique · Sustainable architectural practice · Passive design · Contemporary architecture

---

S. Alikhani (✉)
Huckabee College of Architecture, Texas Tech University, Lubbock, TX, USA
e-mail: sadaf.alikhani@ttu.edu

A. Mehan
Architectural Humanities and Urbanism Lab (AHU_Lab), Huckabee College of Architecture, Texas Tech University, Lubbock, TX, USA

A. Cheshmehzangi and S. Roaf (eds.), *Persian Vernacular Architecture*, Urban Sustainability, https://doi.org/10.1007/978-981-96-1116-4_2

1 Introduction

Given the current environmental context and the increasing frequency of severe weather events, it is essential that we reconsider our design strategies to create architectural principles that protect the mental and physical health, as well as the general well-being of individuals. Across the globe, Vernacular Architecture (VA) covers various approaches aimed at enhancing the quality of living situations. These strategies involve passive design techniques that adapt to the specific climate and weather conditions of the place, the utilisation of locally obtained materials, and the integration of spatial arrangements and protective boundaries to maximise the usage of renewable resources [15, 27]. VA is a good example of intuitive design based on climate, materials, and living patterns [40]. Throughout history, traditional Iranian architecture has been deeply influenced by the local culture, beliefs, and climate [26]. The master architects and builders in Iran, known as Ostad, have devised effective vernacular approaches to tackle the difficulties posed by climate change to improve human comfort. For instance, features such as wind catchers, central courtyards, and mud construction have been refined in response to the arid and scorching desert climate of Iran [14]. Examining vernacular Iranian architecture not only yields historical understanding but also presents opportunities for developing energy-efficient and culturally responsive solutions [6]. Traditional technologies can offer sustainability, cultural preservation, and cost-effectiveness, but may present challenges like "trap" adaptation [36]. To discover the evidence of vernacular architecture and its benefits in today's climate situation in modern architecture, we will be looking at modern buildings to identify the footprint of vernacular architecture and explain how those yardsticks work to be used in the future [17, 19]. This chapter analyses case studies of residential buildings from national heritage registry archives, focusing on Qajar-era Iranian houses [7]. It refines samples by time, climate zone, and less researched dwellings, comparing modern and traditional homes from each climate zone, to investigate how much modern architecture adheres to the values and traits of vernacular architecture and locate the solutions to the subsequent questions:

- What information can be gained from the study of vernacular architecture?
- What is the importance of studying vernacular architectural landmarks?
- Is vernacular architecture important for creating a sense of tranquility within its environment despite technological measures?
- To what extent does vernacular architecture impact contemporary Iranian architectural methods?

In response to these questions, this study evaluates the benefits of using old Iranian VA techniques and principles by modern Iranian architects. This research aims to project the direction of vernacular architecture in the future.

2 Methodology

This study used a qualitative, analytical and comparative research methodology where content analysis and comparison analysis were employed to explore and explain the main themes and patterns in the literature about the VA and its presence in the contemporary architecture of Iran. The aim of this study was to clarify the current body of information in this area, point out the areas that need further investigation about this topic, and analyse the results to develop new approaches. This chapter examines the case studies of residential buildings from the archives of the national heritage registry and refines samples by time, climate zone, and less commonly researched dwellings [24, 25]. All essential information was acquired from the original registration license files of selected Qajar-era Iranian houses registered with the Ministry of Housing and Urban Development [21]. Contemporary houses were selected for their notable qualities and published on Arch Daily and it is worth noting that all these projects were designed by Iranian contemporary architects. In relation to the spatial distribution of old and traditional registered buildings, which are predominantly situated in the center and historic districts, there exist certain regulations governing the construction of new houses within these locations. Contemporary projects are chosen to be developed in the same climatic zone and availability, although in certain cases in different locations.

3 Literature Review

Vernacular Architecture (VA) is associated with traditional cultural practices and makes use of traditional materials, style, and natural concepts to create comfort [39]. It includes imposing elements specific to different climate regions that are commonly found in the built environment and directly contribute to creating tranquility for residents under various climatic conditions. Traditional Iranian architecture involves elements such as geometry, dimensions, proportions, and orientations, which are important design factors for microclimatic conditions and thermal comfort [13]. Native and vernacular Iranian architecture (VIRA) has been well-adapted to its environment for at least two hundred years, and perhaps much more [20]. In this chapter, these elements regarding the climate zone are classified into three categories including all the characteristics of VA: orientation, form, and material. Iranian researchers divided Iran's climates according to the Köppen into four distinct zones in the North, the southern Caspian Sea has a temperate climate due to high rainfall. Hot and humid conditions prevail on the northern shores of the Persian Gulf and Sea of Oman. Iran's mountainous and high plateau regions, which have long winters. These classifications aid architects and urban designers in Iran's range of climates.

3.1 Hot and Arid Climate (HAC)

Iran's climate is mostly arid or semi-arid. However, some regions have special climatic characteristics. This climate is characterised by intense sunlight, strong winds, and minimal precipitation, leading to severe dryness and predictable dust and sandstorms around noon [5]. In this region, traditional Iranian houses are designed with a specific orientation and layout to utilise passive cooling techniques in hot and arid climates, ensuring a comfortable living space for residents during extreme heat.

Orientation in Hot and Arid Climate (OHAC): houses were designed to face north south to minimise the number of walls and windows exposed to sunlight, particularly at the hottest times of the day.

Forms in Hot and Arid Climate (FHAC): The houses in this area exhibit an introverted architectural style, characterised by internal courtyards with two, three, or four-sided façades and seasonal rooms (north-facing in summer, south-facing in winter) [13]. Employ the earth's thermal capacity and a ground level that is lower than the natural one to have basements used for afternoon living for thermal comfort [11]. Cross ventilation and stack cooling with solar chimneys and wind catchers are natural ventilation methods, and night purge ventilation circulates cool night air inside to lower daytime temperatures, benefiting from strategically placed small windows and openings that maximise natural light and ventilation. Well-lit, naturally ventilated spaces improve comfort and reduce artificial lighting and cooling, and a courtyard's shade is another passive cooling solution [28]. Rooftops and balconies are shaded by extended rooflines or wooden lattice screens (Mashrabiya), which allow airflow and privacy.

Material in Hot and Arid Climate (MHAC): Utilising the thermal properties of materials is an important aspect of material selection in this region. Adobe bricks constructed 50–80 cm (about 2.62 ft) thick with a mud and straw mixture, capture, and release heat from the outside during the day and slowly release it at night [13]. Mud-brick (Khesht), a popular construction material, has excellent thermal mass, keeping interiors cool during the day and warm at night [31].

3.2 Cold Climate (CC)

Cold Climate in Cold Climate (CC): The cold climates in the northwest of Iran are distinguished by cold temperatures, long winters, and limited sunlight. Architectural design for cold areas focuses on maximising solar gain, minimising heat loss, and guaranteeing thermal comfort for occupants. Orientation in Cold Climate (OCC): Traditional architecture optimises energy efficiency and indoor comfort by strategically orienting structures to capture winter sunshine and provide shade during the summer [2]. Consequently, the crucial sections of the structure are positioned on the north (northwest to northeast) and south sides [41].

Form in Cold Climate (FCC) The architectural style is focused on energy conservation. It features an introverted design that includes a courtyard with a central pool to allow for the entry of light and air, creating a microclimate. Establishes a semi-open area in the middle of open and close areas that maintains privacy and enhances thermal efficiency [41]. The structure is designed to be squat and compact, with thick walls and a flat roof, which reduces the surface area exposed to the wind and minimises heat loss from the roof. Shading devices such as overhangs, awnings, and vegetation can be used on facades to decrease summer solar heat, increase winter sunshine, and act as decorative objects [2]. The number and size of openings in the house are deliberately limited on the east and west sides to minimise heat loss and the impact of cold winds. Thus, during winter, it's advisable to use south-facing windows to maximise the amount of solar heat obtained.

Material in Cold Climate (MCC): In this region, materials with high thermal mass, such as stone, adobe clay, mud, and brick, play a crucial role. These materials could absorb and store heat during the day, gradually releasing it at night. This helps in regulating indoor temperatures and providing insulation. To protect against cold temperatures, it is recommended to use thick walls made of materials like adobe or brick, ranging from 50 cm (about 1.64 ft) to 85 cm (about 2.62 ft) in thickness. Additionally, utilising the thermal capacity of the ground is also significant in this climate zone.

3.3 Hot and Humid Climate (HHC)

This region is in the south of Iran in a warm and highly humid climate zone and has notable seasonal temperature variations along with exposure to the Oman Sea and Persian Gulf and the prevalence of wet and warm sea winds. The influence of Sudanese and Mediterranean cyclones in autumn and winter [29]. The design strategy in this climate zone is enhanced by natural ventilation [9, 14].

Orientation in Hot and humid climate (OHHC): Buildings are positioned facing the sea to utilise the cool breeze and align with prevailing wind directions to enhance airflow through openings and pathways, encouraging natural ventilation [9, 33].

Form in Hot and humid climate (FHHC): Buildings either have a central courtyard with rooms and halls or a front or rear yard with rooms on one side, surrounded by open, semi-open, and closed spaces [9, 29]. Courtyards are typically constructed on an elevated platform to prevent the building from being affected by humidity. These central open spaces cool buildings and provide natural light and ventilation. Exterior balconies (Shenashir) with wood awnings and louvres from upper floors are an object to shade and control airflow, improving thermal comfort [33]. Traditional southern-style wind catchers known as Badgir are another passive cooling element that captures and directs airflow into buildings for natural ventilation and cooling and helps wind exchangers improve ventilation. By harnessing renewable energy sources like water, wind, and air, these systems have made it possible for people in hot climates to live comfortably without using fossil fuels.

Green spaces with appropriate vegetation are utilised to provide shade in open areas without increasing humidity levels [14]. Building facades with many exterior and interior openings for cross-ventilation and airflow First-floor windows are larger than ground-floor windows to improve airflow and ventilation [9, 33].

Material in Hot and humid climate (MHHC): Local stone (task) is widely used in construction due to its availability and durability. Local plaster (lime) is another popular insulator and climate-friendly material. Wood (chandelier) is widely used in construction because its slow heat transfer helps maintain comfortable indoor temperatures. Wood is also often used for structural elements, interior finishes, and decorative features in traditional buildings. It can contribute to natural ventilation and add aesthetic value to the architecture [9].

3.4 Tempered Climate (TC)

The moderate climate is situated in the northern region of Iran, along the southern shore of the Caspian Sea. The region is surrounded by the Alborz Mountain range to the north and is renowned for its verdant scenery and varied temperature. Weather features, including high humidity levels and significant rainfall.

Orientation in Tempered climate (OTC): Homes are commonly angled to maximise solar gain in winter and limit it in warmer and humid summers. House orientation considers wind directions for natural ventilation and cooling; their orientation also considers landscape views and privacy from adjoining buildings to maximise space utilisation. The building's orientation advantageously depends on the prevailing east and northeastern winds while simultaneously circumventing the discomfort caused by the north-west winter wind. The prevailing orientation in the region is typically north–south, with a slight lean towards the east [38]. Houses are commonly located in the core area of a site, distanced from its outer boundaries [1, 4, 10, 38].

Form in Tempered climate (FTC): typically has an extraverted style. The central arrangement is created to support the inhabitants' daily activities and encourage social contact within the family [10, 38]. Buildings are raised above ground level to avoid moisture collection, and various types of balconies, known as kutams, are semi-open spaces that link indoor sections, shield the building from rain, and offer another living area. Their placement is crucial to optimise ventilation and solar exposure while providing protection from excessive heat and humidity [4]. Building facades with many exterior and interior openings for cross-ventilation and airflow Passageways with double columns all around the building and in front of halls, known as (Qolam Gard, are designed to provide shaded places with proper ventilation. Houses are built with floors erected on a lattice foundation to shield them from elevated humidity levels, and the roofs of buildings typically have a hip roof form. In certain areas, the roof may be lower on one side than the other to deal with the direction of rain [1, 4, 38].

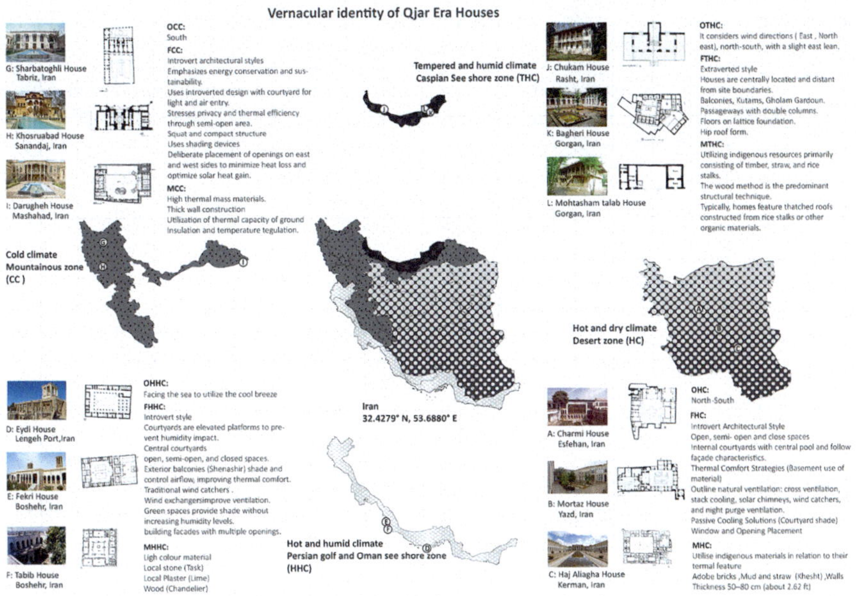

Fig. 1 Explore the vernacular architecture and climate zone of traditional Iranian dwellings. *Created by* authors (2024)

Material in Tempered climate (MTC): Using indigenous materials, including timber, straw, and rice stalks, the wood method is the most used. The thick walls offer thermal mass, which aids in controlling indoor temperatures and improving energy efficiency [10]. Hip roofs are typically thatched with rice stalks (kolush), or other natural materials called gali and offer insulation and protection from the weather, making them a sustainable roofing option ideal for the local climate [34, 37, 38]. Wooden beams, columns, and frames are frequently used to construct the structural framework of buildings, highlighting the plentiful local timber resources in the area (Fig. 1).

4 Contemporary Architectural Project of Iran

Preserving vernacular architecture requires adapting it to modern standards while preserving its original elements. Modernisation has significantly altered urban architectural structures [3, 6, 12, 32]. The environment is another element that can directly influence how architecture is designed, and Iran is not exempt from this influence [18]. It is subject to climate change impacts such as rising temperatures, precipitation changes, and extreme weather [30]. This section examines contemporary projects and their responses to these developments.

Aban House in Esfahan, Iran, designed by USE Studio, showcases a modern interpretation of traditional Persian architecture made use of a 3 × 3 pattern which is the fundamental pattern for many architectural creations in this land is and flexible public spaces. Based on the given arrangement, the courtyard was positioned in the middle and the surrounding spaces were organised accordingly. In this house, the 3 × 3 pattern consists of three courtyards located at the corners. These courtyards are arranged in such a way that each interior space is connected not to just one courtyard in a single direction, but to two courtyards in two different directions. The arrangement of courtyards and open spaces used passive ventilation to support community living; and the entire structure was designed to respond to the climatic conditions of the local area [16]. Aban House consists of spaces that have distinct and precise purposes, while also incorporating open, closed, and semi-open areas that may accommodate a variety of events. The usage of vernacular materials such as local bricks to insulate the house and provide thermal mass is noteworthy.

Yast Khane, a residential building in the city of Yazd, Iran, was undergoing construction when the client approached Mohammad Khavarian + Awe Office to develop a new design strategy for it. The primary focus of this design was to address the project's issues, such as rearranging the site location and resolving problems related to the facade. Additionally, the design aimed to meet the client's specific cultural beliefs, which included incorporating a courtyard and creating a clear distinction between private and public spaces within the home. To address these points, the architectural groups extensively researched climate and vernacular architecture. As a result, they devised a solution that involved constructing tall walls with minimal openings on the west and north sides, in accordance with vernacular principles. This design not only allows for better control of natural light, but also creates a unique entrance passage in a creative manner. They also design an atrium to separate the northern half of the existing structure from the southern part, with the purpose of controlling natural light and creating a division between the public and private areas of the house. It is worth mentioning that they utilise brick as a suitable choice of material due to its high thermal capacity, which is advantageous in this region.

Narboon Villa, designed by Gera Studio Architects, is situated inside the pomegranate orchard of Kerman in Iran. The noteworthy component of this project, aside from incorporating vernacular strategies in its design, is its inspiration drawn from culture and heritage. An evident illustration of this is the preservation of the traditional main clay walls in this garden, with the construction of new walls positioned behind them. Another example is the use of bricks to create the inner facade.

In terms of vernacular strategy, the architects tried to connect the open space with the building's interior by creating a gap between the main mass of the structure. Additionally, they aimed to incorporate the garden into the space, resulting in a design that allows for natural skylight, shadows, ventilation, and the manipulation of privacy using walls that create viewing farms and platforms. The. Their architectural objective was to design a pavilion that drew inspiration from the indigenous culture of the people in that region (Fig. 2).

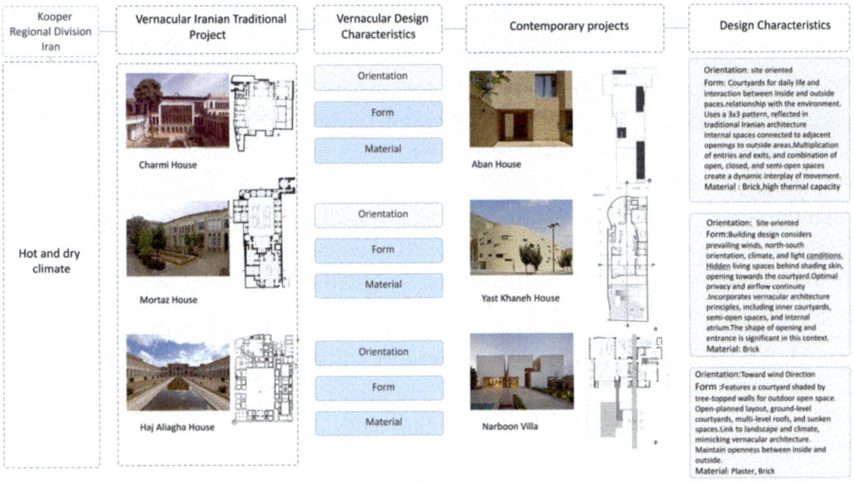

Fig. 2 Looked at contemporary projects in HC and their adherence to VA design features in previous homes. *Created by* authors (2024), *Material Source* Archline (n.d.), https://archline.ir/

4.1 Contemporary Projects in Cold Climate (CC)

The Inside House, Located in Amameh, Iran, within a residential building, The setting consists of a series of underground areas that offer a combination of seclusion and harmonious blending of the structure with the surrounding environment. The Olgooco architects' group is seeking a solution that balances privacy while respecting the surrounding landscape and environment. The building design is aligne with the geometry of the site location, which is a 70 m^2 cave divided into three sections: private, public, and utilities. The interesting aspect of this project lies in the strategic design of incorporating a cave-like structure into the ground, along with a large balcony at the top covered in white natural materials such as soil and trees. This design approach not only provides a vernacular solution for the region by utilising the thermal capacity of the earth but also creates an introverted building that aligns with environmental ideologies. The inner courtyard of the house is situated on a hill, and it slopes towards the west, offering a clear view of the next village. The courtyard is connected to a spacious porch, which allows for a panoramic view of the village through the trees in the garden, regardless of the season. Open and semi-open spaces in this manner take use of the sunlight and unobstructed vistas without compromising privacy, while the project itself safeguards the natural surroundings for nearby residents. Finally, this project's introvert a cave-shaped residence, private–public space separation, and courtyard are all vernacular responses to design in this region.

Hesar Villa, is situated in a mountainous sanctuary in Mashhad, Iran. The most notable feature of this project is the 20-m difference in site elevation, which results in a temperature difference of around 4 °C from the starting point to the endpoint of the site. The design strategy in this project aims to address numerous concerns. Firstly, Afshin Khosravian and Associates designed an introverted dwelling with a vertical slice to achieve natural ventilation and separation between residential and recreational spaces. In addition to natural ventilation, this design approach facilitated optimal viewing angles and established a strong link with the courtyards. The courtyards, with their introvert-extrovert pool, possess a special capacity to regulate temperature and humidity, providing a tranquil environment for its residents. The overall design of this project is characterised by an introverted approach. It incorporates a solution for integrating natural ventilation with the surrounding landscape and use stone as the primary material.

The Man and the House is a housing project situated in the mountainous region of Damavand, Iran. This project by ZAV architects endeavour to develop a design strategy that considers the specific demands and interests of the client which is Mountain and a deep knowledge on industrial construction. They believe that the surrounding setting, particularly the mountain, has a crucial role in shaping the design. The main issue they want to answer is whether the combination of logical and imaginative elements may influence the design of the space. In this regard, architecture teams utilise various colours of materials in a three-story villa, ranging from dark grey stone to a white gypsum sloped roof. This material selection not only represents the transition from fantasy to reality but also caters to the vernacular requirements of this region. Additionally, the design of this building, which is introverted and compact, directly references the vernacular strategy of this region. It also demonstrates how the building seamlessly integrates with the natural surroundings (Fig. 3).

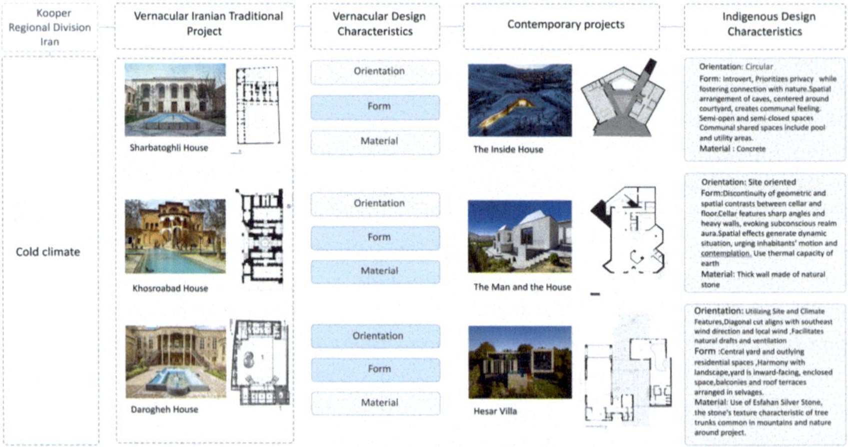

Fig. 3 Looked at contemporary projects in CC and their adherence to VA design features in previous homes. *Created by* authors (2024), *Material Source* Archline (n.d.), https://archline.ir/

4.2 Contemporary Projects in Tempered Climate (TC)

Villa in Darvish Abad, is in northern Iran. The main design strategy in this project aimed to achieve a harmonious blend of tradition and modernity. This was achieved by illustrating the open plan on the first floor, which embodies a modern approach to design. In contrast, the compact plan of the second floor, centred around a lantern-shaped void, represents the traditional concept of planning a house plan. This spatial strategy creates a connection between the sky and earthly realms. In the context of vernacular architecture, the building's distance from the ground level, as well as its open plan and the angle of its openings, adhere to the principles of vernacular methods.

Wicker House, in Mazandaran, Iran aims to design a dwelling that integrates with its surroundings, effectively blending in with its natural environment. Designers believe that the vernacular style of design in this region is derived from traditional village houses, and this project has been designed accordingly. As a result, the project's details uphold important local customs by utilising locally produced natural materials like thatch, clay, and straw to blend in with the surroundings. Additionally, the building's placement maximises views of the charming rural landscape, which is accentuated by the new structure, which represents the development of local architecture from antiquity to the present. The building's roof shape, and material openings are all consistent with the region's vernacular styles.

Netal Weekend Villa, located in the northern region of Iran, is a popular recreational destination for people from all over the country. As a result, it experiences a phenomenon of mixed cultures, which has influenced the traditional architectural style of the area. The project aims to address the challenge of balancing the need for privacy in residential spaces with the surrounding social environment that come from mentioned mixed culture. To fulfil the client's need for privacy and introverted design, while also incorporating the extroverted style of vernacular architecture in the region, this project has been separated into three distinct sections: a private portion, a public section, and a semi-open structure located inside the garden. The design incorporates local construction trends, a sloped roof shape, and openings for natural ventilation and light (Fig. 4).

4.3 Contemporary Projects in Hot and Humid Climate (HHC)

The presence in Hormuz 2 (Majara), designed by ZAV Architects in Iran's Hormuz Island, aims to stimulate the economic and cultural development of the community through the implementation of environmentally sustainable urban planning. The project employs Nader Khalili's super-adobe technique, which involves constructing structures using bags filled with soil, to create small-scale domes using locally

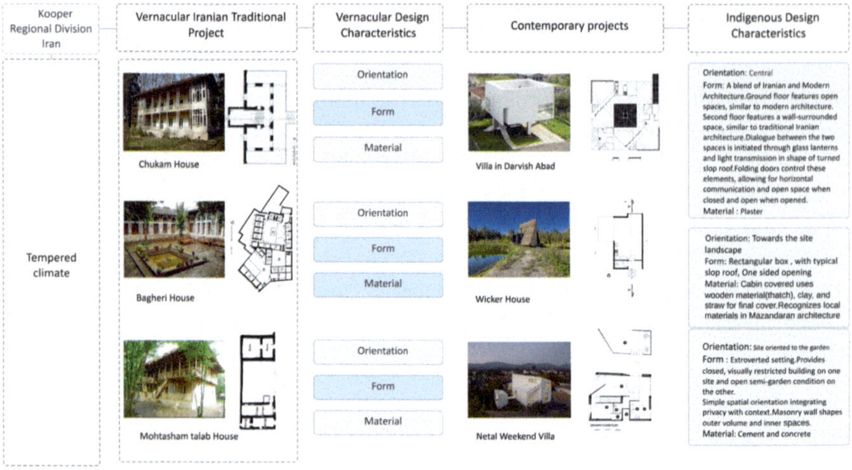

Fig. 4 Looked at contemporary projects in TC and their adherence to VA design features in previous homes. *Created by* authors (2024), *Material Source* Archline (n.d.), https://archline.ir/

sourced materials. This approach allows for the incorporation of the natural essence of the surrounding terrain into the building, while also establishing a harmonious relationship with the traditional ice houses of the region. Majara aspires to promote the local economy and provide versatile spaces that embody the community's essence and the marine life found on the nearby island while considering the convenience for the local workforce and using adaptable design. This project is situated on a site that was previously used for showcasing sand-carpet exhibitions, where Persian carpet designs were created on the sandy shores. The architectural design of the building takes into consideration the surrounding hill and respects its presence. During the project, they made the decision to design it in a way that is noticeable due to the strategic placement of each dome in relation to the environmental topography and inspiration from sand-carpet exhibition. This project was implemented as part of a cultural development initiative in the region, with the aim of transforming it into a tourist hotspot. To achieve this, the decision was made to activate certain nodes within the city. It is believed that by doing so, these nodes will serve as focal points and stimulate activity in other parts of the city. Therefore, this idea is not only a temporary residential complex, but it also lacks any boundaries to resemble a plaza and neighborhood for people, serving as a cultural platform for events. In conclusion, the construction methods, block orientation, and materials used in this project align with the principles of vernacular architecture.

A house looking out to the Persian Gulf, is a new architectural proposal in the fast-rising village on Qeshm Island aims to combine traditional southern architecture with sustainable principles [23]. In this project, Kooshk Office architects want to develop a solution for designing a residential building that adheres to vernacular and traditional architectural principles. One of the inspirations for the design of this

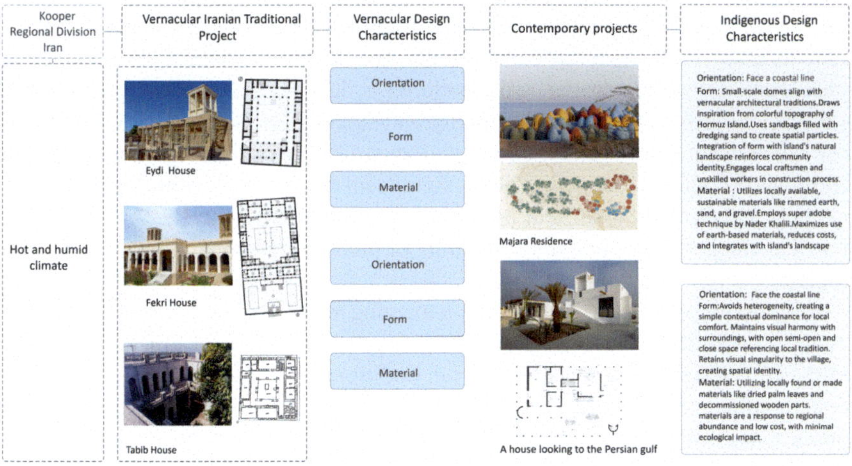

Fig. 5 Looked at contemporary projects in HHC and their adherence to VA design features in previous homes. *Created by* authors (2024), *Material Source* Archline (n.d.), https://archline.ir/

village comes from its natural surroundings, where all the residents are related and wish to preserve the coherence among their houses. Another challenge in this project was ensuring a sustainable energy supply and preserving the 40 square meter existing structure. The design prioritises passive energy-saving strategies that are well-suited to the local environment, such as including balconies, orienting the building towards the sea, and strategically placing openings to capture sea breezes. Additionally, the design aims to retain visual harmony with the surrounding community. The project seeks to achieve self-sufficiency, environmental friendliness, by prioritising the use of local materials such as dried palm leaves and components from dated boats [42]. The ultimate objective is to create a structure that is both ecological and energy-efficient, while also harmonising perfectly with the traditional aesthetic and values of the village (Fig. 5).

5 Discussion and Implications

This research has shown that while architects and clients are incorporating vernacular architecture into modern projects, challenges persist. Adapting vernacular architecture to modern standards requires careful examination and integration of modern materials, techniques, and culture while preserving its original elements. Traditional Iranian architecture emphasises low-energy principles for material performance and resource optimisation in extreme climates. Modern designs could benefit from these principles, incorporating passive cooling systems and adaptive design strategies. Traditional technologies like qanats and badgirs offer valuable lessons for modern sustainable design, especially in regions facing similar environmental challenges

[36]. Additionally, The challenges faced by traditional technologies include adapting to modern standards, material compatibility, and economic viability. Scaling traditional technologies to meet modern building codes and standards is challenging, as it may compromise structural integrity and safety. Material compatibility issues arise when using natural materials in modern structures or different environmental conditions. Economic viability is also a significant challenge, as sourcing traditional materials and skilled labour can be prohibitive in modern contexts. Also, the lack of a clear categorisation and understanding of modern Iranian architecture impedes progress.

In developing countries like Iran, implementing interactive architecture faces limitations. Modernisation has significantly altered urban architectural structures, leading to urban dwellings that often deviate from traditional vernacular architecture due to Western influence, industrial development, and rapid urbanisation. As a result, most houses today follow mandatory orientations due to the city grid or junk spaces, making structures unsuitable for local topography, climate, and cultural contexts.

Iran is subject to climate change impacts such as rising temperatures, precipitation changes, and extreme weather. The country's climate has been categorised into six zones: mild and humid, cool and sub-humid, cold and temperate semi-arid, warm and semi-arid, cool and arid, and warm and hyper-arid. Therefore, vernacular architecture methods may not be applicable everywhere. Also, the transformation of historic residential zones resulted in permanent alterations and a loss of physical identity. However, the current project demonstrates a tendency and hope to use Iranian modern architecture that draws inspiration from local vernacular architecture through climate-friendly design, traditional architectural technology elements, passive cooling strategies, and sustainability principles.

The projects included Aban House in Esfahan, Narboon Villa in Kerman, and Hesar Villa in Mashhad. These projects skillfully utilised vernacular techniques to optimise cooling by employing passive cooling and natural ventilation. The Inside House in Ammameh incorporates underground spaces to store and disperse thermal energy by using the natural thermal mass of the Earth. It offers concrete proof of how applying old methods in a creative way can lead to innovative solutions for achieving thermal comfort in modern architectural projects. Yast Khane in Yazd and The Man and the House in Damavand and A House Looking to the Persian Gulf House highlight the relationship between cultural values and modern architecture, namely in terms of cultural preservation and modernisation. Both buildings highlight the significant importance of vernacular design principles, which lead to architecture that is considerate of cultural values and requirements, while still meeting modern practical needs. The Presence in Hormuz 2 project in Hormuz employs Nader Khalili's super-adobe construction technique to effectively embody the local culture and environment, thereby promoting the advancement of sustainable urbanism [8]. Several other case studies, like Wicker House and A House Looking Out to the Persian Gulf on Qeshm Island, demonstrate a dedicated commitment to sustainability by utilising indigenous materials and passive design strategies [22]. Traditional architecture can provide valuable insights for constructing environmentally friendly buildings. The villa in Darvish Abad Mazandaran and Yask Kahneh integrated ancient construction techniques into the layout, combining them with modern architectural principles. Finally, Traditional architecture can provide valuable insights for constructing environmentally friendly buildings.

6 Conclusions

Exploring Persian vernacular architecture has demonstrated how adaptable and timeless is the design wisdom it embodies, and how relevant to addressing contemporary climate design challenges in a rapidly changing world. A detailed examination of Iran's residential built heritage from the Qajar period and modern times highlights the fact that the traditional forms and practices of vernacular architecture, which are responsive to climate, are not only still as useful for contemporary architecture as they were then but have to offer unique insights into how to design better for a more extreme future. The construction techniques and environmental technologies developed by the Ostads of Iran already provide the bases for low energy, low impact, climatic design principles for contemporary Iranian buildings involving passive cooling, orientation, and materiality.

In today's world, many people believe that modern technology can solve all problems. However, in some cases, technology can actually hinder the situation. It is important to consider the side effects of technology. For example, we are currently facing environmental problems that greatly impact the field of architecture. Therefore, it is crucial to examine traditional building models that were self-sufficient and by studying their traditional and old techniques, we can learn valuable lessons. Given the current water and energy crises parallel to environmental problems, it is crucial to acquire knowledge from vernacular self-sufficient techniques. Although in the past, these techniques were specific to their region and culture and were not exported to other regions around the world due to limited access to knowledge transfer, nowadays, the transformation of knowledge is easily accessible. The application of Iranian Vernacular architecture may effectively handle global environmental challenges, such as extreme weather conditions, while also incorporating cultural elements. This approach can be adapted and modernised to suit individual circumstances.

Iranian vernacular design principles can be adapted to benefit from modern technological advancements within the constraints of modern culture. Current issues that need addressing include how to fit vernacular approaches into modern lifestyles and the reclassification of climate zones in Iran. Following the city's updated climate classifications—mild-humid, cold, temperate semi-arid, warm, and hyper-arid—traditional architectural methods may no longer adequately ensure residents' comfort. Moreover, the rapid urbanisation influenced by Western standards is transforming architectural landscapes and urban settings, making it increasingly difficult to implement vernacular solutions.

The study has underscored the need for new thinking on how best to combines the best of the old approaches to vernacular buildings with contemporary and future social, economic, technological, and climate challenges. The focus is on creating an architectural synthesis that acknowledges Iran's evolving climate and lifestyle changes from traditional to modern times. This approach emphasises the need for designs to harmonise traditional vernacular knowledge with contemporary sustainable practices. By integrating vernacular heritage with modern design, we can ensure resilience and sustainability in the built environment, thereby preserving cultural continuity.

Acknowledgements The author expresses sincere gratitude to the reviewers as well as the guest editors for their insightful comments and valuable feedback. Their constructive critiques have enriched the depth and quality of this work. The authors would like to express gratitude to the Architectural Humanities and Urbanism Lab (AHU_Lab) within Huckabee College of Architecture, Texas Tech University for all the generous support.

References

1. Ahmadi MP, Rostami EY (2023) Façade design recommendations towards visual compatibility of infill buildings with the Qajar architecture of the cities of Mazandaran province. Int J Archit Eng Urban Planning 33(3). http://ijaup.iust.ac.ir/browse.php?a_id=771&sid=1&slc_lang=en&ftxt=1
2. Alitajer S, Shahabi S (2023) Building orientation; climatic adaptation technique in the vernacular architecture of cold regions of Iran (case study: vernacular houses in Hamadan). Pazhoheshha-Ye Bastan Shenasi Iran 13(37):333–361
3. Amini M, Mahdavinejad M, Bemanian M (2019) Future of interactive architecture in developing countries: challenges and opportunities in case of Tehran. J Constr Dev Countries 24(1):163–184
4. Amiri N, Vatandoost MM (2017) The study of the relationship between sustainable architecture and vernacular architecture in the North of Iran. J His Cult Art Res 6(1):436–450
5. Azarbayjani M (2019) Seeking contemporary urban comfort through vernacular architectural principles in hot arid climate. In: Sayigh A (ed) Sustainable vernacular architecture. Springer International Publishing, pp 93–104. https://doi.org/10.1007/978-3-030-06185-2_5
6. Azizi-Bondarabadi H, Haji Sadeghi N (2022) Local seismic culture in Iranian vernacular architecture: evidence from Yazd earthen architecture. Built Heritage 6(1):5. https://doi.org/10.1186/s43238-022-00051-3
7. Boodaghi O, Fanni Z, Mehan A (2022) Regulation and policy-making for urban cultural heritage preservation: a comparison between Iran and Italy. J Cult Heritage Manag Sustain Dev. https://doi.org/10.1108/JCHMSD-08-2021-0138
8. Cheshmehzangi A, Sedrez M, Dawodu A, Heath T, Zhao H, Li T (2023) Urban resilience to future pandemics. In: Resilience vs pandemics: innovations in cities and neighbourhoods. Springer Nature, Singapore, pp 1–12
9. Fani M, Mehdizadeh Saradj F, Yazdanfar SAA (2022) Vernacular characteristics of the built environment in Bushehr city. Int J Environ Stud 79(2):213–244. https://doi.org/10.1080/00207233.2022.2037345
10. Fathi Saqalaksari H, Sahragard Monfared NS, Yazdanfar SA (2023) Assessment of influential components in vernacular housing architecture of villages in the Gilan Plain. Int J Archit Urban Dev 13(3):99–112

11. Foruzanmehr A (2017) Thermal comfort in hot dry climates: traditional dwellings in Iran. Routledge. https://books.google.com/books?hl=en&lr=&id=Hq41DwAAQBAJ&oi=fnd&pg=PT16&dq=Foruzanmehr+A.+Thermal+Comfort+in+Hot+Dry+Climates:+Traditional+Dwellings+in+Iran.+London+(UK):+Routledge%3B+2017.&ots=_F25PBNUsQ&sig=c3LKDQMct5nku9gt_TW16qcE16s
12. Haj Maleki M, Haghir S (2020) Classification of Iranian contemporary architecture, based on trends and challenges. Int J Archit Urban Dev 10(3):47–60
13. Izadpanahi P, Farahani LM, Nikpey R (2021) Lessons from sustainable and vernacular passive cooling strategies used in traditional Iranian houses. J Sustain Res 3(3). https://sustainability.hapres.com/htmls/JSR_1395_Detail.html
14. Khakzand M, Chahardoli S, Niknejad A, Khanijazani T (2023) Comparative study of architectural elements to improve the wind environment in hot and humid climates. J Archit Eng 29(3):04023024. https://doi.org/10.1061/JAEIED.AEENG-1546
15. Khasraghi SS, Mehan A (2023) Glocalization challenges and the contemporary architecture: systematic review of common global indicators in Aga Khan Award's winners. J Archit Urban 47(2):135–145. https://doi.org/10.3846/jau.2023.17176
16. Mehan A, Odour N, Mostafavi S (2023) Socio-spatial micro-networks: building community resilience in Kenya. In: Resilience vs pandemics: innovations in cities and neighbourhoods. Springer Nature, Singapore, pp 141–159. https://doi.org/10.1007/978-981-99-7996-7_9
17. Mehan A (2023) Re-theorizing the collective action to address the climate change challenges: towards resilient and inclusive agenda. Can J Reg Sci 46(1):8–15. https://doi.org/10.7202/1097156ar
18. Mehan A (2023) Gio Ponti and Villa Namazee: (De) listed Modern Heritage. Heritage 6(2):789–801. https://doi.org/10.3390/heritage6020043
19. Mehan A (2022) Tehran: from sacred to radical. Taylor & Francis. https://doi.org/10.4324/9781003140795
20. Mehan A, Lima C, Ng'eno F, Nawratek K (2022) Questioning hegemony within white academia. Field 8(1):47–61
21. Mehan A, Mehan M (2020) Housing as politics: the case of Tehran. In: Spaces for living, spaces for sharing. Lettera Ventidue Edizioni, Siracusa, Italy, pp 56–65
22. Mehan A, Sennema H, Tideman S (2020) Port city heritage: contested pasts, inclusive futures? The port city futures blog. Leiden-Delft-Erasmus (LDE) Initiative
23. Mehan A, Jansen M (2020) Beirut blast: a port city in crisis. In: The Port City Futures Blog. Leiden-Delft-Erasmus (LDE) Initiative
24. Mehan A (2019) The challenges of "comparative urbanism" in post-Fordist cities: the cases of Turin and Detroit. Contour J 4 (Comparing Habitats):1–14. EPFL University, Lausanne, Switzerland
25. Mehan A (2017) In razing its modernist buildings, Iran is erasing its past western influence. The Conversation 1–7. Creative commons license
26. Mehan A (2016) Squares as tools for urban transformation: foundations for designing the Iranian public squares. Revista Brasileira de Planejamento e Desenvolvimento 5(2):246–254
27. Moscoso-García P, Quesada-Molina F (2023) Analysis of passive strategies in traditional vernacular architecture. Buildings 13(8):1984
28. Movehhedi Nia M, Nikpour M, Jahanshahi Javaran E (2024) Evaluation of quality of daylight in a contemporary residential building with a central courtyard in Kerman, Iran. Iranica J Energy Environ 15(3):294–302. https://doi.org/10.5829/ijee.2024.15.03.08
29. Nadoomi R, Sharghi A, Nakhaei S, Azadian R (2023) Regional materials and environmental sustainability in hot and humid climates: a study on Boushehr's vernacular houses. Iran Univ Sci Technol 33(4):1–19
30. Najafi MS, Alizadeh O (2023) Climate zones in Iran. Meteorol Appl 30(5):e2147. https://doi.org/10.1002/met.2147
31. Nakhaee Sharif A, Keshavarz Saleh S, Afzal S, Shoja Razavi N, Fadaei Nasab M, Kadaei S (2022) Evaluating and identifying climatic design features in traditional Iranian architecture for energy saving (case study of residential architecture in northwest of Iran). Complexity 2022. https://www.hindawi.com/journals/complexity/2022/3522883/

32. Palazzo PP (2021) Vernacular patterns in Portugal and Brazil: evolution and adaptations. J Tradit Build Archit Urbanism 2:359–370
33. Parsaee M, Abazari T, Abolhassani SS (2023) Natural ventilation in a traditional city: an exploratory computational study of Bushehr-Iran. Int J Environ Stud 80(3):792–806. https://doi.org/10.1080/00207233.2022.2117479
34. Rahdari A, Mehan A, Malekpourasl B (2018) Sustainable real estate in the middle east: challenges and future trends. Sustain Real Estate Multi Approaches Evolving Syst 403–426. https://doi.org/10.1007/978-3-319-94565-1_16
35. Rahimi J, Laux P, Khalili A (2020) Assessment of climate change over Iran: CMIP5 results and their presentation in terms of Köppen-Geiger climate zones. Theoret Appl Climatol 141(1–2):183–199. https://doi.org/10.1007/s00704-020-03190-8
36. Roaf S (2008) 739: The traditional technology trap (2): more lessons from the windcatchers of Yazd. Retrieved from 6 Aug 2024. From: https://www.researchgate.net/profile/Roaf-S/publication/238691773_739_The_Traditional_Technology_Trap_2_More_lessons_from_the_Windcatchers_of_Yazd/links/54c6470f0cf256ed5a9dcb1d/739-The-Traditional-Technology-Trap-2-More-lessons-from-the-Windcatchers-of-Yazd.pdf
37. Rajendran L, Molki F, Mahdizadeh S, Mehan A (2021) (Re) framing spatiality as a socio-cultural paradigm: examining the Iranian housing culture and processes. J Archit Urbanism 45(1):95–105. https://doi.org/10.3846/jau.2021.14032
38. Salavatian S, Malekjahan FA (2019) Typological analysis of vernacular residential buildings in moderate-humid climate of North Iran. In: Sayigh A (ed) Sustainable vernacular architecture: how the past can enrich the future. Springer International Publishing, pp 115–139. https://doi.org/10.1007/978-3-030-06185-2_7
39. Sayigh A (2019) Sustainable vernacular architecture: how the past can enrich the future. Springer. https://books.google.com/books?hl=en&lr=&id=va6PDwAAQBAJ&oi=fnd&pg=PR5&dq=ali+sayigh+sustainable+vernacular+architecture+&ots=WAkv0NcS3v&sig=riexCywSfIDhITuNbfEKFaWWZu4
40. Shayegani A, Joklová V (2023) Investigating privacy principles' formation in vernacular architecture of arid and semi-arid parts of Iran. Archit Pap Fac Archit Des STU 28(1):3–16. https://doi.org/10.2478/alfa-2023-0002
41. Tamaskani Esfehankalateh A, Farrokhzad M, Tamaskani Esfehankalateh F, Soflaei F (2022) Bioclimatic passive design strategies of traditional houses in cold climate regions. Environ Dev Sustain 24(8):10027–10068. https://doi.org/10.1007/s10668-021-01855-6
42. Tappert S, Mehan A, Tuominen P, Varga Z (2024) Citizen participation, digital agency, and urban development. Urban Plann 9. Article 7810. https://doi.org/10.17645/up.7810

Ab-Anbars of Yazd: A Study of Geographical and Historical Distributions of Cisterns in a Desert City

Hossein Soltan Rahmati, Hadi Torkshavand, and Mojgan Amiribolvardi

Abstract In this time of civilizational crisis, architecture needs to learn from the past how to heal and conserve human civilization by having a holistic vision and a deep understanding of design for extreme conditions. One such good historical example can be found in the Ab Anbars (cisterns) of Yazd. Ab-Anbars are simple yet effective structures that formed a flexible and affordable design to maximise the probability of survival in the most hot and arid climate conditions on the planet by providing clean and fresh water, free of charge, to the communities around them. This vernacular engineering design enabled not only survival in extreme conditions, but more than that spread and sustained the Iranian civilization by helping to build resilient cities and settlements in the desert. In this chapter, the Ab Anbars of Yazd are studied in detail to find out more about how they deeply work. This research has used primary and secondary sources and GIS mapping tools to gather and analyse the data and to answer the main questions about Ab Anbars. This study presents for the first time the geographical and historical distribution maps of Ab-Anbars of Yazd, and by superimposition of the maps of Ab-Anbars and Qanats reveals their underlying relations in the irrigation infrastructure of the city throughout its history. This study also underlines the importance of the revitalisation and restoration of Qanats and cisterns as a tested and reliable infrastructure for the supply and storage of water in case of emergency situations or severe droughts. The architecture of the Ab Anbars of Yazd not only reflected the Indigenous engineering knowledge and expertise, but it also mirrored the very values, beliefs, and aesthetics of Iranian civilization. In a broader sense, this civilizational understanding can be applied to other elements of Persian vernacular architecture to inspire a holistic vision for beautiful and functional architectural innovations to heal and preserve the human civilization.

Keywords Ab-Anbars · Yazd · Persian architecture · Cisterns · Water reservoir · History of architecture

---

H. Soltan Rahmati (✉) · M. Amiribolvardi
Shiraz, Iran
e-mail: Rahmati.x@gmail.com

H. Torkshavand
Yazd, Iran

A. Cheshmehzangi and S. Roaf (eds.), *Persian Vernacular Architecture*, Urban Sustainability, https://doi.org/10.1007/978-981-96-1116-4_3

1 Introduction

The climate of the central regions of the Iranian plateau, where the city of Yazd is located is hot and dry. In Yazd, with an elevation of 1216 m, the temperature in summer may reach or exceed 40 °C, also 97% of the total area of Yazd province is covered with desert [4], and the mean annual precipitation is around 73 mm. The only redeeming natural feature is Mount Shir Kouh (4055 m) and its foothills. Adapting to these extreme climatic conditions to keep civilization alive required architectural solutions to the problem of supplying and storing safe drinking water. The response of traditional Iranian architects (or Ostads) was a combination of indigenous knowledge and tested patterns and innovations that took form, for example, in the enduring companionship of Ab-Anbars and windcatchers (Fig. 1). Although the geographic and climatic conditions of the region have been changed through millennia of the evolution of cisterns, the demand for safe drinking water and the engineering problems associated with it have only increased through time due to various factors such as the construction of new cities, town and villages and development of new routes. Before modern technological advances cisterns were the most important element of keeping civilization, and Ab-Anbars were the beating hearts of Yazd. In this chapter, the Ab Anbars of Yazd are studied in detail to explain how they deeply work. A GIS software (ArcGis 10.8.2) was used to gather and analyse the data, to visualise their historical and geographical distributions and their relation with Qanats. This chapter explains how Ab-Anbars were built, what about them still works and what needs to evolve in order to inspire future architectural solutions for extreme conditions that like Ab-Anbars are both beautiful and functional.

2 Etymology

Cisterns have been called by many names in different regions and dialects throughout their long history of evolution in Iran. The most recognisable of these names is Ab-Anbar that consists of Ab (water) and Anbar/Ambar (reservoir) meaning water reservoir. Other famous names include Howz Anbar, Estakhr (pool), Berkeh (pond) and Masna'a or Manba' (Tank, from Arabic). In Yazdi dialect it is called Ombar that is an abbreviated form of the original Ab-Ambar. In this chapter Ab-Anbar and cistern have been used interchangeably.

3 Historical Background

The story begins a few thousand years ago. The oldest known cistern in Iran is the cistern built near the Elamite city of Durantash (40 km south of Susa in Iran) ca. 1250 B.C. This cistern, made by bricks, is an example of the height of Elamite

Fig. 1 Ab-Anbar of Rostam Giv with four Badgirs. *Source* The Author's own

architectural advancements. There were also several water structures like cisterns and water tunnels discovered in Persepolis from the Achaemenids era ca. 518 B.C. that like most of such structures were built in relation to Qanats or underground connected wells. The tradition continued through the Parthian and the Sassanid empires until the Arab conquest of Iran in seventh century A.D. Before Islam, water was a sacred element in Zoroastrian religion; also to pollute water was a sin, and there was a water festival on the 7th day of the 7th month of the year (Aban). An eyewitness has described a stone cistern of the Sassanid era located by the road from Taft to Yazd that had no tap foot, and one had to go down the stairs enough to reach the level of water. This cistern is ruined today [13: 32]. There is little knowledge and evidence about the locations and number of such structures in Iran after the conquest until nearly three hundred years later. The first appearance of cisterns has been mentioned in Ibn-Balkhi's book. He writes about the Howz (little pool) built by Azedoldula Deilami in the citadel of Estakhr, the ancient capital of Pars province. The ruler built a dam on the valley near the Estakhr citadel that was a seasonal watercourse, and put a roof above it supported by 20 columns, and it was also surrounded by other ponds and reservoirs [10: 156].

The city of Yazd is located in central Iran and near the desert of Dashte Kavir that makes the city vulnerable to severe water tensions, sandstorms and hot winds. If it wasn't for the presence of Qanats and Ab-Anbars there could be no civilized settlements in such extreme conditions. The first written accounts of Ab-Anbars in Yazd date back to the tenth century when Estakhri the author of Masalik-ul Mamalik, a geography book, described the 'rain pools' built by the roads to Khorasan [5]. In the Timurid era (ca. early fifteenth century) more attention was paid to cisterns that were supplied by Qanats, and as witnessed by the author of the oldest history of Yazd, there were 200 cisterns active at the time depending on only one Qanat (Ab-e Taft), and 150 other cisterns depending on Ab-e Shahi. This gives the estimation of 350 cisterns in Yazd at the time that not all of them had the form of Ab-Anbars we know today, and some may have been more similar to open reservoirs or little pools. A lot of cisterns were also built in Safavid, Qajar and Pahlavi periods that were more or less similar in structure to their predecessors. In addition to cisterns, building Chah-e Ab-e Sard (Cold water wells) were common in Yazd. Figure 2 shows a distribution map of Ab-Anbars of Yazd.

It should be noted that the number of existing cisterns from fourteenth century is actually zero, but there were five famous Ab-Anbars of this period mentioned in the historical texts of Yazd, and they were also located along with the existing Ab-Anbars in all the maps. The map shows that the highest concentration of Ab-Anbars is located within the limits of the old city and its periphery. Memarian suggests that of total of 700 cisterns in Iran more than half are located in Yazd province [13: 32]. One of the oldest of these remaining structures that has a building date is the Jennok Ab-Anbar built in 1473, and it takes 25 steps to reach the water level. Many of these historical cisterns have inscriptions that include the year of construction, the benefactor's name, and even some have verses of poetry dedicated to the cistern. Figure 3 shows the Ab-Anbars of Yazd categorised based on the period of construction. The map shows that the majority of the existing cisterns within the current limits of the city have

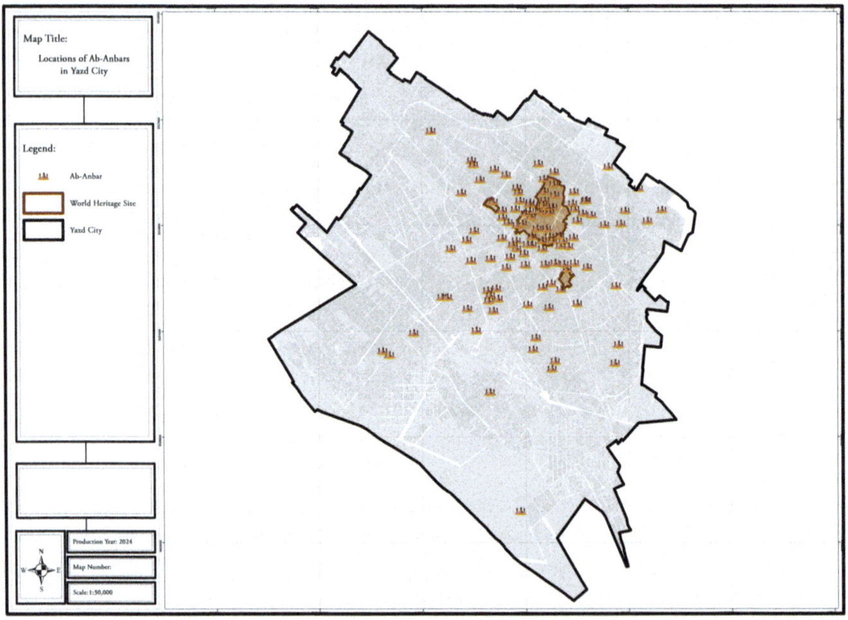

Fig. 2 Geographical distribution of Ab-Anbars of Yazd. *Source* The Author's Own

been built in the twentieth century, but if one looks at the historical world heritage area it can be seen that the majority of the Ab-Anbars are much older and date back to 400–500 years ago. High concentration of Ab-Anbars within the limits of the old city can be interpreted as high population density in this area in a period of 600 years.

By comparing the old cisterns to those built close to recent times, it can be seen that a historical evolution has taken place. The methods of construction evolved as the engineering knowledge advanced and architectural solution to the problems of ventilation, disinfection, cooling and water supply resulted in better functionality and overall performance of these buildings. The peak of this historical evolution can be readily seen in the cisterns built in Qajar and Pahlavi periods (eighteenth–twentieth century). There have been suggestions to keep a number of them for the time of crisis, in case of drought or floods, and to change the rest to accommodate various activities such as sport venues, museums and libraries.

The frequency table (Table 1) shows that most of the existing Ab-Anbars have been built in twentieth century that corresponds to late Qajar and Pahlavi eras. Before this era, and with the exception of the eighteenth century, the number of remaining cisterns is relatively close in other periods.

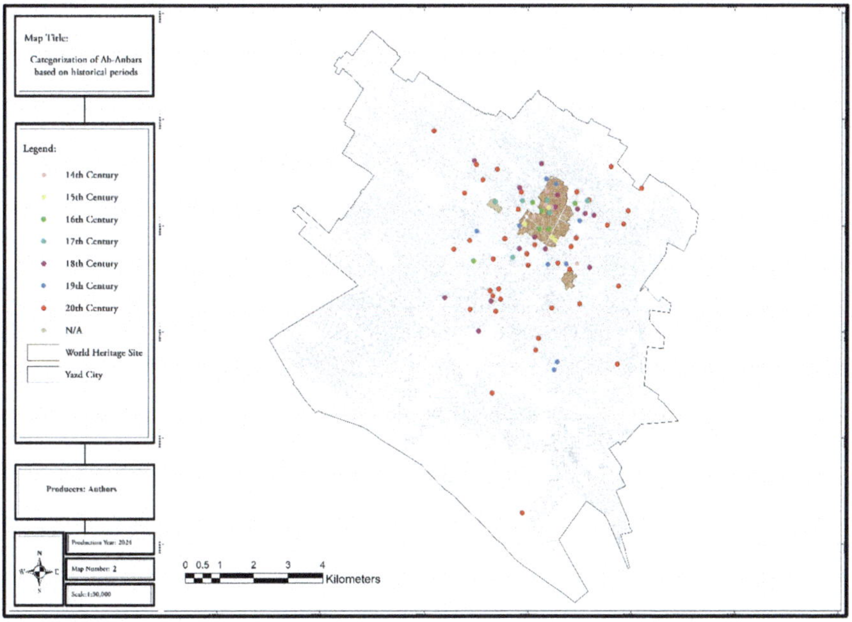

Fig. 3 Historical distribution of Ab-Anbars of Yazd (based on construction dates). *Source* The Author's Own

Table 1 The frequency table of existing Ab-Anbars in Yazd based on the year of construction

Time period	Number of Ab-Anbars
Fifteenth century	5
Sixteenth century	6
Seventeenth century	6
Eighteenth century	15
Nineteenth century	9
Twentieth century	39
N/A	28

4 Typology of Cisterns of Yazd

Cisterns of Yazd can be categorised according to different factors. Ab-Anbars could be public or private, built in the city or in a house or by the roads. They could be supplied by Qanats or rainwater, built with a dome or a flat roof or uncovered. Here, these main characteristics are used to give a better understanding of the cisterns of Yazd and how they operated through centuries.

4.1 Functions

Ab-Anbars can be categorised based on their domain of public usage to public and private cisterns. Public cisterns were available to all, and private ones were designed to service private gardens and households. Although the majority of cisterns of Yazd are roofed and covered structures those built in the villages or in the fields are mostly of the open/uncovered type. Covered cisterns provided access through either a tap foot or a staircase, and open reservoirs provided access either by the staircase, built near the reservoir, or by bucket and hand.

Public. These cisterns form the majority of Ab-Anbars built in Yazd were the main supply of drinking water for all people regardless of their social or economic ranks. These cisterns were built by the kings, the benefactors and the governors, and almost all of such cisterns had dedicated funds allocated by the donners and builders of cisterns. These endowed cisterns make the absolute majority of all cisterns in Yazd, The benefactors even dedicated money for the light inside cisterns. The motives to dedicate such vast amounts of money to public benefit were numerous, but the most important of them all was the good intentions of the benefactors whether Muslim or Zoroastrian to build something that lasts and benefits all. Dedication to public service was such a powerful idea that some cisterns like that of Ali Akbar Rismanian had endowed money dedicated to water bearers to provide water for the elderly who could not use the stairs, and for the big pots of fresh water to be bought for that reason [13: 42]. Money was also reserved to buy salt for disinfection of water. In some cases, the water of the Qanats themselves were also endowed by the owners to the public Ab-Anbars. It can be concluded that in addition to architectural structures needed for the storage of water, a deep sense of dedication and cooperation was the key to survival in desert extreme conditions and survival of civilization. For this reason, in some exceptional cases Ab-Anbars had tall defense towers on their roofs [2]. Another type of public cisterns were the ones built and maintained by the governments in different eras, and the third type of public cisterns were those built and maintained by the residents of neighborhoods themselves. These public Ab-Anbars can be seen in many locations inside and outside the city, along the roads and near Caravanserais.

Private. These cisterns were built commonly inside big gardens like Dowlat-Abad or inside private houses for the sole domestic use of the inhabitants. The sources of water for these rather small cisterns were passing Qanat routes or domestic wells in each house. Hossein Masarrat differentiates between three main types of these private cisterns: (1) Cube/rectangular cube reservoirs with flat or cradle roofs operated by buckets and ropes using man or livestock power (2) Those built right under the residential part of the house that provided access to water through a median space called Pasheer that has the same level as the bottom of the tank and had brass taps, These taps were installed some inches above the bottom level to prevent sludge or other sediments enter the drinking water of the tap. (3) the third form was called Chah-Khaneh (Chah: well and Khaneh: house) and these were built above the floor level and near deep wells. Some of this type of cisterns could provide the needed water of a household for up to a couple of years.

4.2 Applications

Wherever civilization thrived in the central regions of Iran there were cisterns at work for hundreds if not thousands of years. Thus, one can find examples of these marvelous structures all across the region and especially in Yazd which can be safely named the city of cisterns. This widespread use of cisterns made them available in almost all situations.

Cities. In the city itself each neighborhood had at least one Ab-Anbar dedicated to the public. The high number of existing cisterns in Yazd by itself is more than half of all existing cisterns in Iran testifies to this. In villages too if they were small a couple of cisterns could supply enough drinking water for the residents, and if big just like the city each neighborhood could have its own Ab-Anbar. Cisterns had a deciding effect on the formation of urban routes, streets, and alleys and thus on the overall form of the city. The safety of the city depended on these cisterns, and they formed the center of neighborhoods. This was more or less the case for all desert cities of Iran, and after the introduction of tap water, the pattern dissolved. But even in contemporary times in the drought crisis of 2008–2009, Ab-Anbars has proven to be vital to provide some level of water safety, especially for rural areas [13: 13]. Sometimes competition between different neighborhoods was a motive to build ever bigger cisterns, but the core values of doing something for the benefit of the public was still the main motivation and drive to build Ab-Anbars.

Villages. Cisterns built in villages were commonly simpler and smaller than urban cisterns and had fewer details and decorations. On a smaller scale, some village houses and gardens had their own private cisterns that were smaller than the public cisterns for and were called Manba' (from Arabic: source). Compared to urban cisterns, these Ab-Anbars were usually of the open type and access to the water as possible by going down the stairs, and because of their open design, they did not have any Badgirs on them. The common depth for such cisterns was around 3 m, which is rather shallow compared to the city cisterns. It is, however, interesting that some of the most sophisticated Badgir designs can be seen in the village Ab-Anbar of Nasr Abad that has seven Badgirs and two domed reservoir tanks.

Desert. Other types of Ab-Anbars were built in the desert and near farming sites to provide water for agriculture and livestock farming, and some of these even had dedicated spaces for livestock. Like village cisterns, those built in the desert and for farming reasons did not have much depth (Fig. 4). It was common practice to build such cisterns by the intersection of water canals and division hubs administered by the Mirab.

Caravanserais. These buildings had also dedicated cistern inside the building accompanied by a number of cisterns around them. The water was usually supplied by Qanats, and the roofs of most Caravanserais were designed to guide the rainwater into their reservoirs. Supplying water to these buildings was more difficult than urban and rural cisterns because of the sheer distance between the source and the reservoir. In some cases, the distance could be more than 50 km [6: 590, 21: 155]. Their capacity volume could be small (20–30 m^3) or big (84–88 m^3 or more) [20: 244].

En Route. Providing water for the Caravans and passengers demanded for cisterns to be built both near places of rest like Caravanserais and by the roads. Based on anecdotal evidence of the historians and accounts of the travelers the water of these

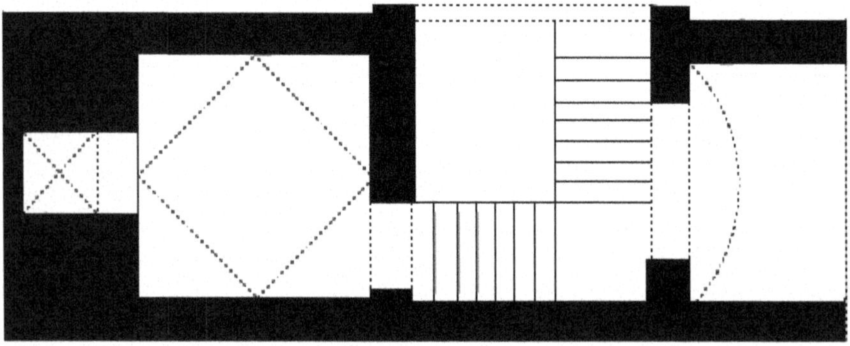

Fig. 4 The plan of a field Ab-Anbar in Mohammad Abad Nodushan. *Source* Redrawn by the authors, adapted from [14: 84]

cisterns could be extremely cold to the point of breaking the teeth [18: 48]. Naser Khosrow writes in his famous travelogue, written in eleventh century, that such reservoirs existed every two Parsangs (~ 14 km) on the road leading to Khor-o Biabanak [12: 168]. Usually, such cisterns had resting places and shelters for the passengers built inside or near them. These cisterns are usually simpler in design and details and also have less capacity compared to other cisterns. Their sources of water were mostly surface water or if accessible nearby Qanats and wells.

From the formal evolution of Ab-Anbars in different locations and comparing the complexity of design it can be concluded that the desert/field and en route cisterns represent the primal or paternal forms that have later evolved into village and later city cisterns. Their flexible pattern language enabled a variety of designs to suit local conditions. The evolved patterns, like the resorts, resting places, Chahar-Taghi (similar to tetrapylons with four barrel vaults and a dome), Pasheer space and Badgirs in turn inspired more complex designs for desert and en route cisterns. Therefore, from the smallest scale of a household to the largest scale of urban areas and by the road drinking water was available to people and livestock alike. Cisterns were also built in citadels and in mountains, but almost all remaining Ab-Anbars of Yazd are either located inside the city, nearby villages or near the old routes.

4.3 Sources of Water

In the desert each and every drop of water is precious, and thus most Ab-Anbars had more than one way of supplying their needed water. In raining seasons the main sources were rainwater or seasonal rivers, and in dry seasons the water was supplied by springs, Qanats or wells. Cisterns that made use of seasonal rivers usually had a bigger capacity than the other types [13: 59]. Among these sources, the springs of the foothills and in mountainous areas were the freshest and cleanest source of all. Rainwater was guided to the cistern's tank either by proper sloping in the proximity of the cistern or by guiding the rainwater from the roofs first to small ponds and then to the tank. This was one of the reasons that Ab-Anbars were built underground. Most of the cisterns that were supplied by rainwater were located outside the city and near the main roads. As an example of this kind of cisterns one can name Ab-Anbar in Kalar Meybod near Yazd (Fig. 5).

Among the sources of water, the main supplier of Ab-Anbars, as mentioned before, were Qanats (Fig. 10). The oldest Qanats in Iran date back to 2720 B.C. [7: 433]. These connected wells were built by the highest engineering sensitivity for thousands of years and without them Ab-Anbars could not provide enough water especially in Summer when other sources were scarce. The most famous of these Qanats were Ab-e Taft va Nasiri that supported over 200 cisterns [11: 178]. Another important Qanat was Zarch that is built over a thousand years ago and is nearly 80 km long and is said to be the longest Qanat in Iran [4: 27]. Ab-e Vaghf Abad was another important source of water. The water could be guided to the reservoir using canals either above or underground, or by use of Payab to tap into the underground canal of Qanats [6: 14].

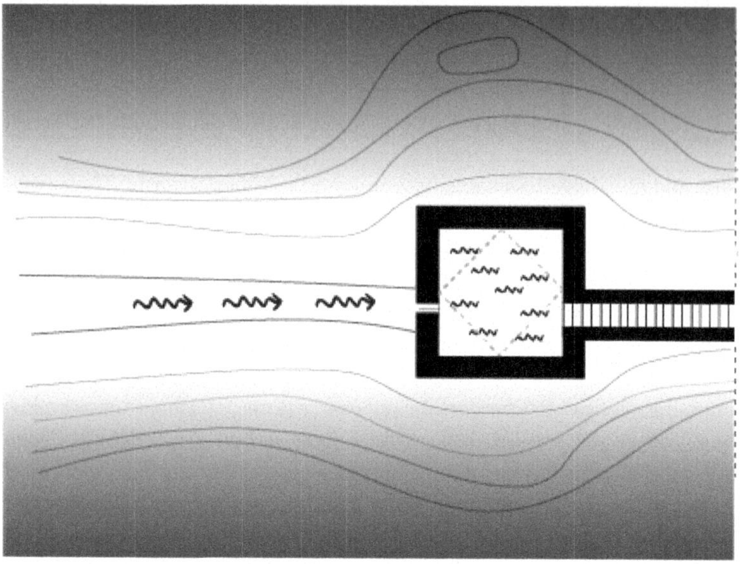

Fig. 5 Ab-Anbars were built according to the slope and direction of surface water. *Source* The Author's own

Wells were another source of water for the cisterns and were used both publicly and privately to fill the Ab-Anbars. These wells were called Chah-e Ab-e Sard (cold water well), and were in use until the recent times were uncontrolled welling and droughts rendered them dry and obsolete. Some cisterns were dependent on nearby wells, and used waterwheels and cows to draw water to the surface. Another recent source of water was tap-water provided by the benefactors for the poor neighborhoods of the city that could not afford subscriptions. Ab-Anbars that had no nearby sources of water were filled 'by hand' using portable tanks. Cisterns could have more than one way of water source, but the most reliable of all these sources, as mentioned before, were Qanats. The existing Qanat water network in the limits of city is 392.95 km long and this number can increase by 45 kms of probable Qanat roots that have been cited in historical texts but not yet exactly identified (Fig. 11).

5 Water Quality of Ab-Anbars

While the majority of anecdotal evidence and historical accounts praise the quality and coldness of water in Ab-Anbars, scientific studies have found a wide range of bacterial contaminations. There were a number of methods for filtering, procuring, and treating water in Ab-Anbars to keep the water fresh and safe to drink. During the physical filtration, any small floating particles would settle at the bottom of the Ab-Anbar once the water became calm. This settling process was accelerated by the addition of clay and lime to the water. The reservoir was dark, in case it was roofed so that water-plants and algae cannot easily grow. Roofed cisterns were also kept the water safe from environmental pollutions, and their water had better quality compared to the uncovered ones [9]. The chemical filtration involves the addition of salt to the water. The salt decomposed, releasing chloride that effectively disinfected the water, and prevented it from putrefaction. The water was kept cool and oxygenised by use of Badgirs, locating the tank underground and building thick walls for proper insulation, and the cisterns were usually filled in winter time with Yakh-Ab (ice-water). The use of fish for getting rid of small water creatures was also common. In recent times, chloride has also been utilised for water disinfection in Ab-Anbars. It was also crucial to clean the reservoir tank of any sledge or other residues before filling it at least once a year or otherwise the Ab-Anbar was due to ruin. Ventilation by Badgirs, Khishkhans, or openings in the ceiling helped to keep the water fresh, oxygenised, and cool. When studying the traditional and vernacular technologies there is always a danger that facts and myth become mixed in the evaluation, and that is what Susan Roaf has conceptualised as the 'traditional technology trap' [17]. Thus, in order to have a realistic understanding of the functionality of Ab-Anbars one should consider the facts regarding scientific testing of the quality of water. According to these studies a vast number of bacteria and even fecal matter were observed in the water of cisterns, and in some cases these pollutions led to widespread pandemics in the past before water-pipes were introduced [4: 96]. Sources of pollution were numerous. Naturally, the surface water that was directed to the reservoir was contaminated with improper or even dangerous materials and pollutants like sewage or animals' excrement. Small animals or birds could sometimes enter the reservoir or die there and pollute the water too. It can be concluded that despite all of the techniques and concerns, the quality of water in cisterns, compared to the contemporary safety standards, is not acceptable in the majority of cases. The quality of water of Ab-Anbars, however, was not all the same and it depended on the source of water, the type of structure and the level of care and hygiene of the people who built and used them.

6 Operation and Maintenance

Ab-Anbars needed constant attention and maintenance to work properly, and some of these structures were used constantly for hundreds of years. It was customary to begin dredging and mud removal at the end of each summer before the raining season arrives. After that, the tank was cleaned and disinfected once again before being filled with fresh water. The drain valve was sealed using special types of wood greased by sheep fat and someone called Ab-Yaar was responsible to clean the canal and also to watch the entire water stream from beginning to the end not allowing anyone to wash anything in the water route, and thus he made sure that the course was clean before any water could enter the tank. For this reason, it was considered good practice to fill the reservoir in nights or near dawn when people were asleep and there was not much action around water. Operating at night also had the advantage to let the water cool down and together with structural properties of Ab-Anbars (i.e. proper ventilation) the water was ice cold even in hot summers of the desert. The canal was once filled with water to wash it clean, and only then it was proper to fill the tank. There was a basket before the entrance to filter out the floating bits of leaves or straws on the surface of water. After the reservoir was filled the valves were sealed by bricks, mud and lime to prevent water or animals from getting into the reservoir. It could take between one to four days to fill a reservoir depending on its size. The amount of water that each Ab-Anbar could take was decided by the donors of Qanat waters. Ab-Yaar and other service providers were also paid from the donations dedicated to the cistern or by the residents of the neighborhood themselves [13: 51]. Building and maintaining Ab-Anbars needed a deep sense of duty and social cooperation or otherwise the city could not survive and thrive in the middle of Kavir.

7 Parts of Ab-Anbars

Ab-Anbar has a simple yet effective and functional structure that has been little changed through time. It consists of a reservoir or tank, Badgirs and ventilators, staircase, tap-foot and entrance. Other patterns or dependent spaces could be added to this core system like resting places, platforms or Sang-abs.

Main Reservoir. The tank was built anywhere between 5 and 18 m underground and usually by bricks. Majority of Ab-Anbars have only one main reservoir, but there are other types built in Yazd with two reservoirs (Gonbade-Sabz and Khajeh Ab-Anbars). This main reservoir could either be cylindrical in form or built like a chamber with columns. Cylindrical reservoirs were relatively cheaper to build. Chamber reservoirs, built often in the form of octagons or semi-octagons, were supported by thick columns and were used for Caravanserais, citadels or in the cities, and could contain much larger volumes of water (over 1000 m^3) [13: 64]. The average capacity of city Ab-Anbars was around 400 m^3. Table 2 shows the estimated dimensions and sizes of a number of Ab-Anbars in Yazd sorted by their maximum capacity. All of these reservoirs were made in cylindrical form. All the top five cisterns have close diameter to depth ratios, and except the Khajeh Ab-Anbar on top the list, the others have been built in twentieth century. This close diameter/depth ratio can also be seen in the bottom of the list in smaller cisterns.

Ventilation. To keep water fresh, oxygenised and cold, and also to prevent it from stagnation almost all covered cisterns have openings to allow proper ventilation. This can be achieved by either having openings in the roof or the supporting walls or by building windcatchers (Badgirs) or Khishkhans to catch even the faintest breezes. Figure 6 shows a diagram of airflow in cisterns with Badgirs. The form and type of Badgirs on cisterns depend on the form and type of the cisterns that they serve. The number of Badgirs can range from one up to seven (mostly two or four) that can have one, two, four, six or eight-directional openings. A modest cistern may not have any Badgirs at all or have a couple of adobe or brick one-directional windcatchers.

Table 2 Various physical characteristics of some Ab-Anbars of Yazd

Ab-Anbar	Estimated capacity (m^3)	Diameter (m)	Depth (m)	Date of construction
Khajeh	4578	18	18	1588
Rismanian	3617	16	18	1921
Rostam Giv	2307	14	15	1944
Afshar	2154	14	14	1955
Ghandehari	2000	14	13	1953
Golshan	2000	N/A	N/A	1958
Shesh Badgiri	1424	12	12.6	1862
Amir Chaghmagh	858	12.5	7	1441
Vazir	731	9	11.5	1717
Shah Abolghasem	662	7.5	15	1671
Tekieh Amir Chaghmagh	600	9	9	1456
Bagh-e Dowlat Abad	502	8	10	1747
Jennok	320	7	8	1473

The number of Badgirs depended most of all on the construction budget, and their arrangement around the reservoir tank were almost always symmetrical or polygonal (Fig. 7).

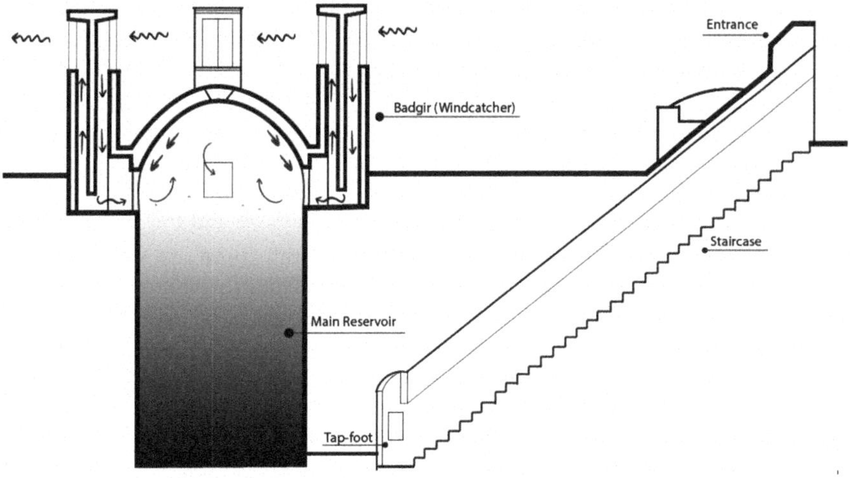

Fig. 6 Airflow and function of Badgirs in cisterns. *Source* The Author's Own

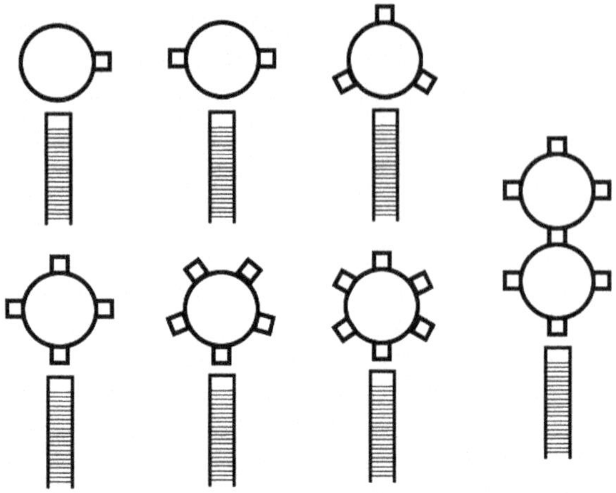

Fig. 7 Arrangement of Badgirs around the reservoir. *Source* Redrawn by the authors, adapted from [14: 56]

Staircase. The main way to access the water in cisterns without tap foots was the staircase, and it was also used for cleansing and maintenance of the cistern. The number of stairs depended on the depth of the reservoir, and was usually between 25 and 70 steps, with 25–30 cm of height each. In open cisterns access to the water was only possible through the stairs one should go down enough to reach the water level. Figure 8 shows the arrangement of the staircase concerning the reservoir tank. If there was enough space it was preferrable to arrange the staircase and the reservoir along the same axis. Other options included arranging the staircase parallel or perpendicular to the main axis or a combination of these arrangements.

Tap Foot. This space provided access to water through big brass, or sometimes lead, taps. Its form was either half-octagonal or square-shaped (Fig. 10). It had sitting platforms on either side and a drainage hole and sometimes a ventilation opening in the roof [14: 50]. The tap was installed slightly above the bottom of the reservoir to prevent sludge and mud getting into the output water. Cisterns with tap feet had the advantage of being cleaner than those cisterns that allowed direct access to water through steps inside the tank. Some Ab-Anbars had two taps and tap-foots one for Muslims and one for Zoroastrians.

Entrance. The entrance was designed according to the size and the stature of the cistern, the location of the cistern (i.e. if it was built near Bazars or mosques), the direction of local winds, and spatial relation with their adjacent buildings and alleys. It was also subjected to the masterful observance of sun and moon paths so that there could be enough light in the staircase. The entrance could be grand or otherwise modest and low profile. Big Ab-Anbars or those built by royalty or wealthy benefactors had inscription stones above the entrance sometimes decorated with poems or verses from the Quran (Fig. 9). Depending on the site and the neighborhood a cistern could have more than one entrance that would provide access from Bazars, schools, mosques or because of the density of local population that one single entrance would

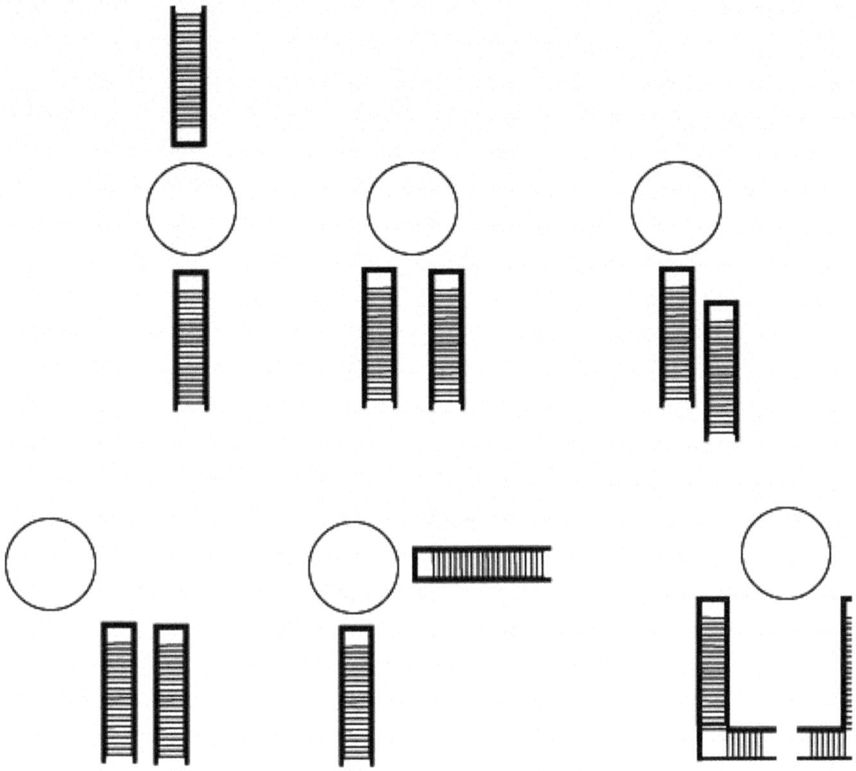

Fig. 8 Location of the staircase in relation to main reservoir. *Source* Redrawn by the authors, adapted from [14: 46]

not suffice. Khajeh Ab-Anbar is an example of this type of cistern that has three entrances.

Ab-Anbars had also dependent and related spaces like Sagha Khaneh, a place for water bearers (Saghas) to bring water to the elderly and the passersby, and/or resting places cooled with the refreshing breeze coming from the reservoir. Some cisterns had Sang-abs that were huge cups carved out of granite for this reason, and if the passenger was lucky or neighborhood was wealthy enough the water in Sang-abs could contain a touch of rosewater for an extra refreshing experience.

8 The Materials and the Construction Process

The history of evolution of cisterns show that the use of materials and the engineering solution evolved through time so that each generation of Ab-Anbars had better functionality compared to its predecessors. Architectural patterns of the cisterns also

Fig. 9 Entrance of Jennok Ab-Anbar, built in the fifteenth century, one of the oldest Ab-Anbars in Yazd. *Source* The Author's Own

evolved and resulted in the construction of some masterpieces of the vernacular and traditional architecture of Iran. Their material was always locally available, and the style of construction reflected the local architectural patterns and traditions.

Materials. The oldest remaining remnants of cisterns in Iran show that they were built with bricks. The brick remained the material of choice for millennia, and it is not surprising that the majority of the cisterns built in and around Yazd were built with bricks. In Yazd, a special type of brick (22 * 22 * 6–7 cm) hardened by extreme heat and pressure at the bottom of the brick kiln was used to build the reservoir tanks [14: 27]. The use of mud-brick was also common in building Ab-Anbars (like in Abu-Alma'ali cistern) because it was the most used material in building houses and other private or public structures. There are cisterns in Yazd that even their domes and Badgirs have been built with adobe and mud-brick. These Ab-Anbars may or may not have a protective layer of bricks or even thatch on the roof. Stone was also the material of choice in foothills and near mountains. As mentioned before, one of oldest Ab-Anbars of the Sassanid era near Yazd was made of stone, but unfortunately it has been ruined, and it further studies on its structure is not possible. Other materials also needed for construction included: brass, lead, wool, egg whites (an expensive

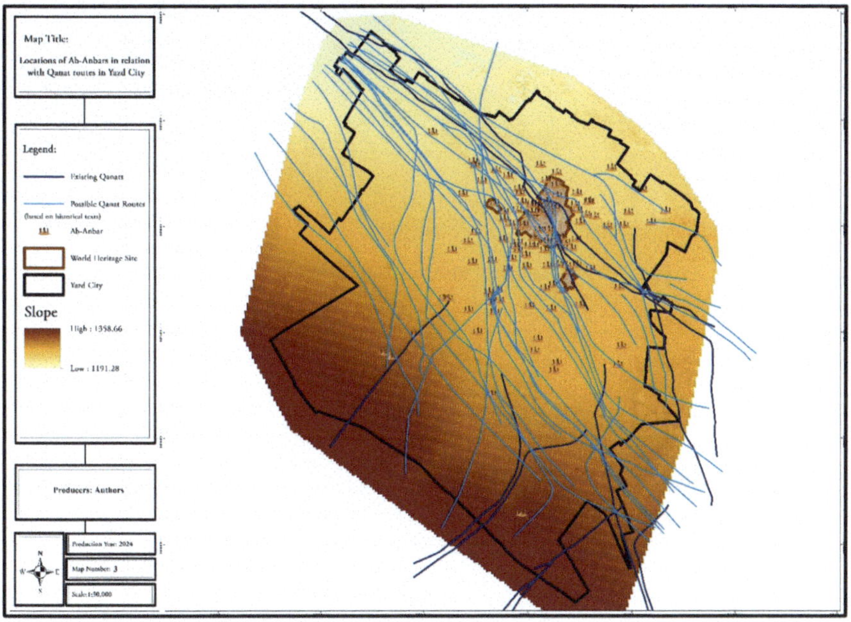

Fig. 10 Locations of Ab-Anbars in relation to Qanats and the slope of the region. *Source* The Author's Own

but effective additive to Saruj), ash, lime, carcass stone, cement, gravels, sand, salt, chalk, tar, tiles, mud, Shuruneh wood (a type of termite-resistant local wood for Badgirs) and wax.

Construction Process. There were three main factors in constructing cisterns namely, proper location in relation to urban/rural centers, status of the building site and site analysis including the slope and proximity to water sources, and finally, the economic situation [13: 66]. Proper location to build an Ab-Anbar in a city or village was the most important factor to be considered before any further decisions. In urban and rural areas, it was both logical and functional to build them as near as possible to Bazars, educational centers, and mosques so that the maximum number of people could use them.

Such cisterns usually had bigger capacity compared to other urban types because of their proximity to high-population neighborhoods and areas [1: 32]. The capacity of the reservoir was decided based on the population of the neighborhood whether it was built in a city or a village. In villages too, there could be located near the mosque or near the farming fields and livestock farming centers. Ab-Anbars were also located and built near the roads and in certain intervals (usually every 20–50 km) or close to Caravanserais or other resting places of the caravans and passengers. It took more than of an ordinary architect or Ostad to build an Ab-Anbar because he should have

Fig. 11 Brickwork of the entrance of Shesh Badgiri Ab-Anbar. *Source* The Author's Own

special knowledge of soil science and land thrust. Ali-Akbar Baghbanzadeh, Ali-Akbar Khorrami, and Mohammad Ali Sadeghies were among the last generation of great Ab-Anbar designers and builders.

The next matter to be considered was the method of supplying water to the reservoir. In most cases, urban cisterns were built near or exactly above the route of Qanats. Figure 10 shows the superimposed map of Qanat routes and the Ab-Anbars built above them. It is evident from the map that the absolute majority of the Ab-Anbars were built exactly above or very close to the Qanat routes. The highest concentration of cisterns can be seen where different streams of Qanats meet each other in the middle of the old city. Another thing to be considered was the slope of the region and of the site. It was both functional and logical to build cisterns in low-slope areas below the level of surface water where the surface water and other sources of supply could easily be directed into the reservoir. Traditional architects and builders had tacit knowledge of soil mechanics and considered clay soil the best choice [13: 68].

They believed that the soil must be hard enough to be impenetrable enough and to bear the weight of the structure and the water it held, but not too hard to make the excavation more difficult than it should be. The situation, however, sometimes demanded that despite the fact a cistern be built on rocky beds, and then gunpowder was used to prepare the ground for construction [9: 307]. Other considerations included avoiding salty soils to prevent the salt alter the taste and quality of water and also avoiding chalky grounds because of their negative impact on the structure by penetrating and applying surplus pressure to the supporting walls. They also avoided dirty places or places near the graveyards and used to get rid of any trees near the construction site because they could penetrate the structure [15: 81–82]. Finally, it is evident that the economic status of the benefactors or the neighborhood was a crucial factor in the scale of the building, its water allowance, and architectural features and details.

Construction Process. Building a relatively large cistern could take up to 2–3 years of hard work. The construction process had two main steps: digging excavation and building the framework or brickwork [14: 28]. They could either excavate the whole space needed for the required functions of the cistern, and then begin to build the structure, or firstly secure the walls of the reservoir tank and then begin digging. In the latter method, they first built the walls of the reservoir by excavating the area between two concentric circles drawn on the ground and filled it with Shefteh (a kind of mortar made of lime, soil, sand, and water) up to the ground level, and this made an impenetrable layer for the walls of the reservoir. Then, they let it sag for a couple of weeks, and only then they began to excavate the area inside the first circle that would eventually become the main reservoir [13: 71]. In some cases, the reservoir had a semi-conical shape that was wider at the bottom for better resistance against hydraulic pressure.

The time required for this phase depended directly on the condition of the soil and the area and depth of excavation, and it could even take a year to complete the process. After the excavation phase was completed, the workers began to dig the waste water wells either in the center or on the periphery of the main reservoir. These wells that were usually connected to the canals of the dry or abandoned Qanats and were used for discharge of polluted water were called Chah-e Harz-Ab (well of waste water). The next phase would be to build the reservoir tank itself, and it could be done in both cylindrical/conical and columned chamber forms. In cylindrical reservoirs of Yazd, the diameter could range from 5 to 18 m, and could contain between 20 and 3000 m^3 of water or even more for the big reservoirs. Chamber reservoirs had much more capacity than the cylindrical ones (in some instances over 100,000 m^3). The most common material to build the reservoir itself was brick. Tested and trusted through millennia, the architects used bricks in building the walls in wavy patterns for increased strength and resistance. In stone cisterns, carcass stones were used to build the walls. The thickness of the walls of a reservoir could be from one to two meters such that nearly half the thickness was brickwork and half was filled with sand and Shefteh for proper insulation [13: 78]. Then, they covered the walls with a plaster made of lime and soft ash that gave the walls an impenetrable yet smooth and polished finish.

Floor construction and applying a sealant layer (from Shefteh or lead) was the next step; and this layer, depending on the diameter of the reservoir, could be one or two meters thick or even more. In case, sealing was done by pouring melted lead to reach a thickness of 10–15 cm. Peernia has suggested that using lead was also important to protect the floor of the tank from deformation [16: 215].

After the construction of the tank and its floor, they proceeded to build the roof of the reservoir that was either flat or domed. Arch-shaped roofs and domes were practical for the reservoirs with diameters of up to 15 m, but nonetheless it was an extremely dangerous and difficult task because of the difficulty in scaffolding and the danger of falling off to the bottom of the tank. The height of the dome was usually close to or equal to the diameter of the reservoir. Domes were a practical choice because of the easy access to local construction materials, that was either brick or adobe, and the form also provided a shadow to keep the water cool like the function of the domes above Yakhchals or ice houses, and last but not the least its aesthetical and architectural beauty especially in case of double-layered domes (like Amir Chaghmagh cistern). Flat roofs were used when other types of buildings like caravanserais or mosques were planned to be built above the cistern, and in any case, the roof was covered and sealed with mortar. Badgirs could be built and finished simultaneously with the roof, or they may be added later in reconstruction. The last step was to complete the architectural details of the entrance.

The process of construction of cisterns varied from one Ostad to the other, but the general principles mentioned here seldom changed. Like other types of vernacular architecture, innovation in construction took many years of test and practice to be accepted and applied in large scales, and thus only the fittest of such patterns evolved and survived.

9 Beauty Matters

Ab-Anbars of Yazd were both beautiful and functional. Depending on the allocated budget for construction, the social status of the benefactor, and proximity to important buildings like mosques and Bazars the cisterns could have different scales of detailed brickwork design and patterns. Figure 11 shows some of the patterns used in the Ab-Anbars of Khajeh and Golshan in Yazd. Some Ab-Anbars have beautiful and carefully carved inscription stones that could contain the name of the benefactor, the year of construction, and even verses of the Quran or poems written exclusively to describe the motive behind the work or simply to adore the beauty and vitality of water or the Ab-Anbar itself.

The use of stuccos was rare, but the use of Karbandi and decorative archs, Muqarnas, decorative columns, and tile work were common in many examples of Ab-Anbars [14: 58]. Ab-Anbars are the best examples of a holistic approach to solving architectural problems in extreme conditions that result in achieving both efficient and aesthetically pleasing structures that are also solid and reliable, and thus, satisfying the Vitruvian principles of Firmitas, Utilitas and Venustas.

10 Back to the Future

The story of the Ab-Anbars of Yazd is not over as some cisterns in rural areas are still in use by the local population, especially farmers. But this is far from what they are capable of providing. The question of the future of cisterns is a multifaceted one and needs to be addressed according to several socio-economic factors in urban and rural strategic plannings. There needs to be a consensus among the policymakers and the scientific community that the importance of these ingenious structures of Iranian architecture, and the infrastructure network that they form with Qanats, go beyond the mere historical value or their tourist attraction capacity. The most important motivation for shaping such a consensus is the problem of scarcity of water that will only get worse in the future due to increase in population and demand for fresh water. Meanwhile, precipitation in Iran is likely to decrease by 15.6% in the coming decades, and this translates to more water tensions and social disputes in the future [8]. Qanats are already categorised as endangered heritage [3] that means there is a danger that this ancient hydraulic technology may lose what has remained of its importance as a water supply solution. Qanats and cisterns need regular maintenance and without proper attention they cease to function as a coherent infrastructure. Therefore, restoration of this ancient water infrastructure can and should be considered as a vital part of future plannings. Qanats can be restored and revitalised, e.g. as the providers of grey water, for the times of need and emergency situations and in case of wars, any grid failures or severe droughts. If restored properly, Qanats and cisterns can provide fresh and safe water with the help of auxiliary modern technologies for proper sanitation and hygiene.

The traditional technology of Qanats and cisterns can also be seen not only as a local solution to water crises in Yazd, but also as a gift from Iranian civilization and architecture to the world—a world that more than ever needs reliable and tested solutions for water supply and storage in the future. There is, however, a danger in imposing the fashionable ideas or models (e.g. embedded energy or carbon footprint) as the ultimate criteria to evaluate these traditional technologies internationally. The danger is two-fold. Firstly, it neglects the strategic importance and the performance criteria of these traditional hydraulic technologies; and secondly, it will repeat the same mistake at the dawn of modernism in imposing an international style and frame of thought everywhere regardless of the climatic and cultural differences in favor of pushing some ideology or quasi-moral agenda [19]. Therefore, we need proper criteria derived from holistic models to take into account the importance of the aforementioned factors. The story of cisterns teaches us what really needs our rigorous attention and focus in order to survive and pass the torch of civilization to future generations.

11 A Reflective Summary

Ab-Anbars (translated to 'water reservoirs' or 'cisterns') were the beating hearts of Yazd for hundreds of years. One could see these covered or open structures in cities, villages, near caravanserais, and by the roads. Their main sources of water were rainwater/surface water, Qanats, and springs with Qanats being the most reliable and accessible of them all. Their simple yet effective combination of vernacular and traditional architectural patterns evolved as a result of an adaptive process that kept civilization alive in the extreme conditions of the desert. These marvelous structures were built mainly by the kings or the benefactors who dedicated a portion of their wealth to the well-being of the public.

The first Ab-Anbars consisted of at least a covered or open reservoir tank, a staircase, and an entrance. Later Ostads (traditional architects) combined other patterns to the flexible architecture of cisterns such as Badgirs for better ventilation and tap-foot for a more hygienic and convenient way of accessing the water. This flexibility in design led to a variety of cisterns that were adapted to the special needs of each region, and made them a necessary part of every neighborhood or urban complex in Yazd. A physical and chemical solution to the problems of filtration and disinfection of water was also developed to provide not only enough water but to make sure that it was safe to consume. Despite these efforts, modern scientific tests of the quality of the water of Ab-Anbars showed significant amounts of bacterial contaminations that make the water improper or even dangerous to use.

Ab-Anbars are still vital in providing water to poor and underdeveloped desert communities, and this requires working towards better designs and more effective sanitation and disinfection solutions. Proper strategic planning is needed to restore this infrastructure for a future that is likely to witness less precipitation and more droughts and social disputes over water. Ab-Anbars of Yazd represent one of the best educational cases for architects to study how beauty and functionality can be achieved in design for extreme conditions. This is how Ab-Anbars of Yazd learnt to evolve and serve human civilization for hundreds of years.

References

1. Abedini M, Saeedi M, Alemzadeh H (1995) Aub-anbar, Islamic great encyclopedia. In: Mousavi Bojnourdi MK (ed), vol 1, 2nd edn. The Center for the Islamic Great Encyclopedia, Tehran
2. Afshar E (1975) Mementos of Yazd, vols 1 and 2. National Works Association, Tehran
3. Branduini P, Zaina F, Zavvari F, Nabati Mazloumi Y (2021) Qanats as an endangered traditional hydraulic heritage. An integrated methodology for documenting, restoring and reusing an ancient Iranian Qanat. In: Proceedings of the joint international event 9th ARQUEOLÓGICA 2.0 & 3rd GEORES, Valencia, Spain, 26–28 Apr 2021. Editorial Universitat Politécnica de Valéncia, pp 142–158
4. Dehghani-Sanij A, Sayigh A (2022) Cisterns: sustainable development, architecture and energy. River Publishers
5. Estakhri AE (1961) Masalek and Mamalek. In: Afshar E (ed) Bongah-e Tarjomeh and Nashr-e Ketab, Tehran

6. Eyvazian S (1995) Trend of the formation of out of town Caravanserais. In: Collection of articles in the conference on the history of architecture and urbanizing in Iran, vol 1. The Organization of Cultural Heritage of Iran, Tehran
7. Farshad M (1997) History of engineering in Iran. Balkh, Tehran
8. Feizi M, Janatabadi NH, Torshizi AS (2019) Rainfall and social disputes in Iran. Water Policy 21(4):880–893
9. Ghobadian V (2008) Climatic analysis of the traditional Iranian building. Tehran University, Tehran
10. Ibn-e Balkhi (1984) Farsnameh. In: Le Strange G, Nicholson RA (eds) Donyai-e Ketab, Tehran
11. Jafari J (2005) The History of Yazd. In: Iraj Afshar (ed), 3rd edn. Elmi va Farhangi Publishers, Tehran
12. Khosrow N (1976) Travel account of Naser Khosrow, attempted by Dabir Siaghi M. National Works Association, Tehran
13. Masarrat H (2010) Research on cisterns of Yazd City. Andishmandan Yazd, Yazd
14. Memarian GH (1993) A Survey of the architecture of Aub-anbars in Yazd. Iran University of Science and Technology, Tehran, Iran
15. Nour Bakhsh H (2002) Water ponds in Persian Gulf. Farhang-e Mardam J 1(2):78–91
16. Peernia MK (1993) An introduction to Iranian Islamic architecture. In: Memarian GH (ed) Iran University of Science and Technology, Tehran
17. Roaf S (2008) The traditional technology trap (2): more lessons from the windcatchers of Yazd. In: Proceedings of the conference on passive and low energy architecture, pp 19–21
18. Safa al-Saltaneh MA (1987) A report of Kavir, travel account of Safa al-Saltaneh Naeinin. In: Galin M (ed) Ettelaat, Tehran
19. Scruton R (2017) The problem of architecture. In: Aesthetics. Routledge, pp 137–141
20. Siroux M (1949) Caravansérails d'Iran et petites constructions routiers. MIFAO, Cairo
21. Varjavand P (1987) Aub-anbar, in Tashayo-e Encyclopedia. In: Javadi A et al (eds), vol 1. Islamic Institution of Taher, Tehran

Flexible Configuration: Selective Connect-Ability in Traditional Iranian Houses

Hassan Estaji

Abstract This chapter explores the concept of flexible configuration of spaces in traditional Iranian houses, considering the challenges posed by changing societies, lifestyles and living conditions. Societies as well as living conditions are subject to continuous change. Societies have to adapt to their environment, and consequently, the living spaces must be changed following new demands. An ideal house should be able to meet the needs of its residents or near-future users throughout their lives. Any changes in the house users, their needs and the physical and cultural environment require a flexible and changeable spatial configuration to be simple enough to be usable and used. The common construction technology in traditional Iranian architecture was the use of load-bearing thick clay walls. In this construction method, it is almost impossible to change the position of the rooms. The use of multiple doors as connections between rooms allows for the alteration of spatial relationships and the creation of diverse configurations. The research utilises space syntax theory and methods to analyse the spatial configuration of traditional houses in Sabzevar, located in Northeast Iran. Space syntax tools are employed to quantify spatial patterns and elucidate the relationships between spaces and house users. Calculations considering potential openings confirm the possibility of creating various spatial configurations in traditional Iranian houses. The presence of numerous openings and potential openings within rooms results in the formation of multiple loops within the system. The number of loops directly correlates with the distributedness of the system. This high degree of flexibility enhances the house's adaptability to future changes and enables it to respond effectively to different scenarios throughout its lifecycle. The findings of this study contribute to a deeper understanding of the design principles and spatial dynamics inherent in traditional Iranian houses. The knowledge gained from this research can inform future architectural practices, facilitating the development of more adaptable and responsive housing designs and helping preserve historical and traditional houses by changing them based on contemporary needs.

Keywords Flexible architecture · Spatial configuration · Connectability · Traditional Iranian houses · Space syntax

---

H. Estaji (✉)
Department of Architecture and Urbanism, Hakim Sabzevari University, Sabzevar, Iran
e-mail: estaji@hsu.ac.ir

A. Cheshmehzangi and S. Roaf (eds.), *Persian Vernacular Architecture*, Urban Sustainability, https://doi.org/10.1007/978-981-96-1116-4_4

1 Introduction

The house as a place for living must accommodate all human growth phases, any changes in the house users and their needs affect the space requirements, but we cannot predict and control these processes. For example, family size and family structure change over time without any fixed patterns. Even if a house cannot necessarily meet all the users' needs at the same time, it must be flexible enough for any changes. The future is unknown; architects and researchers cannot say definitively what kind of demands a building will face in the future. The only way to assess the flexibility of a system is to measure the total number of possible scenarios that may occur in the future.

> Buildings and cities exist for us in two ways: as the physical forms that we build and see, and as the spaces that we use and move through. Bill Hillier [7, p. 2]

The first approach deals with physical form, while the second one is concerned with the relationship between spatial layout and users. The space syntax theory is applied to study the spatial configuration of houses and the relation between spaces and users in detail for each house. This (case) study focuses on traditional houses in Sabzevar (a city in a hot and arid region of Iran) that are registered on the National Iranian Heritage list. Case study research is essentially a mixed-method strategy because here the researcher investigates the cases from different points of view by using various methods. Since the house is a place for human activities; it cannot be viewed independently from its context and users. This research in parallel with qualitative methods to study the human behaviour in the houses uses the Space Syntax theory. The set of Space Syntax tools quantifies the spatial patterns of houses and clarifies the relationships between spaces and the house users.

The time frame of this study is limited to the traditional Iranian architectural styles, before the emergence of Modernism in Iran and, more specifically, in Sabzevar. Since Sabzevar is a small historical city, far away from the capital, the transition from traditional to modern architecture did not take place until the First Pahlavi period (1925–41). It means that the Qajar houses in Sabzevar belong to the last period of traditional Iranian architecture in this region. The fourteen selected houses were built before modern construction materials such as steel columns, steel beams and large glass panes were introduced into Iranian residential architecture. The majority of the selected houses belong to the Qajar and first Pahlavi periods, one was built in the Timurid era, and one in the Safavid era (Fig. 1).

2 Flexibility and Adaptability

We can change the furniture layout of a living room to host a party or make some major changes to divide a flat into two parts for renting. The flexibility and adaptability in architecture follow a fuzzy logic like a tonality between black and white. These

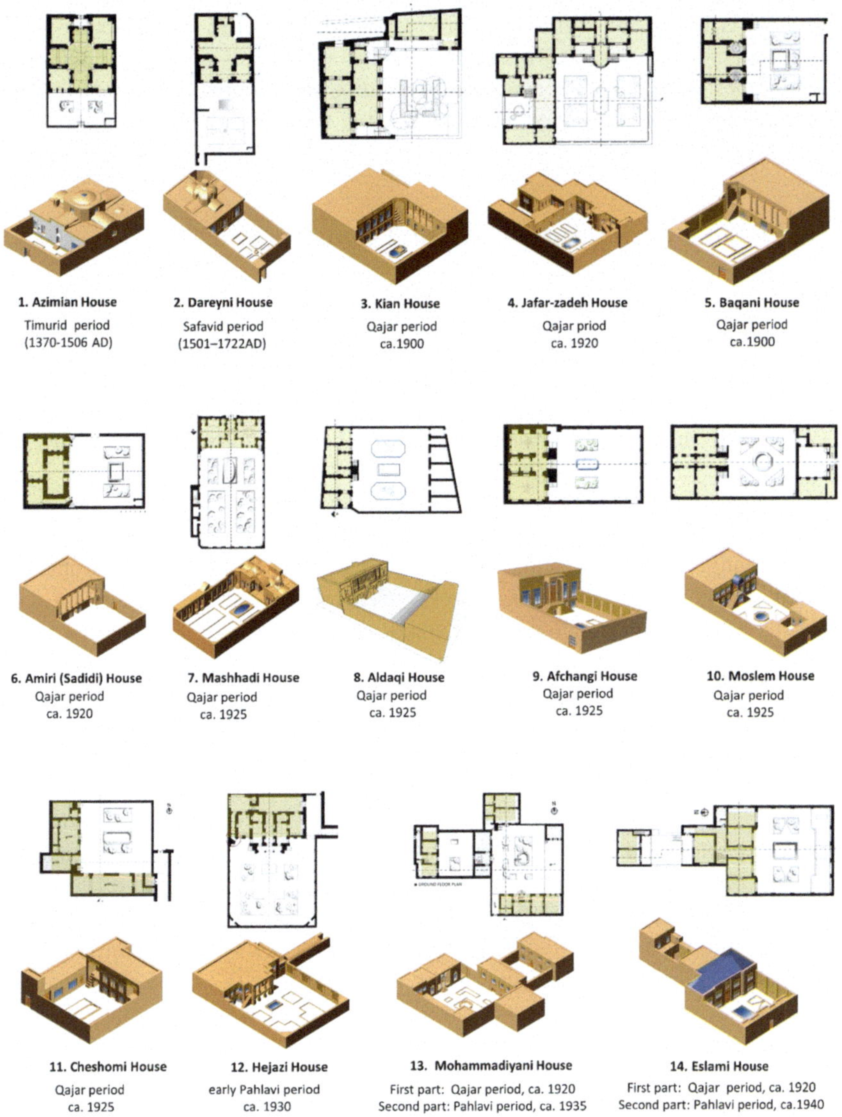

Fig. 1 Traditional houses of Sabzevar (the cases). *Source* The Author

different degrees of adaptability bring a wide variety of definitions and approaches to literature [3]. Robert Schmidt III and his colleagues at Adaptable Futures Research Group, Loughborough University [12], identified four overarching characteristics gathered from their literature review. The first is the capacity for change, "change the size or use of spaces, change its capacity, function, or performance" [1, 12]. A second characteristic is the ability for the building to remain "fit" for purpose. A third topic is value; "maximising its productive use" [6]. The last characteristic is time, which indicates the speed of change. They proposed a comprehensive definition based on these criteria: 'the capacity of a building to accommodate effectively the evolving demands of its context, thus maximising value through life' [12].

2.1 Flexible Configuration in Solid Construction

When we talk about the flexibility in traditional architecture, first of all, we think of traditional Japanese architecture. In traditional Japanese architecture, the sliding doors (Shōji) and walls made from light wooden panels or paper in a wooden frame enable the rooms to be rearranged. Partitionability (divisibility) "is the possibility of splitting up, rearranging or combining different spatial units in a simple way" [5, p. 14].

Unlike the Japanese traditional building, traditional Iranian houses have thick load-bearing brick walls. It is impossible to rearrange the position of room partitions to make changes. However, in these houses the use of multiple doors as a connection between rooms allows the relationship between spaces to be changed and a wide variety of spatial configurations to be created. Connectability "refers back to the traditional system of 'Enfilade', whereby a series of adjacent rooms can be connected through sliding wall panels or doors" [13, p. 190].

A strategy for reducing the amount of required materials in the load-bearing wall was to make niches on both sides of the wall. For this purpose, arches were integrated with the wall (usually 3 or 5) which also provided potential openings and could be turned into niches or doors. This strategy improved the usability of space and above all, increased the building's flexibility to be able to accommodate for future changes. This capability enabled residents to change the spatial configuration of the house with minor constructional modifications. Figure 2 shows various scenarios for a wall between two rooms. Alternative (A) provides three niches for the left room, while the second alternative (B) connects the rooms, and alternative (C) provides two niches for each room and a door between them. In the last alternative, there are two niches for the left room and one for the other room.

3 Research Methodology

A house is not a solid building; it is a system of activities. It has to accommodate a broad range of diurnal and nocturnal activities throughout the entire year and this can best be achieved using a flexible spatial configuration.

This research uses the Space Syntax theory to study the interaction between spatial configuration and human behaviour. The set of Space Syntax tools quantifies the spatial patterns of houses and clarifies the relationships between spaces and the house users.

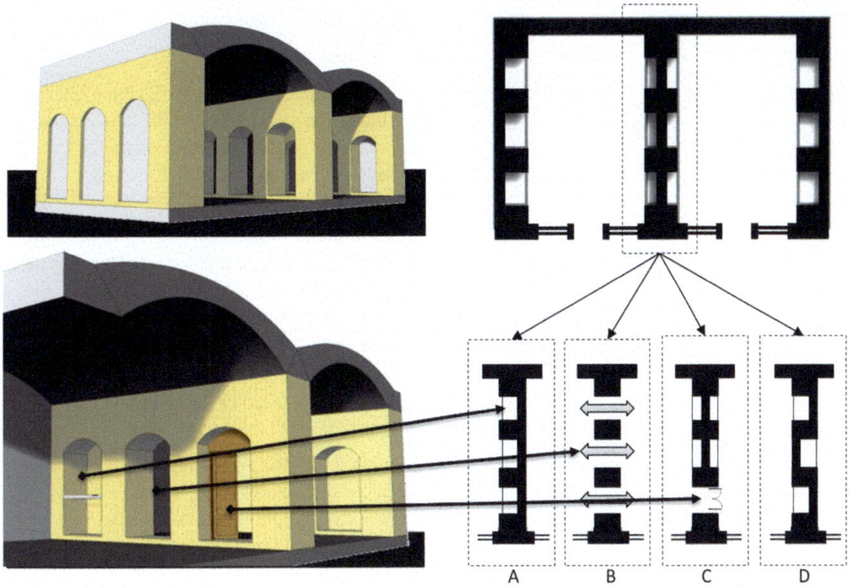

Fig. 2 Different possible scenarios for a load-bearing wall. *Source* The Author

This approach focuses on the study of how the physical layout and organisation of space can influence human movement, interaction, and behaviour within spaces. By mapping and quantifying the spatial properties of a setting, such as connectivity, integration, and accessibility, space syntax provides insights into how people navigate, perceive, and utilise space.

3.1 *Space Syntax*

Space syntax is a theory and a set of methods about space that according to Hillier [7] is based on two ideas:

1. "Space is not a background to activity, but an intrinsic aspect of it" [9, p. 207]. Humans interact with spaces in three ways: moving through space, interacting with other people in space and seeing space from a point in it.
2. Human space is not about one space, but about the inter-relations between the many spaces that make up a whole system. Hiller called it the "*configuration of space.*"

The relationships between spaces help to quantify the properties of spatial arrangements, such as symmetry, asymmetry, distributedness, and non-distributedness. Symmetry/Asymmetry is about the integrating/segregating (less private/more private) effects of space concerning other spaces [11]. The terminology used in space

Table 1 Formula and description of analysis variables in space syntax model

Variable	Formula	Description of variables
Connectivity	Ci = k	Is the direct connections of spaces. the higher the value, the closer the space is to the surrounding space, the stronger the influence on the surrounding space and the better the space permeability [15]
Convex ringiness	CR = I/(2k − 5)	(I) is the number of islands (rings) and (k) is the number of spaces in the system. The ringiness is a number between 0 and 1; 1 indicates the maximum distributedness of a system with respect to itself
Depth	Di = d	Is defined as the least number of syntactic steps in a graph that are needed" to get from one space to another [10, p. 11]
Mean depth	MD = $(\sum D.n)/(k-1)$	Is calculated by assigning a depth value to each space according to how many spaces it is away from the original space, summing these values and dividing by the number of spaces in the system less one (the original space)" ([8], p. 108) Where, $(\sum D.n)$ is the total Depth from the original space and K is the total number of spaces in the system
Relative asymmetry or "integration"	RA = 2(MD − 1)/(k − 2)	Describes the average depth of a space to all other spaces in the system [10]. The highest value indicates the most integrated and the lowest value shows the most segregated space in the system

syntax theory differs from conventional architecture and everyday terminology. In the following paragraphs, these terms are defined.

The "distributedness refers to the degree of control spaces exert on a system as a whole" [14, p. 5], in other words, it is about the possibility of entering space in more than one way [11]. It depends on the number of rings in the system. The ringiness of the convex[1] system is the number of islands (obviously the number of islands and the number of rings is the same) in the system as a proportion of the maximum possible planar rings for that number of spaces [8, p. 100]. "The measure of relative asymmetry [RA] generalises this by comparing how deep the system is from a particular point with how deep or shallow it theoretically could be" ([8], p. 108)

In the following Table 1, some essential definitions and formulas to quantify the relation between spaces are cited.[2]

[1] A 'convex space' is a space which no straight line drawn between any two points goes outside the space. Hillier and Hanson [8]. The social logic of space, Cambridge University Press, Cambridge. pp 97–8.

[2] For more details and calculation methods see: Hillier and Hanson [8]. The social logic of space, Cambridge University Press, Cambridge.

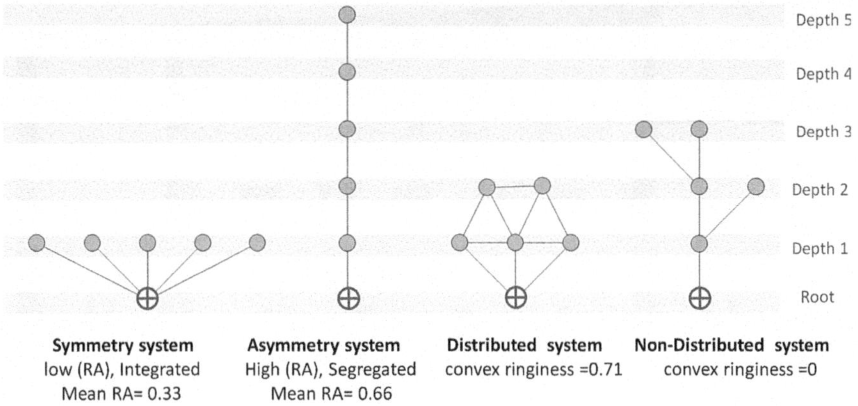

Fig. 3 Various spatial configurations with six spaces. Adapted from Ferguson [4] with modification

The following graphs (Fig. 3) present four different spatial configurations with six spaces that result in different systems with maximum symmetry, asymmetry, distributedness, and non-distributedness.

4 Spatial Configuration of Traditional Houses

Until the Qajar period a household consisted of the extended family that comprised several generations; the grandparents, parents, married sons, and their wives and children. They lived in a large house with a shared kitchen, storage, and service rooms. Any changes in the size and structure of the family (such as marriage) required new spaces. The residents could divide the large houses into smaller units for new members of the extended family by locking some doors or blocking some openings using walls. Multiple entrances to rooms and different movement loops in traditional houses have a direct relationship with the building's flexibility. An example helps to clarify this issue; the first floor of Aldaqi house has a pillared portico, four large rooms, and a connected room. Due to the large number of access doors to each room, this house can change its spatial configuration; some of these possible alternatives are presented in the following graph (Fig. 4).

At first glance, all of them are similar but from a spatial configuration point of view, they are very different. The justified graph of these alternatives reveals these major differences. In this example, the staircase is considered the root of the spatial system (Fig. 5 top). The calculation of the integration value shows the diversity of spatial configuration clearly (Fig. 5 bottom).

These alternatives can respond to different probable scenarios:

A: each room is dedicated to one of the family members or guests.

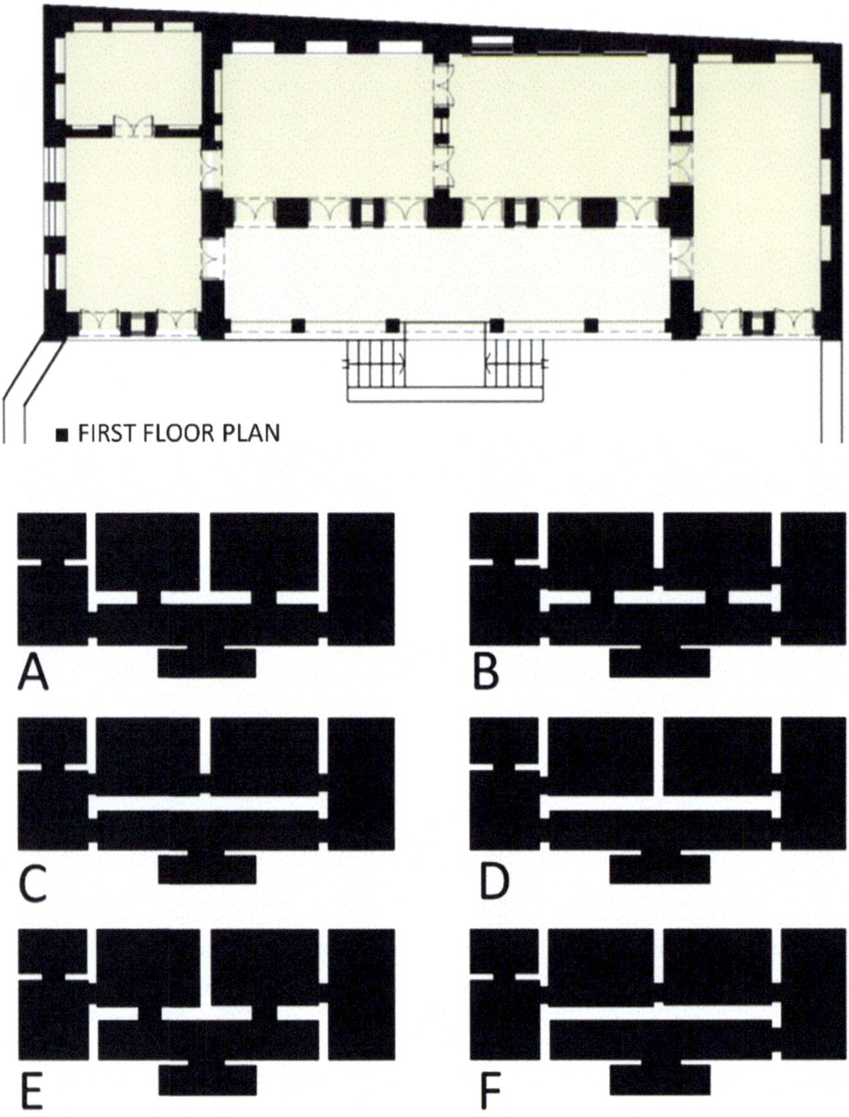

Fig. 4 Some possible scenarios for the first floor of Aldaqi house [2]

B: by opening all doors and connecting the rooms to the portico a large common space is ready for guests in summer.

C: communal space for guests in winter.

D, E: the house is subdivided to create two independent living units. Or the family opts for compact living in hard winter, only heating two rooms and leaving the other rooms unheated [2].

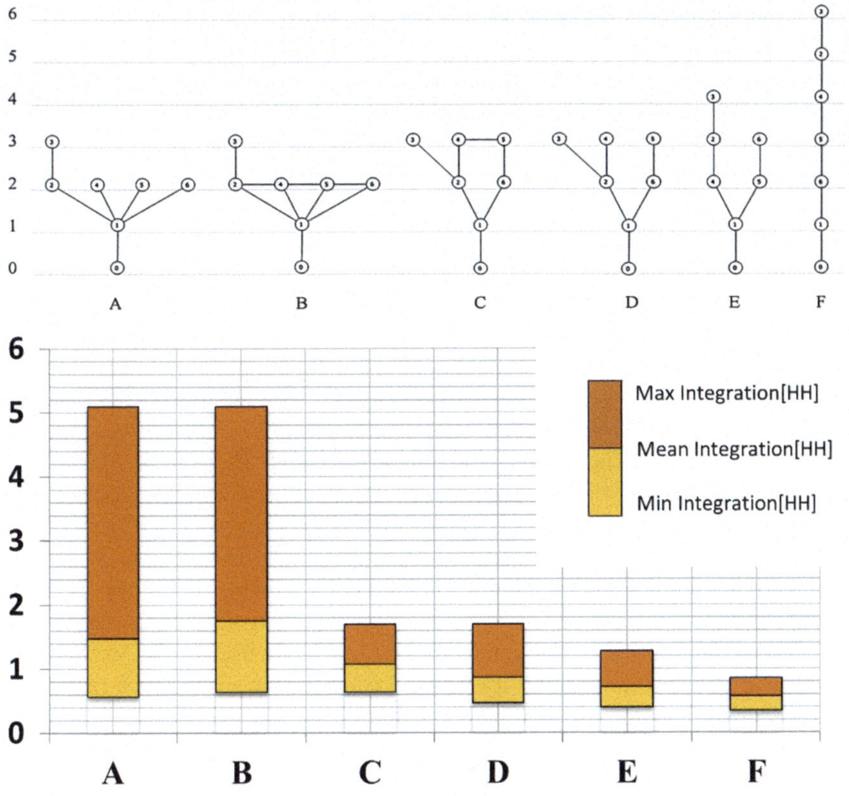

Fig. 5 Justified graph maps and integration values of different alternatives. *Source* The Author

F: it is only an example of spatial variation; otherwise, this scenario is unlikely in practice, because all rooms are converted into a circulation path.

The first indicator for evaluating the flexibility of a system is its size. A house with a large number of spaces, especially multifunctional rooms, can respond better to predictable and unpredictable changes. Table 2 shows the number of spaces and the number of multifunctional rooms for each case. The Mohammadiyani and Jafarzadeh houses with forty-five spaces including the main rooms, service rooms, yards and circulation spaces are the largest houses among the buildings investigated.

The number of openings to the main living rooms is another indicator for evaluating the flexibility of a building. Thanks to the large number of openings (doors) to the rooms the residents can alter the arrangement of the spaces if the need arises. The most integrated spaces with maximum connectivity in all of the cases are the courtyard and the central domes of the Azimian and Dareyni houses.

A house with a large number of loops can better rearrange its configuration by locking some doors (loops) and providing access via another loop. There are nine loops in the Baqani house that it has only sixteen spaces, while two large Pahlavi houses, the Eslami and Mohammadiyani houses, with 45 and 38 spaces have only eight loops (Table 2).

Potential openings increase the number of circulation loops in houses. The increase in the number of loops is directly related to the increase in the Distributedness of houses. Figure 6 shows the impact of potential openings in dramatically increasing the Distributedness of the system. For example, potential openings have been able to transform Darini and Afchangi Houses from a non-distributed building to a completely Distributed one.

For better comparison of the cases the size of houses, the mean connectivity of main rooms, the distributedness of the system (Convex ringiness), and the efficiency of space use[3] are normalised. It helps to group the houses according to these indicators. The following Fig. 7 presents the different house types according to their spatial configurations. The first group is small houses with the maximum efficiency of space use and high distributedness. The Azimian and Dareini houses are located in this group (Fig. 7a). The Baqani, Amiri, and Afchangi houses (group b) are small houses with very high distributedness and acceptable connectivity and efficiency. The group c contains three medium-sized houses that all indicators located in moderate conditions. Two Pahlavi houses, the Mohammadiyani and Eslami houses, and the Jafar-zadeh house (the late Qajar period) are the largest houses among these cases. In terms of space use, they score highly, but their spatial configurations are not flexible at all. The worst houses regarding the efficiency of space use are the Moslem and Mashhadi houses; there are a lot of circulation and service spaces for a few living rooms. The Mashhadi house has the maximum connectivity among the cases and acceptable distributedness, while the spatial configuration of the Moslem house is not flexible. The most flexible spatial configuration according to the above indicators can be found in the Hejazi house (group g); a medium-sized house with high connectivity and great distributedness.

5 Conclusion

The existence of potential openings in traditional Iranian houses increases the potential of the houses to be reconfigured in response to the changes (possible scenarios). There was no fixed furniture in the rooms, so the residents could easily change the function of rooms as needed. This possibility of change at a low cost increases the flexibility of the system for future changes and enables the house to respond to the different scenarios that might happen during its lifecycle.

It is necessary to mention that the basis of this ease of change is the multifunctionality of spaces (rooms) in traditional houses in this region. Traditional Iranian houses are space-based as opposed to furniture-based; there was no fixed furniture in the rooms, so the residents could easily change the function of the rooms as needed. The higher scale of multifunctionality of spaces (rooms) is changing the function of

[3] The ratio of main spaces to total spaces, including service and circulation spaces.

Table 2 Spatial configuration of houses

Name of house	Number of spaces	Number of main rooms	Ratio of living spaces to the entire spaces (%)	Maximum connectivity/ space	Mean connectivity	Mean connectivity of main rooms	Number of current loops	Number of loops-considering the potential openings	Distributedness (convex ringiness) current position	Distributedness (convex ringiness) considering the potential openings
Azimian	16	9	56	Central space (8)	2.5	2.75	5	7	0.18	0.26
Dareyni	14	8	57	Central space (6)	2	2.25	0	5	0	0.22
Kian	25	10	40	Courtyard (7)	2.36	3	5	9	0.11	0.2
Jafar-zadeh	45	16	36	Courtyard (11)	2.47	2.33	10	10	0.12	0.12
Baqani	16	6	38	Courtyard (6)	2.56	3	5	9	0.19	0.33
Amiri (Sadidi)	17	5	29	Courtyard (7)	2.41	3	4	8	0.14	0.28
Aldaqi	26	9	35	Courtyard (11)	2.41	2.42	5	8	0.11	0.17
Mashhadi	14	2	14	Courtyard (8)	2.5	5	4	4	0.17	0.17
Afchangi	16	6	38	Courtyard (5)	2.06	2.5	1	7	0.04	0.26
Moslem	35	6	17	Courtyard (8)	2.43	2.1	8	8	0.12	0.12
Cheshomi	25	8	32	Courtyard (8)	2.04	3	4	7	0.08	0.16
Hejazi	33	12	36	Courtyard (8)	2.82	3.5	15	19	0.25	0.31
Mohammadiyani	45	19	42	Courtyard (8)	2.18	1.63	5	8	0.06	0.09
Eslami	38	15	39	Courtyard (6)	2.31	1.57	5	8	0.07	0.11

Source The Author

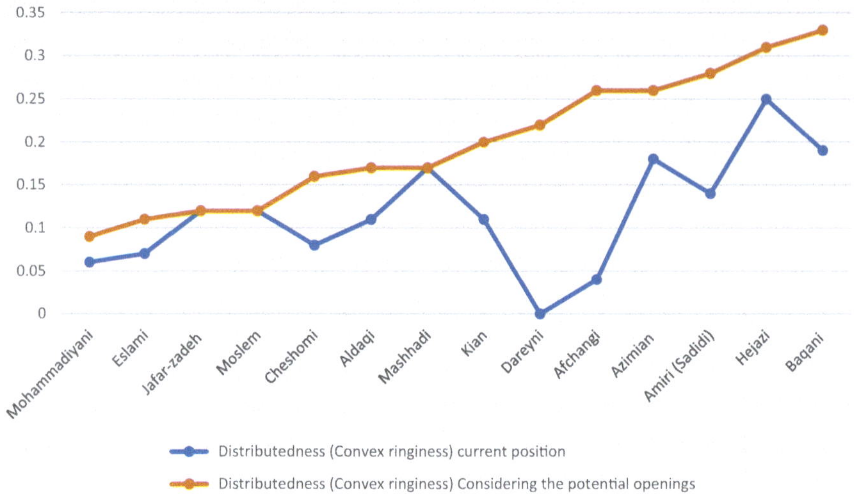

Fig. 6 Distributedness of houses

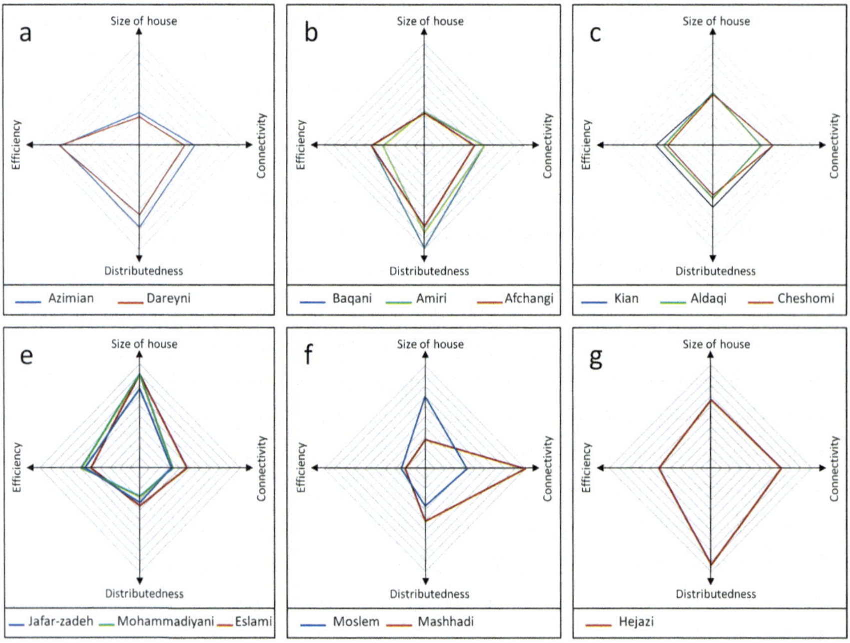

Fig. 7 Different house groups according to the spatial configurations. *Source* The Author

buildings. The flexible configuration of cases especially a large number of rooms and circulation spaces and also a large number of doors and potential openings enable the houses to accommodate new functions In their old structure with minor changes. Some of these houses have experienced these changes in their lifecycles. A wing of the Aldaqi house was converted from storage space into commercial use. The Kian house was transformed into a bank office about a hundred years ago. The functions of the Mashhadi and Eslami houses were changed several times. The multifunctionality in the building scale is an important factor in extending the lifecycle of buildings and saving them from demolition.

As a research finding for designing new projects, anticipating the ability to create multiple entrances (openings/doors) to the architectural spaces and rooms of buildings with minimal structural changes can be a solution to increase the flexibility of buildings.

References

1. Douglas J (2006) Building adaptation. Routledge
2. Estaji H (2014) Flexible spatial configuration in traditional houses, the case of Sabzevar. Int J Contemp Archit "The New Arch" 1(1):26–35
3. Estaji H (2017) A review of flexibility and adaptability in housing design. Int J Contemp Archit "The New Arch" 4(2):37–49
4. Ferguson TJ (1996) Historic Zuni architecture and society: an archaeological application of space syntax, vol 60. University of Arizona Press
5. Geraedts RP (2008) Design for change; flexibility key performance indicators. In: 1st International conference on industrialised, integrated, intelligent construction (I3CON), vol 11
6. Graham P (2005) Design for adaptability—an introduction to the principles and basic strategies. Environ Des Guide 1–9
7. Hillier B (2005) The art of place and the science of space. World Archit 185:96–102
8. Hillier B, Hanson J (1984) The social logic of space. Cambridge University Press
9. Hillier B, Vaughan L (2007) The city as one thing. Prog Plan 67(3):205–230
10. Klarqvist B (2015) A space syntax glossary. NA 6(2)
11. Moser G, Pol E, Bernard Y, Bonnes M, Corraliza JA, Giuliani MV (eds) (2003) People, places, and sustainability. Hogrefe & Huber, Seattle, Washington
12. Schmidt III R, Eguchi T, Austin S, Gibb A (2010) What is the meaning of adaptability in the building industry. In: Open and sustainable building, pp 233–242
13. Schneider T, Till J (2007) Flexible housing. Architectural Press
14. Tipple AG (2000) Extending themselves: user-initiated transformations of government-built housing in developing countries. Liverpool University Press
15. Xu Y, Rollo J, Esteban Y (2021) Evaluating experiential qualities of historical streets in Nanxun Canal Town through a space syntax approach. Buildings 11(11):544

Exploring the Architectural Legacy of Najafabad: Lessons from Iranian Vernacular Architecture and the Safavid Era

Atefeh Karbasi

Abstract Najafabad, a small, traditional city near Isfahan in Iran, has a rich history steeped in excellence in urban design, architectural craftsmanship, and horticulture. This chapter delves into Najafabad's past, describing some of the architectural techniques, cultural narratives, and design principles cherished by Iran's master architects. Historical accounts, both written and oral, highlight historic threads dating back to the Safavid era under Shah Abbas the Great's rule in 1613 AD. These accounts shed light on Najafabad, a thriving city known for its famous almond and grape orchards, and its urban evolution and architectural innovations. Prior research has touched upon Najafabad's urban planning, a substantial gap persists in explaining its architectural legacy that this chapter bridges using detailed fieldwork to provide comprehensive findings gleaned from on-site investigations. These endeavors provide invaluable insights into the intricate design and unique structural features in Najafabad's historic residences include the vaulted roofs, colloquially referred to as "*taq-o cheshmeh*," which have stood as enduring architectural elements that epitomise tradition and resilience since the Safavid era. In many historical houses of Najafabad, a distinct difference in height is observed at the end of the main central axis between the exterior and the interior scale of the primary living space. This difference in height is due to the presence of the storage space of upstairs, which functions as a storage area for garden products, particularly almonds and grapes, and reflects the interweaving of domestic life and agricultural activities in the architectural fabric of Najafabad. This space, situated between the main living area and the roof, acts as a filter above the primary living space due to its specific characteristics in materials, roof shape and thickness, proportions, height, positioning relative to the opposite iwan, and the number, size, and placement of openings. It provides additional climatic comfort, especially against the summer heat, demonstrating how form, function, and climatic considerations come together to create a harmonious space.

A. Karbasi (✉)
Faculty of Architecture and Urban Planning, Shahid Beheshti University, Tehran, Iran
e-mail: a_karbasi@sbu.ac.ir

A. Cheshmehzangi and S. Roaf (eds.), *Persian Vernacular Architecture*, Urban Sustainability, https://doi.org/10.1007/978-981-96-1116-4_5

Keywords Najafabad · Indigenous Iranian architecture · Ṭāq-u Chishmah · Safavid Era · Historical houses · Climatic comfort · Integration of domestic and agricultural life

1 Introduction

Najafabad, a small city near the historic city of Isfahan in Iran, boasts a rich history in urban planning, architecture, and gardening. Historical documents, both written and oral, attest to the significance of Najafabad in urban planning, architecture, and fruit cultivation. The city's renown dates back to the Safavid era, specifically to the year 1613 AD during the reign of Shah Abbas the Great. It is reported that during that time, under the supervision of Sheikh Bahai, the minister of Shah Abbas the Great, Najafabad was established approximately thirty kilometers from the Safavid capital, serving as a means to supply agricultural needs for Isfahan [1, 6, 7, 16]. Previous research on Najafabad has primarily focused on its urban planning, including the predominant irrigation of the city and its orchards through *qanāts* [1], the city streets form an intersecting and geometric network, and the city is surrounded by fruit orchards on all four sides. This image has largely persisted into contemporary times. Until just over a decade ago, no research had been conducted on the introduction of architecture, including the historic houses of this city. As part of a research project [9] and a book [10], the author presented forty historical houses of this city in the form of photographs, maps, and descriptions of life in the spaces of each house in the past, and in an article about the types and evolution of *ayvān* in the houses of this city, it was presented that the central iwan on the north facade, in the oldest houses of Najafabad, is deeper and more prominent compared to other types of iwans in houses of later periods [11]. However, no research has been conducted to analyse the architectural characteristics of these houses based on their internal functional performance. Only a few books have provided limited descriptions, in the form of short sentences, paragraphs, or a single sentence, about some historical buildings in the city, including houses [12, 18]. The present article, based on the author's previous fieldwork, introduces and analyses the architectural heritage of the historical houses of this city, focusing on the interaction of form with the function of the living space, at the end of the central courtyard axis, which is the most important space of the house, in the following structure: a general introduction to the historical background of Najafabad, introduction of fieldwork including data collection methods, field observations, and key findings focusing on the subject of this article; presentation of the architectural features of historical houses in Najafabad; analysis of the relationship between form and function of the living space and the warehouse-like space above it, and finally, the results of this study.

2 The Historical Background of Najafabad

Najafabad is a city located near the center of Iran at 51° and 21 min longitude and 32° and 38 min latitude, in a semi-arid climate with an average precipitation of 120 mm (low rainfall) and a temperature of 38 °C. Much of its land is fertile and under cultivation [1]. This city is situated in a vast and flat plain, with very poor surface water resources [18]. In historical Najafabad, streams flow alongside certain streets of the city. Moreover, access to groundwater is facilitated through *qanāts* and wells in certain areas within and around the city. [14, 18], leading to the existence of numerous productive orchards surrounding the city on all four sides (Fig. 1).

Many refer to Najafabad as the "garden city." In all these orchards, it is rare to find a tree that does not bear fruit and serves solely decorative purposes. Thus, the geography and strategic location of Najafabad have had a significant and astonishing impact on the residents' quality of life and economic prosperity [1]. The irrigation of Najafabad's orchards was done through streams, wells, and *qanāts*; however, it should be noted that this type of irrigation in this climate does not always result in consistent and year-round watering. This has led to the majority of the city's orchard trees being drought-resistant and well-adapted to the hot and dry summer weather. These products include almonds, grapes, and to some extent, pomegranates, mulberries, and walnuts, which continue to thrive. Numerous travelers or writers have

Fig. 1 Aerial photograph of Najafabad city in 1956, orchards surround the city on all four sides. *Source* Surveying Organisation of Iran

visited historical Najafabad in the past and have provided accounts that support the above statements.

Comte de Sercey, the French ambassador, referred to Najafabad as a "large village" in 1839 [4]. Shirvani wrote about the abundant gardens and affordable fruits of Najafabad [17]. Brugsch mentioned that the orchards of Najafabad are extensive: "*We rode through the orchard alleys of this city with our horses for three-quarters of an hour until we finally reached the streets and bazaar of the city*" [3]. The excellent and abundant almonds of Najafabad are emphasised in historical sources.

According to Al-Isfahani, half of Isfahan's firewood comes from pruning the orchards of Najafabad. Najafabad has abundant and excellent almonds, and even all of Isfahan's almonds are supplied from there. The land of Najafabad is filled with grapevines, and Najafabad grapes are famous and unmatched in sweetness and quality. Al-Isfahani [2] Shirvani describes Najafabad almonds as excellent and abundant [17]. The abundance of agricultural products not only met the needs of the people but also led to the export of almonds and other agricultural products to Isfahan, to the extent that a section of Isfahan's grand market belonged to the people of Najafabad, and it even facilitated the export of Najafabad almonds to other countries, especially India [1].

Thus, not only was agriculture one of the most important occupations of the city's residents in the past [12], but also most of the city's inhabitants, regardless of their primary occupation, or better to say, alongside their primary occupation, engaged in fruit orcharding to meet their needs and even sell surplus products. Such practices can still be observed in contemporary times.

Historical sources mention some aspects about the houses of this city, such as the fact that the houses in Najafabad are orderly and built according to urban planning principles [3]. These houses are excellent and rarely seen in other cities [13]. Shirvani mentions about a thousand houses in this city designed with engineering methods [17]. But did the urban and natural characteristics of Najafabad have any influence on the architecture of these houses? Among contemporary sources, only one source mentions in a sentence that in old houses of Najafabad, there were storage areas on the second floor for food and orchard products such as pomegranates and grapes for making raisins [15]. For a precise answer to this question, we delve into field research on the architecture of houses in this city.

3 Detailed Fieldwork and Examination of Najafabad's Architectural Heritage

Examining the oldest available aerial photograph of the city from 1956 provides an interesting insight into the general organisation of historical houses in Najafabad. It indicates that within the intersecting network of the city, each house has a central courtyard in the middle and is surrounded by structures around the courtyard, with curved roofs (Fig. 2).

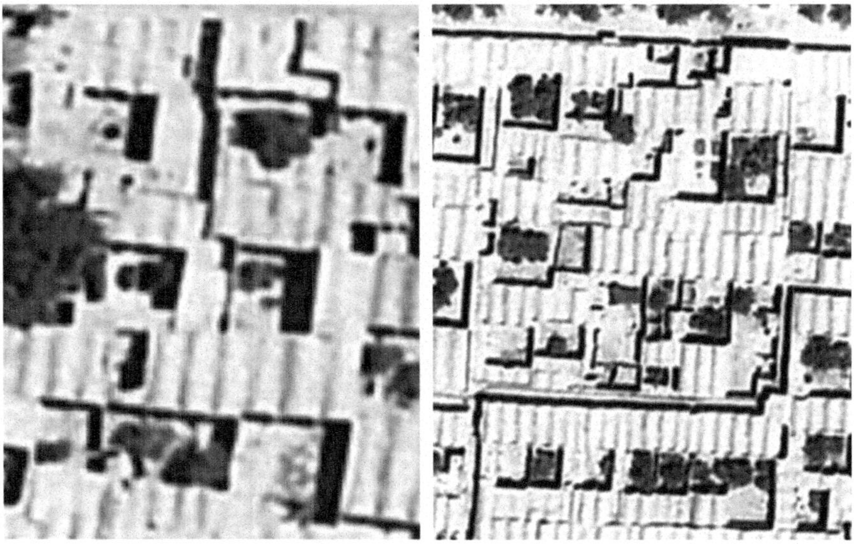

Fig. 2 Aerial photograph showing sections of Najafabad city in 1956. Houses with vaulted roofs (known as "*ṭāq-u chishmah*") are observed, constructed around central courtyards. *Source* Surveying Organisation of Iran

The primary tool for the current research is fieldwork, deemed the most crucial. Over a span of a decade, numerous houses remaining from Najafabad's pre-modern era were meticulously visited and examined. This investigation encompassed houses of various ages and sizes, scattered across different areas of the city and representing diverse social and economic strata. A significant challenge encountered during the detailed examination was the residents' hesitance due to specific traditions. While they were generally accommodating to ordinary visits by a fellow citizen (the author being a resident of Najafabad), they were less enthusiastic about documentation and photography of their homes, often hindering such efforts. Consequently, nearly two hundred houses were visited, with approximately half of them being photographed using either professional cameras or mobile phones. Manual drawings were then meticulously crafted using AutoCAD, followed by their translation into three-dimensional sketches using SketchUp.

It is evident that after examining and documenting some of the houses from the inside, for many others where detailed documentation or internal observation was not possible, the interior layout and spatial organisation could be predicted and understood to a considerable extent based on what was visible from the outside and by comparing it with past and present aerial photographs.

3.1 General Characteristics of the Historical Houses of Najafabad

Limited references from written sources [14, 18], and especially the detailed findings of this documentation and extensive observation, indicate that the oldest historical houses in this city have vaulted roofs. The general shape of these houses is such that a standard vault module, repeated from a minimum of one to a maximum of seven spans, forms the main façade of the house, and in most cases, other facades as well (Figs. 3 and 4).

These houses are known as "vault-roof houses" or "*ṭāq-u chishmah*". Their skyline sometimes resembles waves (Fig. 5) and sometimes, due to the presence of horizontal parapet on the roof, appears flat and horizontal (Fig. 6).

Since the advent of iron in the construction industry in the city, around the year 1966, the traditional construction tradition gradually fades, and contemporary houses are built. Interestingly, the construction of vault-roof houses continued to exist in Najafabad until around 1990. Therefore, this article focuses on this category of

Fig. 3 General appearance of vaulted-roof houses formed by the repetition of vaults. Southern façade. *Source* Photo taken by the author

Fig. 4 General appearance of vaulted-roof houses formed by the repetition of vaults. Northern façade. *Source* Photo taken by the author

Fig. 5 The undulating skyline of the vault-roof houses in the alleyway. *Source* Photo taken by Farshid Nasrabadi

Fig. 6 The smooth and flat skyline of the vault-roof houses in the alleyway due to the presence of parapet around the roofs. *Source* Photo taken by the author

houses, which are prioritised based on their historical significance, prevalence, and even contemporary relevance.

The materials commonly used in building houses include adobe bricks for walls and baked bricks for vaults. These bricks are usually made from the soil found on the site of the house and then fired in a kiln crafted by hand within the house's courtyard. Wooden molds are employed to shape the vaulted arches, and between two such arches, a taller vault, known as "*chishmah*" (fountain), is constructed. Skilled traditional craftsmen create this taller vault manually (Figs. 7 and 8).

In these houses, the proportion of open courtyard space in the center and enclosed building space around it is such that the courtyard is located in the middle, and the building surrounds it on three or four sides (Figs. 9 and 10) or on two sides (Fig. 11).

In some houses, there is a southern front, and because it doesn't face the sun directly, it is more utilised during the warmer seasons, providing a cooler environment. However, in some houses, multiple families, including grandparents, their married children, and their own children, all live together in one house. In this case, each family is assigned a residential room within the house. The main rooms of the northern front are typically allocated to the grandparents, who are considered the

Fig. 7 The facade of the main front of a house, the *ayvān* of the northern front seen from the courtyard side. *Source* Photo taken by Farshid Nasrabadi

Fig. 8 The interior view of one of the chambers with a vaulted ceiling, known as "*ṭāq-u chishmah*". *Source* Photo taken by the author

primary owners of the house. Subsequently, the elder children are allocated rooms in this area, while the other children are accommodated in rooms located in other fronts, including the southern front. The eastern and western fronts may lead to auxiliary rooms or service spaces. Still, the primary service areas such as the kitchen and bathroom are strategically located in the corners where the fronts intersect. In all houses, the northern front, benefiting from direct sunlight, is constructed and serves as the main living space. The space or room located along the courtyard axis is considered significant both in plan and volume, often serving as a central hub for household activities.

3.2 The Relationship Between the Function and the Internal Layout of the Spaces in Tāq-u Chishmah Houses

As mentioned, in *ṭāq-u chishmah* houses, each vault extends from the courtyard edge to the end of the building, forming a unified module. Within this vault, various

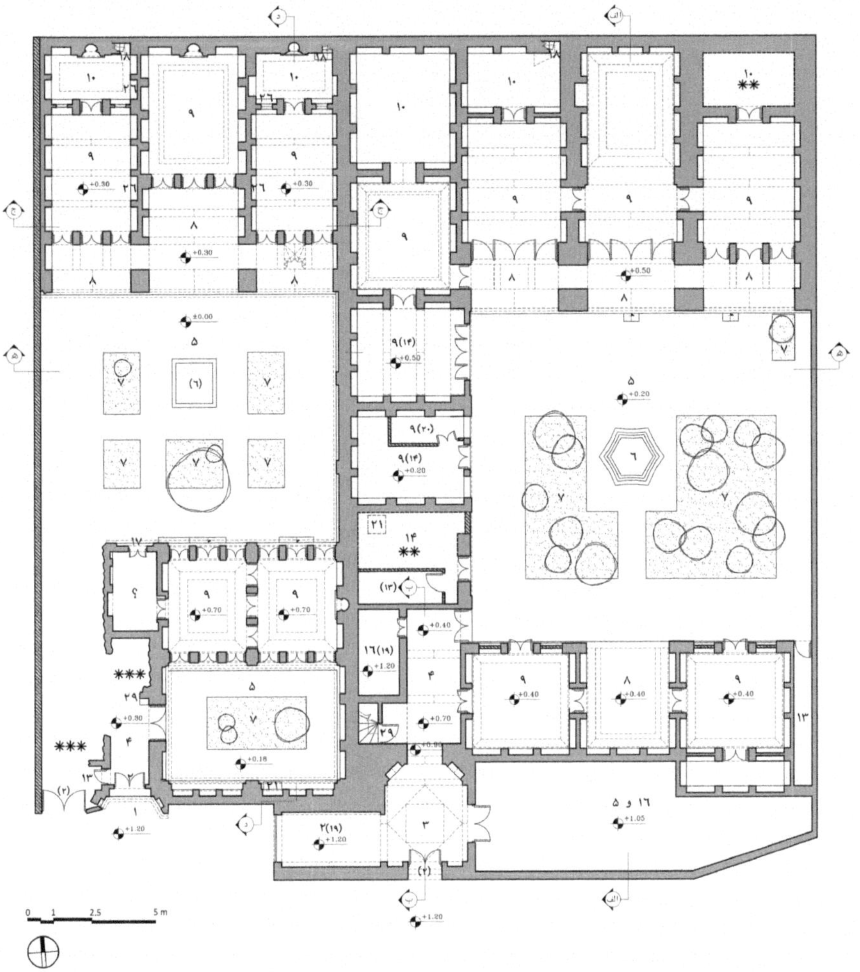

Fig. 9 Floor plan of two adjacent houses, where the northern front is the most important façade. *Source* Plan drawn by the author

spaces are accommodated. Starting from the courtyard, the plan, typically consists of an *ayvān* or semi-open space, followed by the main living room or space, and at the end of the room, there is often a rear room used for storage of additional items, such as bedding. Consequently, three spatial layers are observed beneath this vault in the ground floor plan. In the courtyard elevation and sometimes in the street facade, all vaults are sometimes of equal height, while in other cases, the central vault is taller than the others. This means that the central vault is higher. Up to this point, nothing new has occurred. Many historical houses in Isfahan or historical houses in central regions of Iran follow this pattern, where the central space, located along the axis of

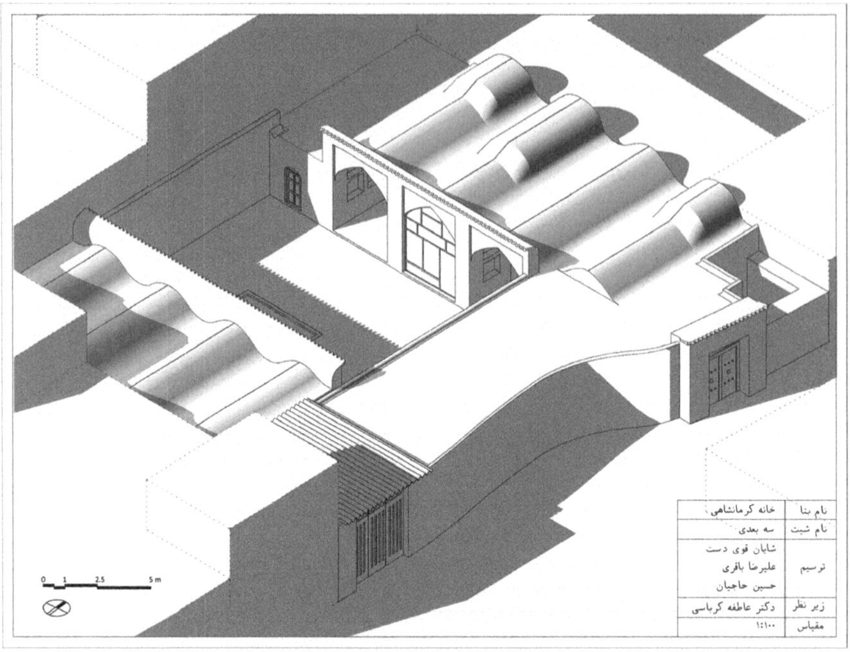

Fig. 10 Three-dimensional rendering of a house constructed on three sides, with the northern and southern fronts featuring vaulted ceilings. *Source* 3D model drawn by the author

the courtyard, has a higher height than the other spaces and stands out prominently [5, 8].

A closer examination of the sections of historical *ṭāq-u chishmah* houses in Najafabad, or upon entering these spaces, reveals interesting findings, as evident in Figs. 12, 13, 14 and 15, which are related to a single house. Whether the central vault is taller than the others or of the same height, and while the opposite iwan to this space have the same height as the other *ayvāns* (Figs. 14 and 15), from within the space, the main living area beneath this vault is noticeably lower in height than the side vaults and appears almost flat (Fig. 12). This phenomenon is most clearly visible in architectural documents (Fig. 13).

In many examples of houses, the central *ayvān*, which is at the end of the courtyard axis, is of the same height as the other *ayvāns* (Figs. 16 and 17). However, the northern facades of the houses reveal that the ceiling of the space behind this *ayvān* is taller than other parts of the roof (Fig. 18). Yet, upon entering the house and entering the northern front, we realise that there is a relatively dark and low room behind this *ayvān*, along the main axis of the house courtyard (Figs. 19 and 20). This architectural phenomenon is well observed in the north–south section of the house (Fig. 21) or the east–west section of the northern front (Figs. 22 and 23).

Fig. 11 Floor plan of a house with eight vaulted chambers, which has been divided into two sections, east and west, by the owners over time. The southern section of the house has been renovated and does not feature vaulted ceilings. *Source* Plan drawn by the author

Fig. 12 The main and central room of the northern front of House A. *Source* Photo taken by the author

Fig. 13 North–South section through the center of the plan of House A. *Source* Section drawn by the author

Fig. 14 The main and central *ayvān* of the northern front of House A, facing the main and central room. *Source* Photo taken by the author

This inconsistency between the interior and exterior arises from placing a storage space above the main living space (Figs. 24 and 25). This storage area, known as upstairs, is used to store agricultural products and is located at a higher level.

3.3 *The Storage Space of Upstairs*

The storage space of upstairs is situated above the main living room of the house, but the only indication of this space visible from the courtyard on a human scale is the hatch above the door of the living room (Figs. 26, 27, 28 and 29).

Thus, the main access to the storage upstairs is through a hatch facing the courtyard, above the door in the main living room. In some examples, the hatch leading

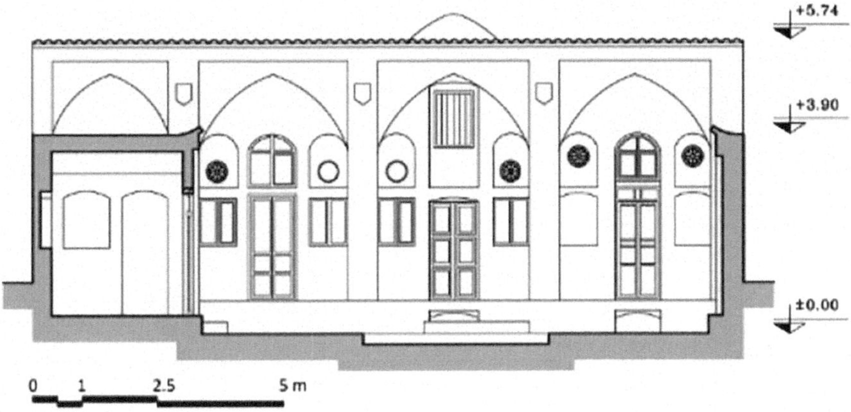

Fig. 15 North facade of House A, showing the main *ayvān* of the northern front located in its center. *Source* Section drawn by the author

Fig. 16 The main and middle *ayvān* of the northern front, which faces the main and middle room of the northern front. *Source* Photo taken by Farshid Nasrabadi

to the upstairs includes several stairs to compensate for its elevated floor, allowing the room behind to have a greater height (Fig. 30). The interviews with local people indicate that access to this hatch is facilitated either by placing a ladder underneath it or by sending children inside it. Children climb up by placing their feet into shelfs in the walls to reach the hatch.

Sometimes, this upstairs storage had secondary access; a steeply sloped staircase in a corner of the rear room, with a small square hatch in the ceiling of the rear room, which was considered an opening in the upstairs floor (Figs. 31, 32 and 33). These challenging accesses confirm the storage-like usage of the space, which is not intended for daily or even weekly use.

Fig. 17 The main and middle *ayvān* of the northern front, which faces the main and middle room of the northern front. The closet seen on the right side of the photo is a contemporary addition, and the *ayvān* has also been converted into a room. *Source* Photo taken by the author

Fig. 18 The northern facade of the houses where in the center or a side of it, the main *ayvān* of the northern front with a raised ceiling behind and above it is visible. *Source* Sections drawn by the author

Local evidence, including interviews and observations, confirms that the primary use of this storage space is for storing products such as almonds and grapes. These kinds of dried fruits are considered essential and valuable nutrition for the people during the cold winters of the region. When almonds or walnuts are harvested, their green shells need to be removed first, and then they are spread out in the sun to dry.

Fig. 19 The main room (living room) of the northern front with a low and flat ceiling, located along the courtyard axis and behind the central *ayvān* of the northern front. *Source* Photo taken by the author

Fig. 20 The main room (living room) of the northern front with a low and flat ceiling, located along the courtyard axis and behind the central *ayvān* of the northern front. *Source* Photo taken by the author

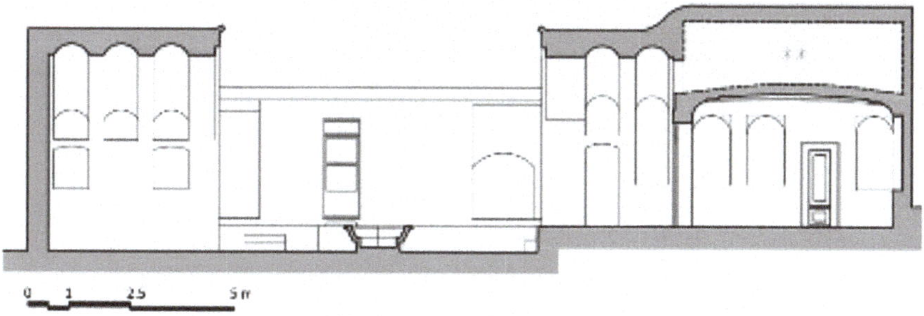

Fig. 21 South-North section through the center of the plan, where on the right side, the main room (living room) of the northern front with a low and flat ceiling is visible, and there is a space above it. *Source* Section drawn by the author

Fig. 22 Eastern–Western cross-section of the northern front showing three rooms with low, flat ceilings, and a space elevated above them, with its roof higher than the rest of the roofs. *Source* Section drawn by the author

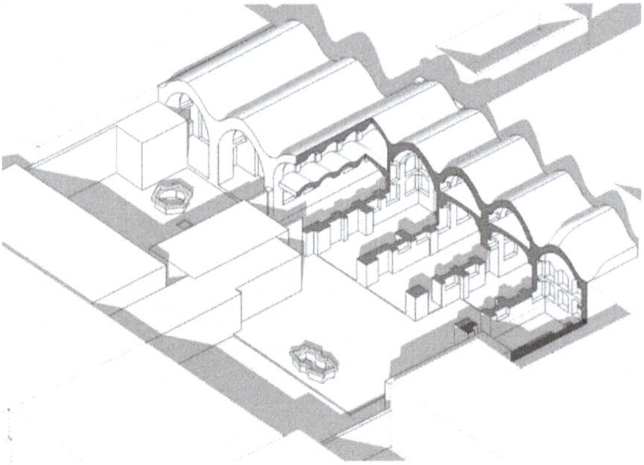

Fig. 23 Perspective-section of the house corresponding to the section depicted in Fig. 22. *Source* 3D model drawn by the author

Fig. 24 A view from inside an upstairs, where even the floor, which serves as the ceiling of the room below, is not entirely leveled. This image demonstrates that despite its size and central location within the house, this space has never been used for human habitation. *Source* Photo taken by Farshid Nasrabadi

Fig. 25 The main *ayvān* of the house, where even today, some people engage in activities related to the preparation and storage of almonds during the harvest season (such as almond drying and using a scale for weighing) can be seen in this courtyard. Above the entrance to the main room, the entrance to the upstairs storage area is visible. *Source* Photo taken by the author

Fig. 26 The space visible in the photo, which is now the living room, used to be a semi-open space or an *ayvān* in the past. The hatch above the door of the room provides access to the storage upstairs. *Source* Photo taken by Farshid Nasrabadi

Afterward, these products are ready for sale. However, some individuals store these dried fruits for sale in later seasons, rather than during times when the market is saturated with freshly harvested nuts. Others keep them for their annual needs. For storing almonds, a space is needed that is protected from humidity, cool, and away from sunlight. It seems that the upstairs space possesses all of these characteristics.

Fig. 27 The front space, now depicted as the living room in the photo, used to be a semi-open space or an *ayvān* in the past, with a large door separating it from the rear space, which had a lower ceiling. The hatch above the entrance door to the room provided access to the storage upstairs in the and is now blocked off. *Source* Photo taken by Farshid Nasrabadi

Fig. 28 An example without alteration, showing access to the storage upstairs through the hatch above the entrance door to the room, maintaining the original condition of the *ayvān* and the room. *Source* Photo taken by the author

The storage space of upstairs is situated higher than ground level and, unlike basements, which typically absorb moisture from the soil, it remains unaffected by any moisture. It lacks any significant openings to the outside and is often dark. Sometimes, there are divisions on the floor of this space to separate different types of almonds or dried fruits (Fig. 34). Grapes are another well-known product of the city, especially when dried and sold as raisins. To make raisins, grapes are hung from

Fig. 29 The hatch above the entrance of the living room, located in the axis of the courtyard, is the only indication visible from the courtyard that leads to the upstairs on a human scale. *Source* Photo taken by the author

Fig. 30 The hatch above the entrance of the living room provides access to the upstairs. *Source* Photo taken by the author

ropes attached to the ceiling in a dark and cool place. Thus, the attic serves as the ideal space for producing and storing raisins. Occasionally, in the ceilings of attics, loop-shaped brackets were permanently installed between the mud bricks forming the ceiling. These brackets were used to tie grape ropes to them and suspend them from the ceiling (Figs. 35 and 36).

Typically, there is a small opening in the shape of a window at the end of the attic, opposite the entrance door from the courtyard side.

Fig. 31 A hatch in the ceiling of the rear room for accessing the upstairs. The stairs are very few in number. *Source* Photo taken by the author

Fig. 32 The stairs at the end of the rear room and a hatch in the ceiling of the rear room for accessing the upstairs. *Source* Photo taken by the author

Fig. 33 The hatch in the ceiling of the rear room for accessing the upstairs. There are no stairs, and one must place their feet into the wall ledges and pull themselves up. *Source* Photo taken by the author

Fig. 34 The division of the upstairs floor for segregating different types of dried fruits. *Source* Photo taken by Farshid Nasrabadi

Fig. 35 Small brackets placed beneath the mortar of the roof bricks in the upstairs to tie and dry grapes for making raisins. *Source* Photo taken by the author

Fig. 36 A dimly lit view of the upstairs from inside through the hatch in the ceiling of the rear room. Two opposing air vents, creating cross-ventilation, are partially visible. *Source* Photo taken by the author

4 Conclusion

The study of the architectural heritage of the "*ṭāq-u chishmah*" houses in Najafabad reveals that functional, formal, and climatic factors have played a role in defining and characterising the storage space of upstairs. The upper space, located on the northern or sun-facing side, plays a key role in storing dried fruits or in producing raisins from grapes. Its formal significance is clear due to its placement at the end of the courtyard axis and the deeper central *ayvān*. This importance is sometimes accentuated by an increase in height in the façade and volume. Thus, the visual identity of this space is prominent in both the courtyard façades and the adjacent streets. However, from a human perspective and in graphical representations, the main living room beneath the upper room has a lower height compared to the side rooms. While the upper floor's reduced ceiling height might compromise the quality of the living space, the unique architecture of the upper room meets the practical needs of the homeowner for storing and processing garden products while also playing a crucial role in enhancing the climatic comfort of the main living space below.

The depth of the *ayvān* in front of the storage space of upstairs protects it from sunlight. The curvature of the vault and the high thermal mass of the adobe walls and ceiling of the upper room ensure that external heat does not significantly penetrate through conduction or radiation. The proportions of the space, or its geometry, are elongated from north to south with a narrow span from east to west. This depth and minimal exposure to sunlight contribute to cooling the space. The entrance vent of the upper room, which leads to the iwan, along with the opening at the end of the interior space of the upper room, facilitates airflow, prevents air stagnation, and helps cool the space during the hot season. Ultimately, the storage space of upstairs minimises the impact of external heat, whether through conduction or radiation.

Thus, the storage space of upstairs serves both as a suitable storage space for foodstuffs and as an insulating layer above the ceiling of the main living room, aiding in regulating the temperature inside the living room and preventing the ceiling from coming into direct contact with outside air. It prevents the living room's ceiling from heating up in summer and from direct exposure to cold outside air in winter, making the central living room cooler and more comfortable compared to similar spaces and neighboring rooms on the northern façade of the building.

Therefore, a space initially intended for storing garden produce and processing raisins and preserving almonds and walnuts has transformed into a protective layer for the main room of the house against the harsh climate of the region. This choice serves as a revealing and motivational example for modern architectural methods and decision-making processes. Preserving and restoring these houses, encouraging their renovation for new uses, and introducing them on a national and global scale to create similar examples could be both educational and inspiring.

References

1. Akhzari A (2024) Tārīkh-i Najafābād az Āghāz tā 1343 [The History of Najafabad from the Beginning to 1964] [Unpublished Manuscript] (in Persian)
2. Al-Isfahani MM (1989) Niṣf-i Jahān fi Ta'rīf-i al-Iṣfahān [Half of the World, in Description od Isfahan] (Sotoudeh M, Ed.). Amīrkabīr, Tehran (in Persian) (Original work published in 1897)
3. Brugsch HK (1988) Safarī bi Darbār-i Sultān-i Sāhib-Qirān: 1859–1861 [Reise Der K. Preussischen Gesandtschaft nach Persien 1860 und 1861] (vol 2) (Kordbacheh MH, Trans.). Iṭṭilā'āt, Tehran (in Persian). (Original work published in 1863)
4. Comte de Sercey FE (1983) Īrān dar 1839–1840 m.: Sifārat-i Fawq al-'Ādah-yi Kunt du Sirsī [Une Ambassade Extraordinaire: La Perse en 1839–1840] (Eshraqi E Trans.). Markaz-i Nashr-i Dānishgāhī, Tehran (in Persian). (Original work published in 1928)
5. Diba D et al (2011) Khānah-hā-yi Iṣfahān [Maisons d'Ispahan] (trans: Qasem Sichani M et al). Islamic Azad University, Isfahan (in Persian). (Original work published in 2001)
6. Esmaeili M (2006) Āmūzish va Parvarish-i Najafābād dar Bastar-i Tārīkh [The Education System of Najafabad in the Context of History]. Ghazal, Isfahan (in Persian)
7. Jaberi Ansari MH (1999) Tārīkh-ī Iṣfahān [The History of Isfahan] (Mazaheri J, Ed.). Mash'al, Isfahan (in Persian)

8. Jebel Ameli A (1996) Khānah-hā-yi Mu'āṣir-i Iṣfahān [The Contemporary Houses of Isfahan]. In: Majmū'ah Maqālāt-i Kungirah-i Tārīkh-i Mi'māri va Shahrsāzī-yi Īrān [The collection of Articles in the History of Architecture and Urbanism of Iran Congress], vol 4. Cultural Heritage and Tourism Research Organization, Tehran, pp 99–136 (in Persian)
9. Karbasi A (2018) Mustanad-Sāzī-yi Kālbadī-Ijtimā'ī-yi Chihil Khānah-yi Tārīkhī-yi Shahr-i Najafābād [The Socio-Physical Documentation of Fourty Historical Houses of the City of Najafabad] [Unpublished Research Project]. Historio-Cultural Buildings and Contexts Research Department, Cultural Heritage and Tourism Research Organization, Tehran (in Persian)
10. Karbasi A (2020) Chihil Khānah, Kālbad va Zindigī dar Chihil Khanah-yi Tārīkhī-yi Najafābād [Forty Houses, the Body of the Building and Life in Fourty Historical Houses of Najafabad]. Cultural Heritage and Tourism Research Organization, Tehran (in Persian)
11. Karbasi A (2023) Gūnah-shināsī-yi Ayvān dar Khānah-hā-yi Tārīkhī-yi Najafābād [Typology of Terrace of Historic Houses in Najafabad]. In: MMI vol 13, no 35, pp 19–36 (in Persian)
12. Khalili F (2015) Muqaddamah-i bar Tārīkh va Farhang-i Mardum-i Najafābād [An Introduction to the History and Culture of the People of Najafabad] (3rd edn). Mihr-i Zahrā, Najafabad (in Persian)
13. Mohandes MH, Mohandes A (2019) Safarnamah-yi Tihran bi Shiraz [The Travelogue of Tehran to Shiraz] (Abrishami F, Trans.). Khanah-yi Tarikh va Taṣvir-i Abrishami, Tehran (in Persian). (Original work published in 1890)
14. Mortazavi M (2017) Khātirāt-i Hashtād-Sālah-yi Duktur Siyyid Mustafā Murtaẓavī [The Eighty-Year Memories of Doctor Seyyed Morteza Mortazavi]. Mihr-i Zahrā, Najafabad (in Persian)
15. Motahhari M, Noormohammadi Najafabadi 'A (2013) Munugirafī-yi Shahristan-i Najafābād az Dīdgah-i Mudīrīyat va Barnamah-Rīzī [The Monograph of the City of Najafabad from the Management and Planning Viewpoint]. Mihdī Muṭahharī, Najafabad (in Persian)
16. Savory R (2010) Īrān-i 'Aṣr-i Ṣafavī [Iran Under the Safavids] (trans: Azizi K). Markaz, Tehran (in Persian). (Original work published in 1980)
17. Shirvani Z (1832). Bustān al-Sīyāhah [The Garden of Explore] (Mostowfi A, Ed.). Sanāeī, Tehran (in Persian)
18. Yazdani Najafabadi A (2017) Risālah-yi Dīyārīah-yi Khurākūshk [The Treatise on the Land of Khurakhushk]. Mihr-i Zahrā, Najafabad (in Persian)

Brick Decorations in Persian Vernacular Architecture: Case Study of Dezful

Amir Tayyebi and Kourosh Momeni

Abstract The traditional architecture of the historical cities of Iran (post-Islam era) has always expressed the Iranian-Islamic culture of the native people. Architecture is a unique art that has always become more complete and enriched due to the evolution of the experiences of masters in different periods of history. The old architects have used various methods in the building to respond to the climatic, cultural and social needs of the residents. Islam's emphasis on respect for hierarchies and emphasis on privacy of spaces in Muslim architecture and urban planning, including the city of Dezful located in Khuzestan province of Iran, has caused this feature to affect all social and cultural aspects. The houses of the Qajar era in Dezful city are a valuable example of contemporary Persian architecture, which have been well adapted to the special climatic conditions of Khuzestan. On the other hand, in the specific architecture of the underground levels of traditional houses in the warm and semi-humid climate of Iran, including in the city of Dezful, one can see solutions of climate design and accordingly meeting human needs in order to provide comfort. The main element of the old body of Dezful city is brick. The use of brick includes all the components of the structure to the decorations of the old texture of Dezful. With the skill of architects and old masters, many beautiful and diverse decorative combinations of bricks have been formed. Knowing the solutions of architects in the old context of Dezful city, especially the houses of the Qajar dynasty period, can open the way for contemporary Muslim architecture. In this chapter of the book, with studies and field observations of the houses of Dezful city; Eight prominent traditional houses are introduced and their architectural features such as the interior and exterior, the public and private zoning of the house spaces, the use of the covered porch, the middle and intermediate corridors between the spaces, the way of controlling the openings, documenting and introducing the types of Brick motifs of Dezful texture are investigated.

Keywords Brick decorations · Vernacular architecture · Brick · Qajar architecture · Dezful

A. Tayyebi (✉) · K. Momeni
Architecture and Planning Department, Jundi-Shapur University of Technology, Dezful, Iran
e-mail: a_tayyebi@jsu.ac.ir

A. Cheshmehzangi and S. Roaf (eds.), *Persian Vernacular Architecture*, Urban Sustainability, https://doi.org/10.1007/978-981-96-1116-4_6

1 Introduction

According to Yavari [17: 12] "*Ancient Iranians believed that "art" consists of two words "ho" and "nare" and it means a good person, a good man and a good woman, so the work of art is goodness, beauty and beauty*". Hegel says about this: "*Art makes the perceptible spiritual and the spiritual perceptible*" [17: 13]. This word has a conceptual meaning and can be examined and pondered in many fields. It is one of the important aspects of art and its flourishing in architecture, whose peak creativity and beauty can be seen in decorations which conveys concepts to the viewer not in the language of speech, but in the language of role and design, and overcomes the world of matter.

Architectural decorations are a type of art innovation that is an integral part of Iranian architecture, and its use has been common throughout the history of Iranian architecture and has always had a tremendous impact on the audience. Human always seeks perfection and beauty and tries to beautify his place of living, and to achieve this goal, he uses various motifs and motifs. According to Rahnavard [15, pp. 77–78] "*decoration in Islamic art is to express the holy atmosphere. Ornament, which is evaluated as one of the visual foundations of Islamic art, is a visual means or expression to honor the "material": surface, color, line, volume, brick, mud, plaster, tile, etc. to the highest horizons. They are exalted and find color and semantic identity and finally a supernatural personality and become spiritual and divine*". Decoration is one of the important factors in Iranian architecture, which makes the facade of the building out of the dry and soulless state and gives it a special identity and character.

According to Zarrin Kamari [18: 8] "Iranian architectural decorations are not only used for beautification, but they are components that have a multiple use, like the finishing piece of a puzzle". The artistic value of the decorative arrays and arrangement of the buildings can be considered to be influenced by the interaction of different cultures and ethnicities from a special aspect and it can be said that "artists have been able to express many cultural and religious concepts with the symbolic language of designs and patterns" [5]. The types of decorations in the Islamic architecture of Iran include: plastering, tiling, bricklayin, carving, mirror work, painting plant and geometric designs, and calligraphy and divided into several categories. These divisions include human, animal, plant, geometric and linear motifs (calligraphy), also the effective factors in the process of architectural decorations can be categorised as follows: expert manpower, materials and materials, the role of governments, religion and Scientific, cultural and artistic centers [9]. Decorations are the result of cultural and social developments of their time and it can be claimed that bricklayin is one of the most beautiful and artistic masterpieces of their time.

The art of laying bricks, in order to create decorative facades appropriate to the shape and use of the building, is called bricklayin, which some like Sir Edward Lutis introduces it with the attribute of Iranian magic [13]. Brick is one of the indigenous and efficient materials in the architecture of Iran. As a result, due to its easy access, high resistance to environmental conditions and the possibility of creating beauty in small and medium-sized buildings, it was used as the main building material.

Since the beginning of work and since the use of brick, which is considered as the primary building material, Iranian architects have created the most beautiful patterns and designs when constructing walls and covering domes and creating earrings, muqrans and arches, and in Its evolutionary process has created unique masterpieces with Gereh-chini,[1] Gol-andazi[2] Khofte-Rafteh[3] [7: 7]. Although this element is the smallest part of a huge and high building, it plays an essential role in design, shape, structure and decorations and can be widely used in all spaces and architectural components. He created innovative things, so that many words have been recorded in the field of this artistic method: "*Don't talk about Iranian bricklayin, but talk about the magic of Iranian bricks*" [14].

In general, bricks can be used in architectural decorations in two ways: 1. Plain and unglazed bricks that are ocher and pea colors 2. Glazed bricks that can be seen in different colors. In the field of brick decorations, Iranian architects and artists, as evidenced by the works left from the 4th century AH onwards, have tried various methods that can be divided into several groups: 1. Rag-chini[4] 2. Gol-andazi 3. Gereh-chini 4. Colored Gereh-chini 5. Khofteh-rafteh or Patkin[5] 6. Khavoon-chini[6] 7. Bricklaying with decorative patterned or stamped bricks" [7, p. 63]. Bricklaying, as an integral part of Iranian architecture, stands out as a beautiful and artistic masterpiece, reflecting the magic of Iranian craftsmanship.

2 Methodology

In this study, the methodology employed involves a detailed investigation and analysis of architectural features, specifically focusing on the brick decorations in the traditional houses of Dezful city, with a special emphasis on the Qajar period. The following steps were taken to conduct the research:

1. Historical Context Examination: The historical context of Iranian architecture, with a focus on the post-Islam era and the specific characteristics of Dezful city, was explored to understand the cultural and social influences on architectural styles.
2. Literature Review: Existing literature on Islamic art, Iranian architectural decorations, and specific bricklaying techniques in historical Iranian architecture were reviewed to provide a theoretical background for the study.

---

[1] Creating a beautiful surface by intertwining geometric shapes.

[2] Decorative bricklaying on the facade of the building, for example, by creating a flower pattern.

[3] After and before arranging the bricks one by one in the wall face.

[4] A row of bricks, clay, tiles or other than that they are stacked on top or next of each other.

[5] A method of corner construction in the outer corner of the building that used in Khuzestan and in the alleys of Dezful.

[6] This method of bricklaying is used in Shushtar and Dezful.

3. Case Study Selection: Eight prominent traditional houses in Dezful city, representing the Qajar architecture period, were selected as case studies for in-depth analysis.
4. Field Observations: On-site field observations were conducted to study the architectural features of the selected houses, focusing on interior and exterior spaces, public and private zoning, covered porch usage, corridors, openings control, and documenting brick motifs.
5. Identification of Brick Decoration Types: Different types of brick decorations prevalent in the Qajar architecture period, such as Rag-chini, Gol-andazi, Gereh-chini, Colored Gereh-chini, Khofteh-rafteh, Khavoon-chini, and Bricklaying with patterned decorative bricks were identified and documented.
6. Analysis of Architectural Elements: Architectural elements unique to Qajar houses in Dezful, such as Jolokhan and Hashti, Iwan and room entrance areas, covered Iwans, and privacy considerations in spatial design, were analysed across the case study houses.
7. Comparison and Interpretation: Findings from the field observations and analysis were compared and interpreted to understand the architectural principles, design choices, and privacy features integrated into the traditional houses of Dezful during the Qajar period.

3 Historical Context

Architecture, as a point of view or a desire that changes and transforms or descends over time, has its roots in culture, which also never accepts a fixed form. Based on this, how to want architecture and how to decorate it are two separate moments of a single phenomenon. The first is considered as a cultural moment and the second as a technological moment, expressing architectural culture, groups, and societies.

Many factors affect artistic works and architectural spaces, which are part of culture. This effect is sometimes clear and sometimes subtle and not obvious. Such as the effect of the cross in the design of historical churches in an obvious way and the hierarchy between all types of spaces, including the elements of the entrance space of the Iranian Grand Mosque and its cultural meaning and content, which remains hidden from the observer's view in the first stage. Some of the characteristics of the culture do not have the same effect on the architectural space, some are easy and some are difficult to identify. One of the effects of culture in arts can be seen as the importance of some numbers and shapes in decorations. One of the reasons for paying attention to numbers is related to man's understanding of the order of the world in the past, when he received it in the most tangible form, i.e. numerical order. Many of these decorations and lines were specially used in constitutional activities and ceremonies, arts, schools of thought, etc. The use of any number or shape in any of the activities or architectural arts, especially decorations, on the one hand depends on the visual and counting features of that activity and on the other hand, on the features and characteristics of that shape or number. Therefore, shapes and numbers were considered and used in architectural art that are compatible with the mathematical, geometric, quantitative and tangible features of this art and also with the general culture of this land. Such as the number four, characteristics of square and circle, common characteristics of circle and square, cube, spherical volumes, and Chahar-taghi.[7]

The widespread use of bricks in Iranian buildings is rooted in the events and developments of Iranian society. The most important building element, both structurally and decoratively, is the brick element, which, while simple, has given the building a special charm. Bricks can create thousands of combinations due to the variety of geometric shapes and sizes. Mainly, bricks are used as an arrangement on the same level, such as Khofte-va-Rasteh, or by placing bricks back and forth to create shadow and volume, such as Khavoon-chini, and to create prominent volumes such as a prominent Ghatar-bandi[8]" [9: 119]. Iranian architects have tried to preserve the organic state of this material in relation to the building by widely using brick in various types of decoration.

3.1 Brick Decorations in the Qajar Period

During the Qajar dynasty, unfortunate events happened in Iran and the country suffered weakness in various fields, also the Isfahani style went into decline during this period. At this time, when Iran's relations with the West increase, Iranians experience self-loss and are influenced by the culture and industry of Western civilization. On the other hand, due to the propaganda that was carried out by some people inside the country; Following western style and standards intensified.

According to Kiani [6] in the architectural landscape of Iran following the Safavid period, a fascinating interplay of communication, interaction, and imitation with the West led to the emergence of fusions and eclecticisms within architectural styles. During the Qajar era, this dynamic continued as ambassadors, merchants, and court travelers facilitated cross-cultural exchanges. Notably, the semi-circular Roman arch gained prominence in Iranian architecture, gradually evolving into a distinctive style. Initially influenced by past architectural traditions, it eventually acquired a unique character in contemporary architecture—a hallmark of Qajar design. However, as time progressed, this stylistic evolution shifted toward pure imitation or complete acceptance.

Although architecture in the Qajar period was considered weak compared to the previous periods such as the Safavid period, the greater connection with the West caused the standards of Iranian architecture to mix with the elements of Western architecture and create beautiful and pleasing works and innovations from the artistic aspect. Among the innovations of this period was the creation of space and giving

[7] A building with a domed ceiling that is built on four columns and its four openings are raised and open.

[8] Standing bricklaying around the edge of the roof of the building.

importance to the facade and exterior spaces in such a way that introversion gave way to extroversion and the art of bricklaying with various motifs to beautify the exterior of buildings flourished.

The art of bricklaying of the Qajar period does not reach the flourishing period of bricklaying of the Seljuq period, nor the scale of the period of the very diverse brick art creations of the Safavid period [19], however, in the facade of the building, the borders of the frames, the scrolls of inscriptions on the Sardare-bana[9] tiles and Espar[10] with Ajore-tarash[11] and Pish-bor,[12] bricklaying and more, with simple or patterned combinations, have been used a lot, as it was said in this period due to the importance of finding the outside space, bricklaying is used in a variety of ways and with differences compared to the past for decoration. It has been taken [9] in such a way that the art of ax flourished and the bricklayers showed their original tastes, and creativity in creating patterns became one of the characteristics of this period.

The bricks were different in different periods. This difference was visible in the dimensions, color, geometric shape, and the way the brick was cut. In the Qajar period, the bricks had their own characteristics, so the bricks were mostly square (the most common sizes are 4 * 19.5 * 19.5), as well as molded and turned bricks, which are in size And various geometric shapes, especially round ones, are used and belong to this period. Koore-malat[13] brick is another example of bricklaying of this period, which had a great impact on building facades [7].

3.2 *Brick Decorations in Dezful*

Brick is the main material due to the limitation of using wood in these areas of the country and the problem of using stone due to heat transfer in tropical regions, as well as being economical and finally easy to use and the high ability to apply and use brick in all sectors. The buildings, especially their great efficiency in erecting wide and tall covers, were noticed in the architecture and art of architecture in all the vast areas of Iran. But the wide use of brick shows two important technical points regarding the climate of this area:

1. The coefficient of expansion and contraction of brick against the cold is such that it prevents the creation of cracks in the building.
2. Compared to other materials, brick has the ability to store thermal energy, and as a result, the transfer of thermal fluctuations of the external environment to the

[9] The upper part at the threshold of the house. Ornaments above the door of the house.

[10] A wall of the building, which is not load- bearing. A part of the facade of the building that does not have doors and windows.

[11] A brick taht at the work to the sizes of various planes are cut and ready to work with and each of its pieces has a special name.

[12] A type of brick that used in columns and ornamental strips.

[13] Vertical seams in briclaying.

inside through a brick wall is not severe, and without the intervention of heat-generating and cooling devices, more pleasant conditions and atmosphere can be provided.

The oldest bricks used in Dezful can be seen in the oldest building of this city, the old Sassanid bridge, which are square bricks from the Sassanid period. The course of changes in brick dimensions in Dezful after the Sassanid bridge in the tomb of Sheikh Ismail Qasri with bricks of 30 * 30 cm and 6 cm thickness and finally in the Safavid period to 18.5 * 18.5 cm and 5.5 cm thickness. 3 cm can be seen, the bricks of the Safavid era are also used with some minor changes in the Qajar period, and the Zargar house, as one of the most significant buildings left from the Qajar era, has the same type of brick, and this is the main material. It is the constituent and filling element of this building (see Fig. 1). On the outside of the door of this house, you can see the beautiful decorations of brick tiles. Regular Gholve-chini[14] along with fine brick Gereh on the ceiling and floor of the Hashti[15] are other features of this building. The yard is completely paved with traditional bricks and there are small crescent-shaped Iwans[16] around the yard.

Iranian architects (according to the surviving works) have tried various methods in the field of bricklaying from the fourth century AH onwards, which include:

1. **Rag-chini**

"Bricklaying by combining bricks of the same color and creating different designs and patterns on a smooth surface, which is also called the Rag-chini. There are many different designs that depend on how the bricks are placed in the facade of the building, and some designs are named after the position of the bricks in the facade" [7: 314]. If the bricks are placed on top of each other in a simple and horizontal way, they are called Rag-chin [11: 70].

2. **Gol-andazi**

In this type of facade construction, the bricks are combined in such a way that each brick goes back one Kalook[17] compared to the previous row to create the desired pattern in the middle [11: 72]. In this type of bricklayin, the arrangement of bricks is such that it creates a pattern. In this way, the bricks are combined in the form of five lines, seven lines and nine lines, and it is called Gol-andaz. Such designs are sometimes "Hamroo" or flat, or "Hashtogir[18]" or raised and locked" [3: 85].

---

[14] Work with river stones.

[15] A variant of the entrance door of the building, the connecting organ between the inside and the outside, is located in the old vestibule buildings after passing the threshold of the door and before the corridor.

[16] One of the oldest fundamental organs in Architecture of Iran.

[17] A piece of brick equal to 1/8 of that brick.

[18] Bricklaying bond between two walls that are built next to each other or two walls that are perpendicular to each other.

Fig. 1 A view of the front of the Zargar house. *Source* Photo taken by the authors

3. **Gereh-chini or Gereh-sazi**

Gereh-sazi, a delicate and productive technique of decorative bricklaying, involves assembling variously sized cut and axed bricks. These bricks are meticulously combined to form intricate geometric patterns, including triangles, rhombuses, squares, rectangles, and trapezoids. By skillfully interlocking these elements, craftsmen create captivating designs such as polygons and star shapes [7].

4. **Colored bricklaying or colored Gereh-chini**

"In this way, different designs are created with bricks of different colors and in a flat manner. The use of colored bricks in the beta facade dates back to the Ilkhanian period. In the historical buildings of Shiraz, we see the use of various examples of this bricklaying method over several centuries" [7: 315].

5. **Khofteh-rafteh bricklaying or (Patkin)**

According to Bozorgmehri [3: 85] Bricklaying occasionally results in a raised and interlocking design known as 'Hashtogir.' Octopuses exhibit significant diversity. The renowned Hashtogir style, exemplified by Hashtogire-shirazi, involves repeating depressions and prominences across six Bands[19] and five rows (Raj or Rag). This brickwork technique is referred to as 'Shesh-bande-Shirazi'.

6. **Khavoon-chini**

According to Kiani [7] in the cities of Shushtar and Dezful, a unique bricklaying method known as 'Khavoon-chini' was widely used. This technique involved creating small mosaic-like patterns called 'Ajore-tarash' (3 × 3 × 3 cm) on the surfaces of brick buildings. By strategically placing these patterns, builders could utilise shaded hollows to reduce heat absorption into the structures. Additionally, Khavoon-chini paintings drew inspiration from the 'Chalipa[20]' pattern, which was creatively applied in various combinations. Naema [10] says Khavoon-chini involves using brick as a decorative element, utilising its thickness to create various decorative patterns on building facades. Essentially, Khavoon-chini arranges the building's design in the form of Amood.[21] It is a kind of decorative pattern with bricks and sometimes with simple blue bricks and tiles, which is made above the edge of the entrance facades and entrances to create a light shadow, and it was especially common in the southern regions of the country [3: 87]. Aminzadeh [2] also says Khavoon-chini, a vertical form of bricklaying, saw significant development during the Safavid era. It persisted until the end of the Qajar period and into the early Pahlavi period.

7. **Bricklaying with patterned decorative bricks (Mehri[22])**

To prepare Mehri bricks, mud is poured into a mold with the desired design on the bottom. This design is usually raised and sunken, and after drying and shaping the mud, it is taken to the oven. Mehri bricks are sometimes placed between two rows of simple bricks or combined to create various motifs [3: 84].

4 How to Use Bricks

Due to the special and unique use of bricks in buildings, Dezful architecture includes two parts: physical use and decorative use of bricks, which is applied by appropriate use of the number and combined shape of bricks, as well as measurements, distances, and arrangements.

Fakhro-Madin[23]

Bozorgmehri [3] says in this specific bricklaying technique, various Gereh motifs are created using bricks. When the brickwork on the facade serves as a source of

[19] The distance between two bricks.

[20] A kind of Iranian decorative pattern.

[21] A layer of brick or tile or plaster that covers the roughness of the buildings.

[22] A kind of bricklaying that may be: Four bricks side by side. Single brick among the simple bricks ...

[23] Bricklaying to make a grid of full and empty.

illumination, knot motifs are crafted in a hollow manner Fakhro-Madin is one of the types of bricklaying that creates a mesh space. In addition to the passage of light and ventilation, it creates unique shadows in the interior. According to Pirnia [12] At times, garden and religious enclosures feature mesh fences, allowing outsiders to glimpse their beautiful interiors. This design was evident in the walls surrounding Isfahan's four gardens and the Iwans of the bazaars near Naqshe Jahan. These walls, constructed from fired clay pieces, incorporated both geometric and non-geometric shapes. The deliberate arrangement of holes in these structures aimed to enhance their aesthetic appeal, and they were referred to as Fakhro-Madin networks.

Using bricks to create a parapet on the roof (Riya[24])

Naema [10] says Riya refers to the use of brick mesh walls as roof parapets. These walls are one height high and serve to create a private and secluded yard, protecting the roof. In Dezful, residents utilise the roof as a functional area and as a leisure space during certain months.

Blade use

According to Naema [10] Dargah[25]s who opted not to use them employed a brick blade resembling a Kalagh-par,[26] referred to as Hejabe-Dargah. If this blade was utilised, it would subsequently be destroyed. Locally, the term for straticulate brick is "Kalagh-par." Additionally, the Ghenas[27] blade used in the Suzangar house is visible in the walls and Mahtabi[28]s of the second floor.

Fereize[29]

According to Naema [10], the brick Feriez is typically arranged in two rows and can be removed by using an ax and soaking it with water. Additionally, Bozorgmehri [3] notes that the Feriez is positioned in front of the surface of the Khavoons, serving as protection against precipitation and preventing it from falling directly onto the Khavoons.

The use of bricks in the creation of stairs

According to Naema [10], brick staircases typically consist of twelve to fourteen steps, connecting one floor to another. Notably, the initial step's height is intentionally greater than the subsequent steps to prevent children from climbing it. These staircases are constructed using bricklaying techniques and are commonly found in houses, taking the form of Herreh (Ghadnama[30]) design.

---

[24] The wall of the parapet of the roof in the language of the people of Dezful.

[25] Doorway.

[26] Piece of three-sided brick. A method of bricklaying the floor of the courtyard or room.

[27] A type of brick that is cut into triangle.

[28] Terrace. It is called Mahtabi (moonlight) because when the air is warming in the spring, they sleep in it at night and the moonlight.

[29] Ornamental frame on the facade of the building and above the head of the door maked by brick.

[30] A brick that used from the side and the length standing in the façade.

The use of bricks in the construction of columns

Naema [10] says in Suzangar house, when constructing a column, they start by shaping the bricks into a round form. These rounded bricks are then stacked on top of each other. If they need to create a window adjacent to the column, they transform it into a side column, which is essentially a 1/4 column. At the column's end, you can observe various brick patterns known locally as "Kartelal".

5 Case Study of Dezful City

Knowing the solutions of the old architects of Dezful City, especially in the houses of the Qajar era, can open the way for modern Islamic architecture. Eight important traditional houses in Dezful city and their architectural features such as spaces and brick decorations, interior and exterior areas, public and private spaces, etc. are introduced (Table 1).

5.1 Privacy in the Architecture of Dezful Houses

The standards of creating space in an Iranian house are based on respect for privacy. Harim, which is a spatial reflection of respect, regulates the relationship between the individual and society. Harim is the limit of the private distance of the most secluded person in the house to the most public gathering of family members and relatives in the house in the form of intermediate and successive spaces, the criteria that shape the spatial levels of the house [4].

The religion and culture of Dezful City create the need for introversion in people and houses. Therefore, privacy has been one of the most important factors in the formation of Dezful houses. In the traditional Dezful residence, the rooms of the house can be located on any front, depending on the location of the land and the way of entering the house. The room and the Iwan are often located next to each other and each has a common side with the yard. In the more advanced traditional houses (belonging to the Safavid period), there was a communication space between the room and the Iwan called Gholamgardesh which is a corridor or a room between the other two rooms that have access to both (Table 2).

In a traditional residence of Iran, there is a separation between the place of activities related to the outside relations of the man of the family and the place of family activities, which is in the form of separation either on one level or between two different levels. In carrying out activities related to the family, no space is assigned to a specific activity and each space is used multiple times at different times of the day. In Dezful houses, the spaces of the residential unit can be divided into four categories: living spaces, intermediary spaces, service spaces, and open spaces [1].

Table 1 Characteristics of case study houses in Dezful

Name	Description	Plan	Picture
Qalambar house	One of the special features of this building is the direct connection of its Shavadan[31] to the Shavadans of nearby houses. Also, its ground floor is 30 cm higher than the yard floor by three or four steps. According to the residents of the neighborhood, the first reeds of Dezful were planted in this house, in terms of architectural and historical value, it is included in the number of national works and monuments	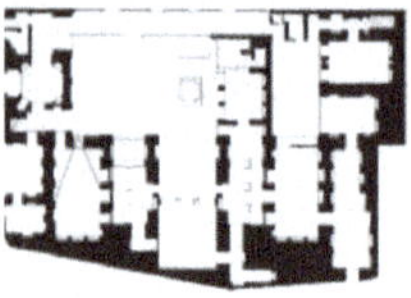	
Abdi house	The mentioned building is located at the end of an alley in Sakian neighborhood. Its entrance was once located under a Sabat[32] called Abdi Sabat. Unfortunately, this Sabat was destroyed by unknowing human factors and without obtaining the necessary legal permission in 2005. The building entrance has a Sardar with beautiful brick decorations and an inscription with the date 1353 AH	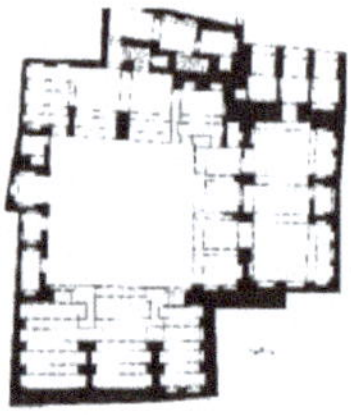	
Suzangar house	Suzangar house is one of the valuable buildings of Dezful, which is located in a cul de sac. Among the features of this house, we can mention the small dome of the Hashti with Pabarik[33] and brick Shaparak[34]'s resting on thick columns, also the floor of the Hashti with regular and delicate cobblestone and the floor of the courtyard is decorated with bricks	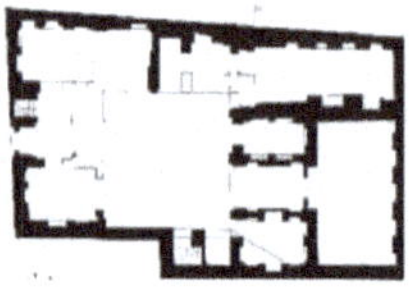	

(continued)

[31] The deep basement in the old houses of Dezful and Shushtar, which is very cool and can be used during the day when it is hot.

[32] The covered part in the old alleys.

[33] A part of Karbandi that has shape of a rhombus with two elongated legs.

[34] One parts of Karbandi and Karbandi is kind of ceiling in Iranian architecture.

Table 1 (continued)

Name	Description	Plan	Picture
Shamsodin (Sayyed Sadr) house	This building has a small and beautiful Sardar with simple bricklaying decorations, which stands out with an old wooden door in a neighborhood called Katkatan. The small and beautiful Hashti of the building has created a suitable space for creating privacy, and with a low staircase, the privacy of the building is fully defined	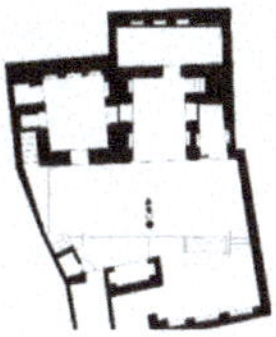	
Firouzi house	The House is located at the highest point of the historical site overlooking the Dez River. One of the special features of this house is the variety of its bricklaying, more than 25 types of bricklaying are used in its facade decorations. The location of this house in the vicinity of monuments such as Sabat Azizipour and Pella Bachilon and the untouched urban fabric of the area has given it a special place	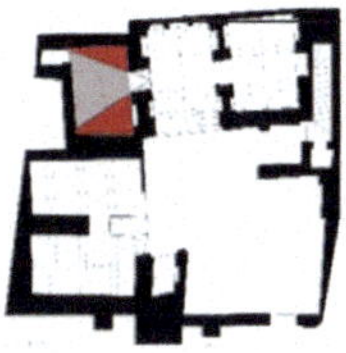	
Tizno house	This construction style is exactly suited to the climate and lifestyle in this region. This reason has led to the division of the buildings into two internal and external parts. Also, preserving privacy for the people of the house is one of the special features of these buildings. Kalil,[35] Tagho-Tavizeh[36] are decorative elements used in the building	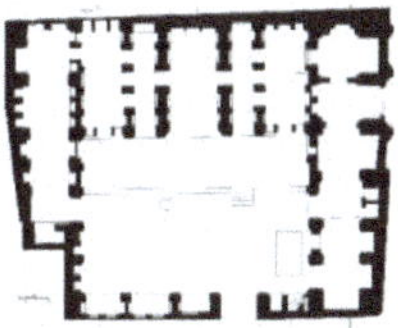	

(continued)

[35] A type of Arc.

[36] A kind of combination of arch and opening.

Table 1 (continued)

Name	Description	Plan	Picture
Delvar house	This house has a large Hashti in proportion to the surface of the building, which makes up an area of about a quarter of the entire surface of the building. This building has two residential Iwan and another Iwan that opens to the entrance Hashti. According to the proportion of the entrance Hashti, whose occupancy level is more than the yard of the house, the importance of privacy in the presence of residential spaces is determined		
Sohrabi house	The historical house of Sohrabi belongs to the Khan[37] of Siahposhan neighborhood named Jafar Khan Sohrabi. The large front of this house with beautiful and unique bricklaying decorations and a beautiful Hashti is one of the architectural elements that are present in most of the old houses of Dezful. This building in an exceptional way enchants the eyes of every viewer with its beauty. Khavoon-chini decoration, which is typical of the beautiful and original traditional architecture of Dezful, is shown in a special way in this building		

Source Photos taken by the authors

Living spaces are among the spaces that have multiple functions and are used depending on their proportions. Intermediary spaces (as their name suggests) are intermediaries and interfaces between two or more spaces, and sometimes these spaces also take on a biological role. Among these spaces, we can mention the Iwan. Only service spaces such as kitchens and toilets or Pastoo[38] and warehouses have specific and defined activities. Other spaces are not assigned to a specific activity and are used multiple times at different hours (Table 3).

[37] A title that expresses nobility.

[38] A small storage behind the room or next to it in the house for placing the bedding items and …

Table 2 Spatial circulation of Dezful house plans

House name	Spatial circulation	House name	Spatial circulation
Qalambar	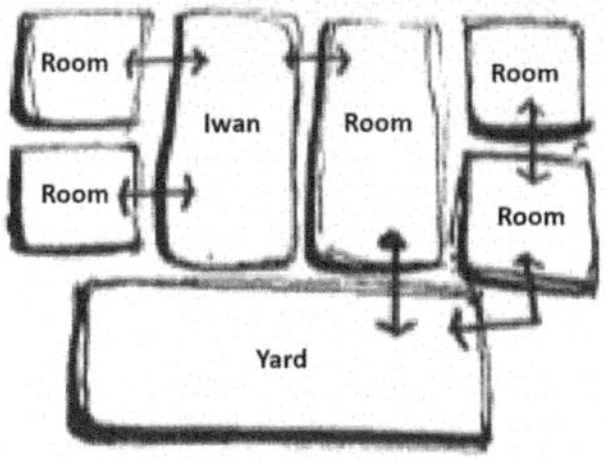	Firouzi	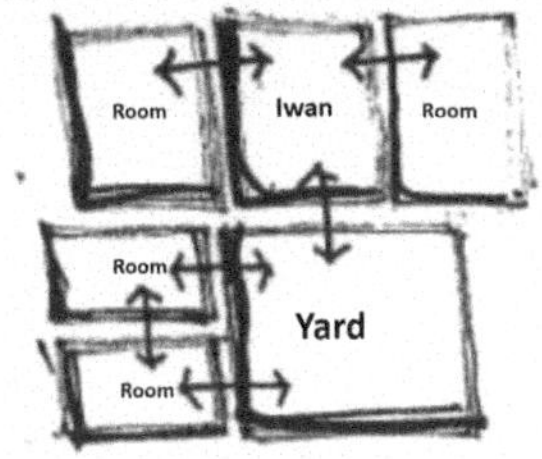
Abdi	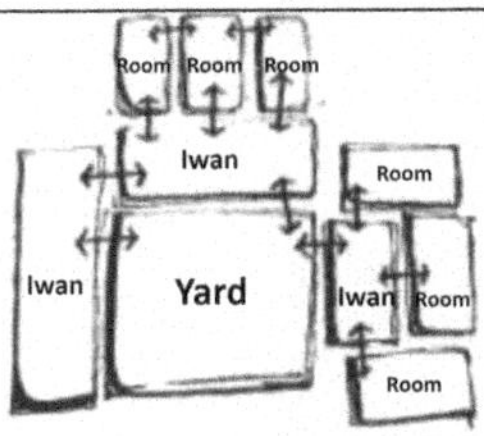	Tizno	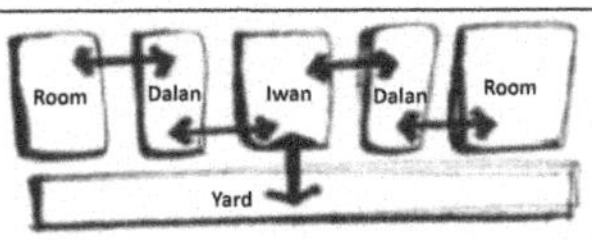
Suzangar	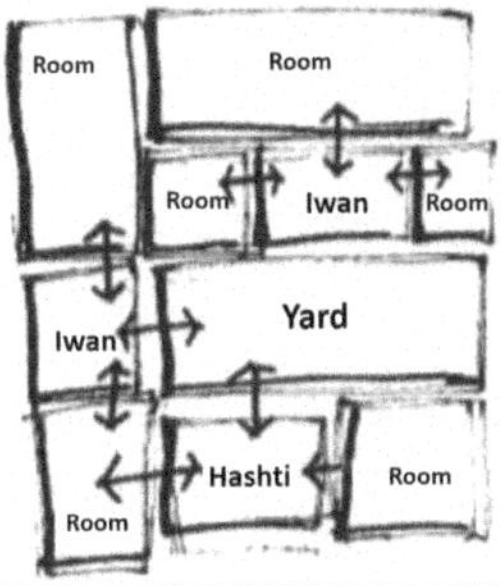	Delvar	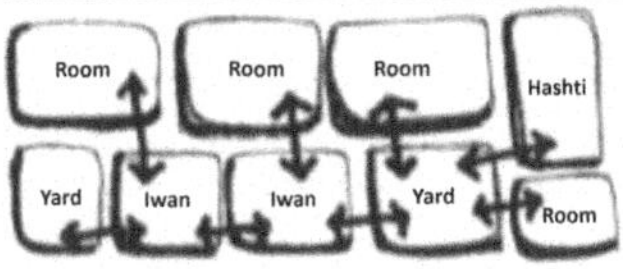
Shamsodin (Sayyed Sadr)	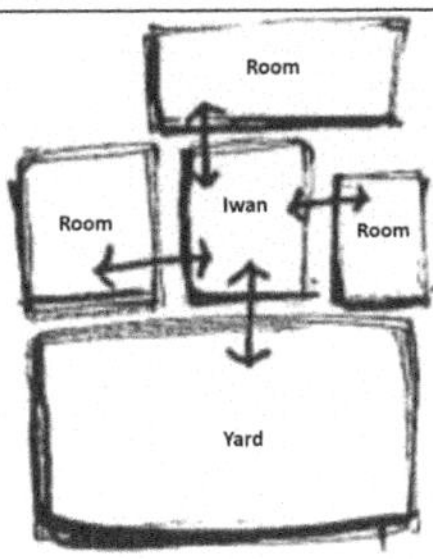	Sohrabi	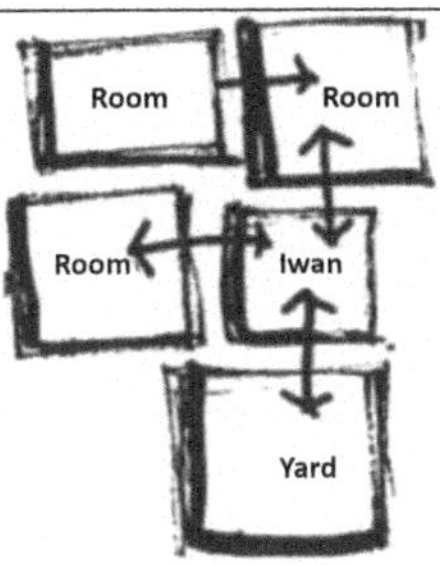

5.2 *Some Basic Architectural Elements of the Qajar Houses of Dezful*

After the visit, general review and analysis, elements were found in the architecture of the Qajar houses of Dezful, which can be used as models for the architecture of

Table 3 Public and private areas in Dezful houses

Abdi	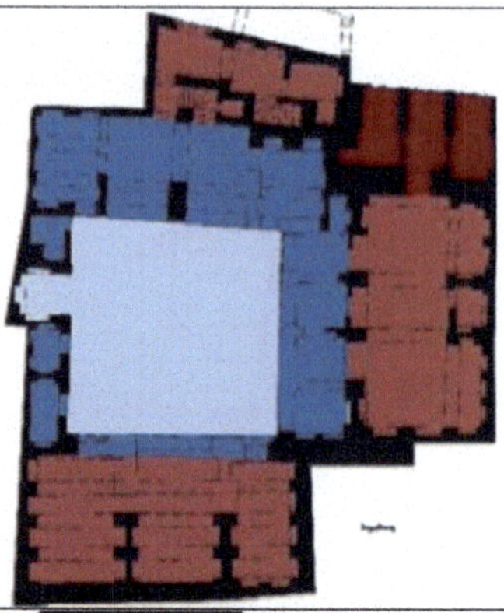
Qalambar	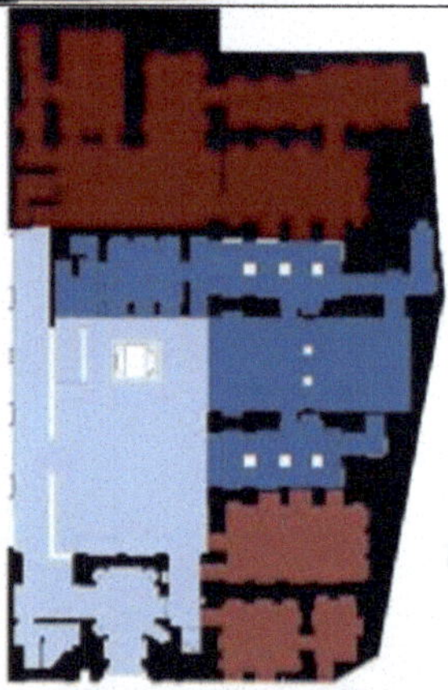
Suzangar	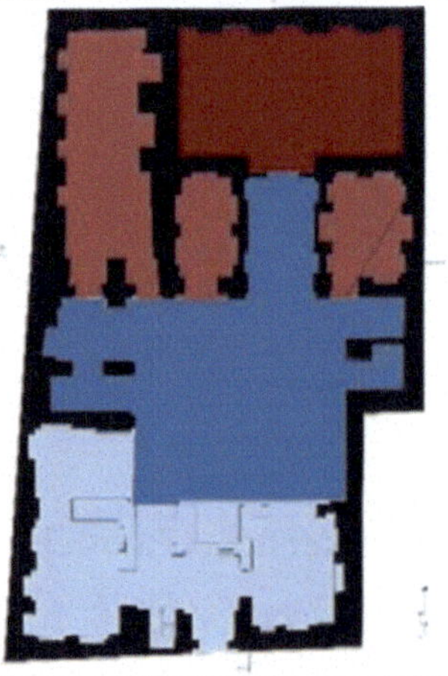
Public Public-Private Private Completely Private	

Table 4 Jolokhan and Hashti in the Qajar houses of Dezful

Abdi	Suzangar	Shamsodin (Sayyed Sadr)	Firouzi

Source Photos taken by the authors

the traditional houses of Dezful in this period by repeating them in other houses. They will be explained below.

5.2.1 Jolokhan[39] and Hashti

This architectural element, despite its small scale in the Qajar houses of Dezful, has various roles in social relations, explaining privacy, defining the connection between inside and outside, as well as respecting the role and position of the addressees. On both sides of the Jolokhan, there are usually platforms (called Pirneshin[40]) designed for pedestrians to rest and visitors to wait for the owner of the house, which plays an effective role in defining the sanctity and privacy of the houses (Table 4).

This space is often placed immediately after the entrance space and one of its functions is to divide the entrance path into two or more directions. The Hashti of the Qajar houses of Dezful is classified into intermediate spaces and depending on the surrounding space and spatial relationships, it may lead to spaces other than the yard. For example, in Suzangar house, the Hashti is connected to the roof of the house with a staircase.

5.2.2 Iwan and the Entrance Area of the Room

In traditional houses of Dezful, the need for privacy is provided in the form of connected spaces, connected to the boundaries or separate territories from each other. Privacy inside the house is related to the organisation and arrangement of spaces. In the public domain, it is necessary to have a space for receiving and receiving guests, and any audio and visual access between the space of the guest house and the main

[39] In front of the house.

[40] In traditional Iranian architecture is a platform that a place on both sides of the entrance door, for resting while waiting to enter or talking with neighbors.

Table 5 The Iwan and entrance space of the rooms of the Qajar houses of Dezful

Qalambar	Suzangar	Firouzi	Tizno

Source Photos taken by the authors

private rooms of the family, such as bedrooms, etc., must be completely preserved [16].

The entrance space here fulfills its mission well in separating the private area of the house from the public area by determining the sanctuary and adding an intermediate space. The Iwan and entrance vestibule (which is considered at the beginning of the room) prevents a person from seeing all the parts of the room by being in this place and limits the view of the person entering in a way (Table 5).

5.2.3 Covered Iwan

The Covered Iwan in Dezful Qajar houses is one of the main functional spaces, especially in the summer season, and has the following uses:

- In summer and spring as a room for sitting and resting (in the evening and at night), using the south Iwan in winter;
- Night sleep
- House working [8]

Therefore, it is important to maintain privacy and limit the view to the Iwan of these houses. This space uses openings to control the radiation and ventilation of the spaces. In addition, the view of this space is controlled and the privacy of the spaces behind it is maintained (Table 6).

6 Conclusions

A fine combination of regional cultural values and popular brick decoration, Dezful Qajar houses use their orientation towards geographical paradise lies to introduce privacy characteristics with optimum climatic design relevance in the architectural landscape. Although the principle of privacy is as important, it becomes way less

Table 6 Covered Iwan in Dezful Qajar houses

Qalambar	Abdi	Tizno	Delvar

Source Photos taken by the authors

interesting than how brickwork and inventive patterns in laying bricks stand out for these traditional buildings that really shows not only utilitarian usage but also stories from artistry human histories.

The use of brick in large masses is not only an aesthetically pleasing solution but also fulfills the basic requirement since it helps maintain temperature efficiently and promotes practical living in hot conditions, meeting the prerequisites for sustainability. Qajar era architects decorated their buildings in a variety of techniques such as Rag-chini, Gol-andazi, Gereh- chini and Khavoon -chini using the most simple element available (brick) to create designs which brought depth and character on facade.

Through these findings, it is recognised that the decoration of Dezful's traditional houses and spaces are not just decorations; they include values in various ways to effectuate given responses on a cultural level and practical responses towards natural conditions. The preservation and adaptation of these ancient brick-laying methods highlight how the present can learn from the past, encouraging contemporary architecture that reflects its traditional culture while still contributing to addressing modern requirements. This investigation on Dezful brick decorations could provide a value for enriching Contemporary Islamic Architecture. A synergy between old techniques and new methodologies of architectural practices can help us in creating a built environment that is not just visually appealing but also more grounded to its cultural roots as well.

References

1. Ahari Z, Amini Sh (1992) Building pattern of houses in cities of Khuzestan provinces. Road and Housing Development Research Center, Tehran (in Persian)
2. Aminzadeh B (2010) Evaluating the beauty and identity of a place. Hoviyate Shahr 7: 6–22(in Persian)

3. Bozorgmehri Z (2011) Iranian Amoods (recognition, pathology and restoration). Soroushe Danesh, Tehran (in Persian)
4. Haeri Mazandarani MR (2009) House in culture and nature of Iran. Urban and Architecture Research Center, Tehran (in Persian)
5. Khazae M (2002) A thousand inscription. Islamic Art Studies Institute, Tehran (in Persian)
6. Kiani M (2013) The place of decorative bricklayin art in the architecture of the first Pahlavi period. J Fine Arts Archit Urban 18:15–28 (in Persian)
7. Kiani MY (1997) Decorations related to the architecture of Iran in the Islamic period. Nader, Tehran (in Persian)
8. Mahmoudi A (2005) A review on importance of porch in traditional houses (case study: Bam). HONAR-HA-YE-ZIBA 22:53–62 (in Persian)
9. Makinejad M (2008) History of Iranian art in the Islamic period. Ministry of Culture and Islamic Guidance, Tehran (in Persian)
10. Naema Gh (1997) Dezful: city of brick. Cultural Heritage Organization of the country, Tehran (in Persian)
11. Pirnia MK (2002) Building materials (Ajhand, Andoud, Amood). Cultural Heritage Organization of, Tehran (in Persian)
12. Pirnia MK (2011) Introduction to Islamic architecture of Iran. Soroush, Tehran (in Persian)
13. Pope AU (1981) A survey of Persian art: from Prehistoric times to the present. Oxford University
14. Pope AU (1965) Persian architecture: the triumph of form and color. George Braziller
15. Rahnavard Z (2018) The islamic art wisdom. Samt, Tehran (in Persian)
16. Valizadeh Oughani MB (2014) Moral thoughts and principles in the spatial structure of Iranian traditional houses. Pazhuhesh-e Honar 7:47–60 (in Persian)
17. Yavari H (2009) A survey in Iranian art and architecture. Simaye Danesh, Tehran (in Persian)
18. Zarrin Kamari H (1997) General patterns of Iranian mosques. Abadi 54:8–17 (in Persian)
19. Zomorshidi H (1988) Tagh and Ghos in Iranian architecture. Keyhan, Tehran (in Persian)

Contemporary Design Interventions

Dances with Shadows: A Retractable Wood Cover of a Courtyard in Historic District of Yazd

Azadeh Khaki Ghasr and Hossein Pourmehdi Ghaemmaghami

Abstract The significance of shade in sweltering climates for sustainable living is underscored by the vernacular architecture of Yazd. Through a central water body, a tree canopy, and structures that encircle the courtyard, courtyards offer shade within the urban fabric of Yazd. These courtyards generate positive pressure, which propels cool air into interior spaces, thereby establishing a cool microclimate. Cool air from the courtyard replaces warm air extracted through high ceilings, windcatchers, and other exhaust systems. This design strategy reduces interior temperatures by implementing a natural ventilation system in arid regions. A Yazd courtyard's underground substructure, however, compromises its ability to generate shade and cool air. The retractable wood cover over Yazd University's courtyard named 'Ellieen' is the subject of this chapter's description and analysis of a recent design-build project. The cover's design generates dynamic shadows on the courtyard surfaces, adding a new aesthetic element to the context. The project's main purpose is to address the increasing need for improved comfort in traditional courtyards during hot seasons, driven by climate change and the need to adapt existing structures. Additional considerations are shade optimisation during the day by its changeability; material selection for the cover should prioritise heat reflectance, ventilation, and durability. Integration with architecture is important too. The courtyard cover should seamlessly integrate with the overall architectural design, complementing the aesthetics and functionality of the historical building. The concept of temporarily covering open spaces and yards in Yazd dates back to the past, with fabric covers called 'Poosh' used to shade religious places. To achieve the goal, the project attempts to shade the courtyard the courtyard via a cover, known as one of the open spaces of the Art and Architecture School. The courtyard is 10 m $\times$ 13.80 m^2 in size. There is no tree since the courtyard has a subterranean level under it as a library. Around the courtyard, there have been various activities, including a buffet for students and a prayer room. In addition, this courtyard serves as a link between other courtyards of the faculty complex houses. The design problem involves lightweight, responsive courtyard coverings that retain the original mass and space while allowing for a variety of activities in the

A. Khaki Ghasr (✉) · H. Pourmehdi Ghaemmaghami
Department of Architecture, Faculty of Architecture and Urban Planning, Shahid Beheshti University, Tehran, Iran
e-mail: a_khakighasr@sbu.ac.ir

A. Cheshmehzangi and S. Roaf (eds.), *Persian Vernacular Architecture*, Urban Sustainability, https://doi.org/10.1007/978-981-96-1116-4_7

courtyard while matching the historic district's skyline. The design-build process of the prototype resulted in a telescopic cover that can collapse into a convex shape. It features wooden beams in the larger courtyard and metal rails on the smaller sides. The mechanism opens the upper half of the cover with the help of an electric motor, exposing half of the courtyard surface. A shorter half-cover is then fitted inside. The two sections move simultaneously and completely out of the courtyard's opening, repeating the process in the opposite direction. By using only wood, the design spans a distance of 13.8 m, demonstrating the structural capabilities of glued laminated timbers. The involvement of students in the construction of the shade structure at Yazd University's Faculty of Art and Architecture went beyond mere labor contributions. It had a profound impact on both the students' sense of belonging to their academic community and the overall project management process. The decision to engage students in the construction of the shade structure aligns with a cornerstone of vernacular architecture, which is leveraging local labor with preservation skills and knowledge.

Keywords Shading · Glulam · Courtyard · Wood architecture · Moveable cover · Learning by doing · Prototyping

1 Introduction

Buildings' comfort conditions, such as their open spaces and courtyards, can affect their occupancy. The microclimate of courtyards is influenced by their geometry, orientation, opening size, materials, and landscape [27]. Nowadays, there are various design strategies and techniques that can be used to create shaded courtyards, enhancing their comfort and usability. One such strategy is shading, which can improve thermal performance and reduce heat stress in outdoor spaces [12, 15]. Shading in outdoor spaces is emphasised, and practical tips for incorporating shade elements into courtyard design are provided. These tips include the use of vegetation, hybrid shading, and careful consideration of the local climate and environment [5, 9, 24, 26].

In Yazd, houses' courtyards are a typical example of spaces that provide a microclimate condition. Traditionally, Yazd's courtyards were shaded using natural elements and tree cover, providing shadow and reducing ambient temperatures. The "poosh" was particularly important in public open spaces within religious buildings, such as mosques, where open spaces often lacked tree shade. In recent years, various approaches have been implemented to enhance the utilisation of open spaces within the existing structures of Yazd's historic districts. These strategies primarily involve the implementation of fixed lightweight, semi-heavy, and heavyweight coverings. While a few of these interventions have proven successful in upgrading, optimising, and increasing the flexibility of open spaces, others have resulted in detrimental alterations to the fabric of the historic districts.

This project aims to create a shading structure for the courtyard of the Faculty of Art and Architecture in Yazd, Iran (Fig. 1). The courtyard's three walls have a flat design with minimal gaps due to various anatomical interventions at different periods. This has reduced the formation of shadows along the surfaces. Additionally, planting trees is impossible due to an underground structure beneath the courtyard. Therefore, the project aims to explore alternative strategies for creating shade without relying on vegetation.

There are two types of approaches to covering open spaces. One is to create a fixed cover, while the second is to create a mobile cover that can be quickly stabilised. The movable cover can be fully assembled in some seasons or designed to be folded or collapsed. According to Wolfe [25], in the past, structures were designed to remain fixed, and under normal conditions, none of the members of the structures would move or rotate. However, there are many situations where the mobility of the components allows a space to be open during the day and closed during the night, which has made it necessary to change the structures themselves. Nowadays, retractable or deployable roofs are preferred because they are highly flexible and allow different uses to occur in the same space. However, it should be noted that the answer for each

Fig. 1 The courtyard at the faculty of art and architecture in Yazd is called Ellieen prior to the installation of its retractable cover. *Source* Photo taken by Sanaat_e_Sakhtan office

specific building will be different and suitable for its purpose, according to Smith [23].

Otto [18] made significant contributions to the field of movable roofing systems. He came up with an innovative matrix that included various patterns, such as parallel, central, circular, and perimeter patterns. Wolfe [25] has addressed common types of movable roofing structures, which include bellows, folding, telescopic, and scissor structures. Patidar and Gandhe [19] categorised the typology of retractable roof structures based on the frequency of opening and closing. Atlamaz and Akgun [1] conducted a systematic critical review of retractable roofs, providing referable categories based on their material, mechanism, and geometry. The design of retractable structures, ranging from large-scale deployable roofs to smaller retractable canopies, requires a thorough understanding of structural forces. This knowledge is essential in determining the optimal geometry and deployment mechanisms, ensuring both functionality and feasibility [22]. Designers of open-and-close roofs must carefully consider the impact of precipitation loads in addition to wind, snow, and self-weight [7].

2 Background Knowledge

The choice of material for a retractable roof depends on various factors, including the desired aesthetic, budget, climate, and intended use of the covered space for each specific project. Besides the commonly considered factors for material selection in retractable roofs, several additional parameters gain priority when selecting lightweight roofing for a historical building's courtyard as a unique challenge: compatibility with historical context, reversibility and minimal impact, durability and weather resistance, lightweight and load-bearing capacity, a balance of structural integrity, aesthetics and visual harmony, sustainability and environmental considerations, maintenance requirements and life expectancy, compliance with regulations and building codes, community involvement and public perception, professional consultation and expertise [2, 4, 8].

Retractable roofs are usually made of aluminum, wood, or a combination of both. Aluminum is lightweight and durable, making it a popular choice. Modern aluminum roofs often use intelligent mechanisms for opening and closing. However, aluminum is generally more expensive than wood, which is also lightweight and has natural origins. Wood is often a preferred choice for parasite architecture due to its lightweight and sustainable properties [21]. Glue-laminated timber (GLT), instead of traditional wood timber, was developed in the 1990s to enhance the structural strength of wood [6]. GLT is a material composed of thin wood veneers that are bonded together with strong adhesives. Metal connectors are commonly used to secure the ends of GLT beams and connect them to the structural framing [17]. GLT and aluminum are commonly used in combination for light roofing structures [14]. Polycarbonate, a lightweight, transparent plastic with high impact resistance and UV protection, can be used as a cover in retractable roof designs, but not as a structural element. Additionally, fabric roofs are lightweight and affordable, but they may not offer the same level of durability and weather protection as aluminum or wood roofs.

Investigating the Yazd context focusing on courtyard covers is essential after considering the various factors that influence design and material suitability in a global scope. There are three types of courtyard coverings in Yazd: summer, winter, and all-season patterns [11]. The summer pattern provides maximum shade during the summer and is removed when not in use. During the cold seasons, the winter pattern is installed in the courtyard and is waterproof with an airlock to keep it warm. The all-season layout ensures year-round user comfort by offering shade and air circulation in warm and cold seasons. After considering various factors, it appears that the summer pattern is the most appropriate. Additionally, curved structures that complement the skyline of Yazd can be constructed using selected technology, such as GLT technology. However, before drawing any conclusions in the design-build process, it is essential to consider a few methodological steps.

3 Methodology

3.1 The Scope of Work and Performance Expectations

To address the lack of shade in the Ellieen courtyard and the limitations preventing the use of natural vegetation for thermal comfort, we proposed designing an extension to serve as a canopy for generating shade when needed. This action research focuses on designing and building a retractable roof. We used mixed methods with a focus on qualitative approaches. The steps taken alongside the design and building process included:

(a) Providing open-ended questionnaires to Yazd Faculty of Architecture faculty members to get their feedback on the kind of roofing and the possibility of student collaboration throughout the construction phase.
(b) Conducting a literature review of the subject in global and local scope.
(c) Analysing practical experiences of add-on roofing Yazd courtyards in the past and contemporary era.

In Step A, the design principles, recommendations, and considerations were identified based on the findings. The primary requirement for the addition is that it should be identifiable as an addition. The response could range from complete contrast to complete harmony with the context. After considering various patterns, the summer pattern was found to be the most appropriate for the cover as it provides ample shade. The proposed covering should be retractable, not fixed, to accommodate the diverse activities in the courtyard without disrupting them. It is preferable to have a covering that does not require a supporting structure within the courtyard, and mechanical control is preferable over manual control for ease of use. This project takes place within the indigenous context of Yazd, and utilising student labor during the construction phase can engage collective effort and provide hands-on learning outcomes for students interested in this field. In Step B, rules and recommendations were determined by reviewing the literature on adding extensions to historical buildings. Table 1 partially includes the achievements of Step B.

During step C, an on-site investigation was conducted in Yazd to examine the existing examples of add-on structures for courtyards. It was observed that most of

Table 1 Aspect of the design guide for the movable cover

Properties	Description
Climatic	The ability of the covering to provide shade, resistance to detachment in the face of seasonal winds, the possibility of airflow and prevention of greenhouse effect, water and sunlight resistance of the covering itself
Functional	The proposed covering should not hinder but enhance existing activities and functions in the courtyard above the library. If possible, it should provide a more suitable framework for their realisation in terms of physical capabilities
Structural	The structure covering the area must be light to prevent roof instability during earthquakes. The design balances connecting and separating the structures, as they are movable. Controlling vibrations during movement is a significant challenge, and measures must be taken to ensure stability during earthquakes
Feasibility	The designed covering should be feasible for construction and implementation within the country. More importantly, it should be constructed using local labor and expertise familiar with the execution context. This is because of localising this experience and enabling its generalisation to other spaces with similar needs
Economic	It is necessary to complete the project within a reasonable budget and strive for cost-effectiveness
Aesthetic and meaningful considerations	Preserving the presence of the sky in the courtyard is of paramount importance, as the courtyard seeks to connect with the vastness of the sky. Alignment with the Qibla direction and alignment with the sun for optimal sunlight exposure from the other side are essential considerations
Construction methods and material selection	It is essential to use the right building materials while taking the structural system into account during the design phase. This entails carefully evaluating the material's attributes to make sure it is appropriate for the intended use
In terms of use	The courtyard aims to become a new university destination as a result of adding the cover. It will improve practicality and offer cover while fostering an environment that is favorable for particular purposes
Modifying historic districts	The cover design should seamlessly blend with the courtyard's existing character and surroundings, without obstructing the sky view, altering its spatial composition, overshadowing the skyline, disrupting visual harmony, or depriving surrounding spaces of natural light

these structures were heavy or semi-heavy. However, some lighter styles were also found, such as tent-like structures that were replicas of traditional woven covers. Matini [16] analysed the current samples of lightweight courtyard roofs for historical buildings, focusing on Yazd. He proposed design guide models for erected and removed or convertible roofs, mainly focusing on the geometrical, mechanical, and technical aspects, as well as the main construction elements of these lightweight roofs that adapt to Iranian courtyards. Based on the written documentation and practical experiences, the various aspects of the design guide for the movable cover in the present project are listed in Table 1.

3.2 Design-Build Process

The process of selecting an appropriate pattern for the segmented roof system was a critical step in the design process. After careful consideration, it was decided that parallel patterns were the most suitable choice for the given context, due to their balance of structural integrity, material efficiency, ease of assembly, and aesthetic appeal. Within the realm of parallel patterns for movable roofing systems, according to Otto's matrix [18], two primary approaches were considered: overlapping system and folding system. Based on these two approaches, various alternatives were designed and outlined in detail in Table 2.

Table 2 presents three alternatives for the shade structure. Alternative 1 is a fabric cover and crescent-shaped wooden beams, while Alternatives 2 and 3 are made entirely of wood. Each of the latter options consisted of four separate pieces that could be assembled two by two and parked on either side of the courtyard roof. After consideration of design guidelines, constraints, and opportunities, Alternative 3 was deemed the most appropriate choice. Rooftop levels that could accommodate parking for the movable roof panels were taken into account in making this decision. Subsequently, alternative 3 was optimised to produce the final solution, as shown in Fig. 2. In the optimised option, the number of blades has been reduced from four to two while maintaining the orientation towards the sun and the Qibla for an open view of the sky. At the same time, the requirements for adequate shade casting, ventilation, and other necessities have been met.

3.3 Student Involvement in the Construction Process

Student involvement in the construction process presents a valuable opportunity to experience and bridge the gap between academia and industry, fostering a more dynamic and innovative construction sector. By organising a student involvement workshop focused on construction planning, the construction process has been started. To equip the enthusiastic students with the necessary knowledge and skills for participation in the construction workshop, a comprehensive introductory session was

Table 2 Three alternatives are displayed as potential responses

Type	Sketch
Foldable system	 Alt. 1
overlapping system	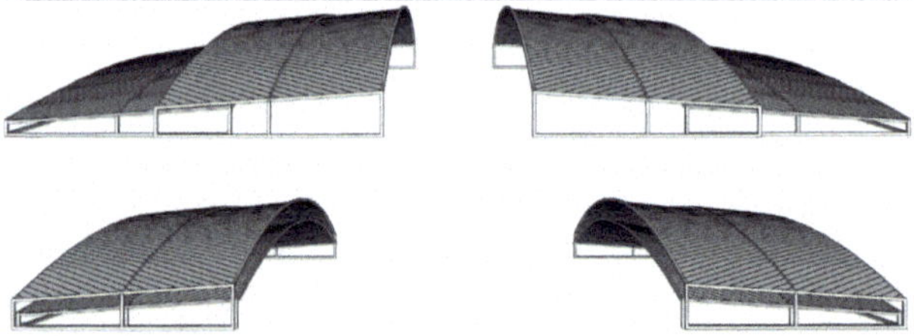 Alt. 2
	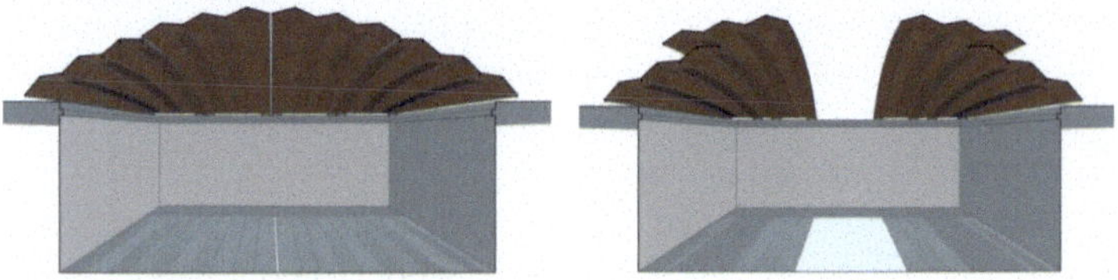 Alt. 3

organised. This informative gathering brought together 60 highly motivated students eager to engage in the construction process. The volunteer students were mostly from the field of architecture, but there were also students from the fields of urban planning and painting. The applicants were from various academic levels, with the majority being undergraduate students, followed by graduate students, and a small number of doctoral students. The introductory session for the construction workshop effectively combined theoretical instruction with practical demonstrations and hands-on experience, ensuring a comprehensive learning experience for the participating students. The construction of the shade structure required a well-structured approach to ensure efficient execution and effective coordination among the student volunteers. Two key implementation strategies played a pivotal role in the process: regular weekly meetings on weekends and a dedicated virtual communication group. The weekly meetings served as a valuable platform for reviewing progress, planning, maintaining communication, and construction on a one-by-one scale (Figs. 3 and 4). Key functionalities of the virtual group included weekly progress reports, volunteer coordination, and real-time updates.

The shade structure's construction involved a well-coordinated effort. The skilled mechanics tackled the metal parts (rails, motors, and wheels) while the students built the wooden components simultaneously. This approach ensured both elements

Fig. 2 The optimised alternative for the cover answer. *Source* Photo taken by Sanaat_e_Sakhtan office

progressed smoothly. Once the wooden frame, metal connectors, and movable roof panels were fabricated, a thorough testing phase began. First, manual operation confirmed the smooth opening and closing of the panels. This was followed by testing the electric motor drive system. Upon the completion of all tests, the structure became operational. For a visual representation of the product, refer to Figs. 5 and 6.

Fig. 3 Samples of workshop pictures: students making multi-layer wooden beams. **a** Formwork layout on ground; **b** Adding glue between wood layers and clamping for initial hold; and **c** Securing beams with additional clamps for final tightening. *Source* Photo taken by Sanaat_e_Sakhtan Office

4 Discussion and Implications

The retractable wooden cover based on the summer pattern generates shade and allows natural air circulation and ventilation within the courtyard. This feature promotes the flow of cool air and improves thermal comfort conditions. To determine the strength of retractable structures, there are two methods used: simulation and computational methods, or experimental testing [13]. For our design-build project, our team used an experimental load-testing approach. We started by constructing three beams and then applied the estimated wind load and self-weight of the structure to determine the load-bearing capacity of the beams. GLT technology enhances the cover's structural properties in several ways, as Hayashi and Gondo [10] and Buchanan et al. [3] stated this technology offers increased strength, improved

Fig. 4 Samples of workshop pictures: students putting wood pieces on beams. *Source* Photo taken by Sanaat_e_Sakhtan office

Fig. 5 The usage after deployment and testing of the movable roof structure. *Source* Photo taken by Sanaat_e_Sakhtan office

Fig. 6 An example of diurnal shadow dynamics. *Source* Photo taken by Sanaat_e_Sakhtan office

dimensional stability, design flexibility, and economic advantages by reducing material consumption and weight, utilising wood waste in production, and offering an extended lifespan due to its durability and resistance to moisture damage, thereby reducing frequent replacements and maintenance costs.

The practice of action research has led to several notable achievements: improved design-build skills among students, a deeper understanding of the local context, the empowerment of the local community, and the sharing of knowledge. The construction of the shade structure serves as an example of the transformative impact of human-centered project management. By embracing trust, empowerment, and alignment with stakeholder needs, the project team was able to tap into the collective energy and expertise of students and faculty, resulting in a tangible improvement to the university's infrastructure.

After using the retractable wooden cover, POE research has been done to know about the range of responsiveness of the courtyard after the cover add-on. Social impact assessment has been considered through evaluating the impact of the courtyard cover on user behavior and fostering social interaction within the space. Pourmehdi Ghaemmaghami et al. [20] analysed that 'the existence of the cover, at first, has helped to evolve the collective role of the courtyard, and at the second level, sometimes, this role has been changed into being a destination'. They also addressed the positive correlation between courtyard sociability and student collaboration (direct or indirect) to improve the quality of the educational space for the students.

This project could serve as a model for future endeavors, demonstrating the power of collaborative and student-centered approaches to achieve remarkable outcomes. The research highlights the potential for sustainable, culturally sensitive, and locally driven design solutions that can address diverse needs and contribute to the country's development goals.

5 Conclusion

The project focuses on a new approach that involves distinct phases to address various aspects of the courtyard cover implementation. The first phase involves researching the suitability of the chosen materials for the project's specific needs. The project explores the potential of utilising easily available materials in the construction process. Furthermore, the involvement of students in the fabrication process fosters hands-on learning and engagement with the project. In conclusion, this project demonstrates a connection to the values of vernacular architecture through the use of vernacular design strategies such as shading, material assessment, practicality, local resources, community participation, and the creation of social space. The project aims to provide an innovative solution that respects and reflects the wisdom of traditional hot-arid architecture.

The present action research has demonstrated the effectiveness of a localised approach to design education and practice. By emphasising the growth of local human capital, the research has contributed to the creation of more sustainable, culturally appropriate, and community-driven designs. This approach offers a model for seeking to leverage local resources and expertise to address context-sensitive development challenges. Beyond these specific applications, action research has broader implications for the construction industry in Iran.

The approach promotes sustainable design focused on shade design strategies in hot-arid regions to achieve comfort conditions. It is a lesson we learned from vernacular Iranian architecture that can be applied to similar climates around the world. Other advantages include local expertise, cultural heritage preservation, and innovation in the construction industry. It emphasises renewable materials and adaptable structures, encouraging eco-friendly practices. The research also focuses on local capacity building and knowledge dissemination, fostering a skilled workforce, and promoting self-reliance. The technology is applied to traditional structures and historic sites, contributing to Iran's rich cultural heritage. The research also showcases the potential for innovative solutions using readily available materials, encouraging further exploration and development.

Acknowledgements The authors express their gratitude to Yazd University's faculty members and officials for their invaluable contributions to the project. They thank the students at the Faculty of Art and Architecture for their enthusiasm and dedication, Dr. Ali Akbar Sharifimehrjerdi for his unwavering support, and the faculty members for their time and expertise in interviews and questionnaires. They also thank Dr. Mansour Bidoki and Engineer Amirabbas Aalamdar, officials of the Office of Construction Projects, for their trust in the research team and financial support, which enabled the project's successful completion.

References

1. Atlamaz B, Akgün Y (2023) A critical review and novel classification proposal for kinetic roof structures. J Archit Sci Appl 8(2):587–608
2. Barozzi M (2015) Design novel covering system for archaeological areas. In: Textiles composites and inflatable structures VII: proceedings of the 7th international conference on textile composites and inflatable structures, Barcelona, Spain, 19–21 Oct 2015. CIMNE, pp 105–114
3. Buchanan A, Carradine D, Jordan J (2011) Performance of engineered timber structures in the Canterbury earthquakes. Bull N Z Soc Earthq Eng 44(4):394–401
4. Cabello-Briones C, Viles HA (2017) Evaluating the effects of open shelters on limestone deterioration at archaeological sites in different climatic locations. Int J Archit Heritage 11(6):816–828
5. Callejas IJA, Durante LC, Diz-Mellado E, Galán-Marín C (2020) Thermal sensation in courtyards: potentialities as a passive strategy in tropical climates. Sustain Times 12(15)
6. Canfjorden R (2022) Folding wood: investigating architectural opportunities of folded plate principles in timber structures
7. Cao Z, Liu M, Wu P (2019) Experiment investigation and numerical simulation of snowdrift on a typical large-span retractable roof. Complexity 2019:1–14
8. Cetin NFY (2013) Architectural design characteristics of protective structures at archaeological sites and their impact on conservation of remains. Izmir Institute of Technology (Turkey)
9. Elgheznawy D, Eltarabily S (2021) The impact of sun sail-shading strategy on the thermal comfort in school courtyards. Build Environ 202:108046
10. Hayashi S, Gondo T (2021) Analysis of the construction of a reinforced-concrete free-form roof formwork and the development of a unit-construction method. J Build Eng 34:101924
11. Khaki Ghasr A (2016) Cover patterns strategies for exposed spaces in the historic districts of Yazd. J Archit Hot Dry Clim 4(4):81–103
12. Lin TP, Matzarakis A, Hwang RL (2010) Shading effect on long-term outdoor thermal comfort. Build Environ 45(1):213–221
13. López-Cabeza VP, Diz-Mellado E, Rivera-Gómez CA, Galán-Marín C (2022) Shade and thermal comfort in courtyards: experimental versus simulation results. Buildings 12(11):1961
14. Marker P, Jirasek R, Schmidt T, Bleicher A (2023) Development, realization, and experimental validation of an active hybrid roof structure based on elastic kinetic and rigid-body transformation. Int J Space Struct 38(2):83–100
15. Mahmoud RMA, Abdallah ASH (2022) Assessment of outdoor shading strategies to improve outdoor thermal comfort in school courtyards in hot and arid climates. Sustain Cities Soc 86:104147
16. Matini MR (2014) Courtyard lightweight roofs for historical buildings in Iran; analyzing the current samples and proposing design guide models. J Fine Arts: Archit Urban Plann 19(3):15–28
17. Moggio N (2012) Glued laminated timber space truss systems
18. Otto F (1972) Wandelbare Dächer/ *Convertible Roofs*. Institute for Lightweight Structures (IL), University of Stuttgart, Stuttgart
19. Patidar S, Gandhe V (2017) Typology of retractable roof structure. Int J Adv Res Ideas Innovations Techno 3(3):556–560

20. Pourmehdi Ghaemmaghami H, Khaki Ghasr A, Seddighian A (2022) Evaluating the sociability of "Ellieen Yard" in the Yazd University's art and architecture school after applying its retractable wooden cover. J Archit Hot Dry Clim 10(15):77–93
21. Rinaldi S, Frunzio G, Guadagnuolo M, DI Gennaro L, Massaro L (2021) A sustainable material for sustainable architecture: wood in parasite architecture. In: Congresso Internacional Sobre Patologia e Reabilitação das Construções. Universidade Federal do Ceará, pp 481–488
22. Shen X, Zhang Q, Lee DSH, Cai J, Feng J (2021) Static behavior of a retractable suspen-dome structure. Symmetry 13(7):1105
23. Smith JK (2003) Current technologies and trends of retractable roofs. Doctoral dissertation, Massachusetts Institute of Technology
24. Soflaei F, Shokouhian M, Abraveshdar H, Alipour A (2017) The impact of courtyard design variants on shading performance in hot-arid climates of Iran. Energy Build 143:71–83
25. Wolfe MH (2013) Analysis of deployable strut roof structures. Doctoral dissertation, Massachusetts Institute of Technology
26. Yang SR, Lin TP (2016) An integrated outdoor spaces design procedure to relieve heat stress in hot and humid regions. Build Environ 99:149–160
27. Zhu J, Feng J, Lu J, Chen Y, Li W, Lian P, Zhao X (2023) A review of the influence of courtyard geometry and orientation on microclimate. Build Environ 236:110269

A Semiotic Analysis of Cultural Continuity Between Traditional Courtyard Houses and Contemporary Apartments in Iran

Hirou Karimi, Mohammad Anvar Adibhesami, Maryam Ghasemi, Peyman Bahrami Doost, Fatemeh Pajani, and Mohsen Riahi

Abstract Housing design inherently reflects cultural perspectives, yet modern, homogeneous apartment complexes often overlook the distinct local identities embedded in traditional residences. This study employs a semiotic analysis to compare key cultural symbols expressed in Iran's introverted courtyard houses with those in contemporary outward-facing apartments, aiming to uncover meanings and assess continuity amid architectural shifts. The ornate 18th-century Abbasian house and the modernist 1977 Ekbatan complex serve as case studies, representing the contrasting eras of traditional introversion and contemporary extroversion through their morphology, ornamentation, and hosting rituals, all of which hold deep cultural significance. The analysis applies Peirce's semiotic theory to classify architectural meanings as icons, indexes, or symbols. The findings reveal significant shifts from sheltering, climate-responsive, and community-oriented courtyard houses to exposed, technology-driven apartments designed for external visibility and aligned

H. Karimi · M. Ghasemi
Department of Architecture, Faculty of Architecture, Eastern Mediterranean University, North Cyprus Via Mersin 10, Famagusta, Turkey
e-mail: maryam.ghasemi@merysland.com

M. Anvar Adibhesami (✉)
Department of Architecture, School of Architecture and Environmental, Iran University of Science and Technology, Tehran, Iran
e-mail: hirou.karimi@emu.edu.tr

P. Bahrami Doost
Department of Architecture and Urban Planning, Shahid Rajaee Teacher Training University, Tehran, Iran

F. Pajani
Department of Urban Planning and Design, Faculty of Arts and Architecture, Tarbiat Modares University, Tehran, Iran

M. Riahi
Department of Architecture, School of Art and Architecture, University of Kurdistan, Sanandaj, Iran

A. Cheshmehzangi and S. Roaf (eds.), *Persian Vernacular Architecture*, Urban Sustainability, https://doi.org/10.1007/978-981-96-1116-4_8

with global lifestyles, thereby diminishing localised distinctions. While contemporary housing forms offer upgraded amenities, the discontinuity in embedded sociocultural associations risks weakening inhabitants' sense of identity and affiliation. The study highlights strategies for balancing cultural continuity with the necessary evolution of housing, advocating for policy changes, community participation, and a re-engagement with vernacular traditions. By tracing the fading cultural codes, semiotic frameworks provide crucial insights into reconciling heritage with modern progress in the ongoing transformation of housing design.

Keywords Housing · Traditional · Contemporary · Privacy · Semiotic

1 Introduction

Symbols exist all around us, carrying cultural meanings that are interpreted by individuals based on their backgrounds [12]. In architecture and designed spaces, buildings and spatial configurations can act as signs that communicate identity and values [6]. As Broadbent [6] states, "the built environment abounds in cultural codes which signify socially agreed-upon meanings." The layout, aesthetics, and functions of a house relate to semiotics as they signify deeper cultural concepts to residents and visitors.

Housing holds particular importance as an embodiment of cultural perspectives. As Sadalla et al. [51] argue, "The house reflects and affects the psychology of its inhabitants because it is an icon, repository, and mediator of individual and household meaning." Domestic spaces contain symbols of privacy, relationships, values, and more [29, 51]. However, modern homogenous housing lacks connections to traditional cultural symbols, resulting in dissonance [28].

This chapter conducts a comparative semiotic analysis of traditional Persian homes versus contemporary apartments in Iran to reveal cultural symbols and evaluate preservation of identity. Two case studies are examined: the Abbasian traditional courtyard home and the modern Ekbatan apartment complex. Spatial configurations and architectural elements are analysed as sign systems, interpreting their underlying meanings related to Iranian culture. The analysis utilises Peirce's conceptual categorisation of signs as icons, indexes or symbols (see Table 1). The meanings revealed through the semiotic analysis provide insights into maintaining cultural identity in Iran's evolving domestic architectures.

Table 1 Peirce's semiotic sign types with examples from domestic architecture analysis

Sign type	Definition	Architectural example
Icon sign	Sign that resembles or imitates the object	Ornate Orosi doors as icons of beauty and extravagance
Index	Sign with an actual connection to the object	Courtyard cooling system as index of desert climate
Symbol	Sign with arbitrary, culturally-assigned meaning	Guest rooms symbolising hospitality

2 Literature Review

2.1 Overview of Semiotics and Meaning in Architecture

Semiotics, or the study of signs and symbols, has increasingly been applied in architecture to analyse how spaces and structures communicate meaning culturally [7]. Studies have examined architecture using semiotic theory and concepts to reveal the significations and messages embodied in design elements at both micro and macro scales.

The investigation of semiotics in architecture becomes especially pertinent in the context of the essay on the cultural continuity between contemporary apartments in Iran and traditional courtyard houses. By examining the semiotic aspects in Iran, researchers can uncover the underlying cultural meanings and values that have persisted or evolved over time [44]. The overview of semiotics and meaning in architecture may offer insights into how these architectural styles communicate cultural continuity, reflecting the societal changes and influences that have shaped the built environment in Iran [27]. The language of architecture is a special form of symbolic art meaning the process of granting the symbolic value of aesthetic phenomena caused by the cultural and historical context [33]. Through the application of semiotic analysis, scholars can interpret the symbolic meanings sent by architectural features and spatial arrangements in Iran's comparison of modern apartments and traditional courtyard houses. This method broadens our understanding of architectural beauty while also strengthening our respect for the cultural importance ingrained in the built environment.

2.2 Theoretical Basis for Architectural Semiotics Foundational

Semiotic theories provide a basis for interpreting sign systems within the built environment. Swiss linguist Ferdinand de Saussure differentiated between the signifier (physical form) and the signified (underlying concept or meaning) as the key

Table 2 Assessment of cultural sign continuity between traditional and contemporary housing cases

Type of cultural sign	Assessment of continuity
Icons (forms resembling ideas)	Low—courtyard morphology lost in extroverted towers
	High—some ornamentation sustained
Indexes (signals tied to cultural realities)	Low—climate-responsive features replaced by mechanical systems
	High—certain access coding preserved
Symbols (culturally-assigned meanings)	Mixed—some hospitality rituals reinterpreted, but overall meaning diluted
	Low—the significance of the home as a sanctuary has decreased

components of a sign [8]. American philosopher Charles Sanders Peirce further characterised signs into categories of icons, indexes, and symbols based on their relationships to represented objects [45] (see Table 2).

In Table 2, cultural signs from traditional courtyard houses and modern apartments are compared, reflecting their continuity or loss. While some ornamental features persist as visual icons, the fundamental spatial and climatic functions—such as courtyard designs—have largely disappeared. Certain cultural symbols and rituals remain but are often transformed, contributing to a diminished sense of sanctuary and cultural identity in contemporary housing designs.

2.3 Applying Semiotic Theory in Architectural Analysis Studies

Applying semiotic theory in architectural analysis studies have applied semiotic perspectives to analyse meaning in architecture at both micro and macro levels. At an intricate scale, semiotics can decode details like façade ornamentation and layouts of homes [14]. Broadly, urban plans have underlying cultural syntax to be interpreted, as streets, squares and monuments create "discourses" [10, 17, 22] (Table 3).

The traditional Abbasi House is characterised by an introspective courtyard-focused morphology, with blocked sightlines to the outside world and indirect access sequences. This is further emphasised by the presence of screened balconies, known as Shabestan, and muted lighting, creating a private, inward-looking atmosphere. In stark contrast, the contemporary Ekbatan Apartments exhibit outward-facing extroverted forms, with floor-to-ceiling glazing and balconies fully open to exterior views. This design approach facilitates continuous visibility from the street and promotes a bright, open-plan lighting scheme, fostering a spacious and welcoming atmosphere [41].

Table 3 Key cultural themes encoded in traditional (Abbasi) and contemporary (Ekbatan) case study dwellings

Cultural theme	Traditional Abbasi House	Contemporary Ekbatan apartments
Introversion versus extroversion	Courtyard-focused introspective morphology	Outward-facing extroverted forms
	Blocked sightlines to outside	Floor to ceiling glazing
	Indirect access sequences	Balconies fully open to exterior views
	Screened balconies (shabestan)	Continuous visibility from street
	Muted lighting	Bright open plan lighting
Hospitality	Stepped access filters insider versus outsider	Television-oriented furniture arrangements
	Ornate guest reception rooms	Enmeshed family/entertaining zones
	Discrete service areas	Sacrificed privacy
	Differentiated finishes by access	
	Open plan dissolves host/guest thresholds	
Environment	Internalised courtyard gardens	Terraces connect residents to distant views
	Water features and trees indoors	Mechanical heating, cooling, lighting
	Windcatchers, screens filter air/light	Severance from local climate contexts
	Protective, climate-moderated	
Concept of home	Refuge for family life to unfold	Spaces showcase modern lifestyles
	Personalised rooms by generation/role	Visibility signifies attainment
	Inward focus prioritises security	
	Performative display outwards	

2.3.1 Hospitality

In traditional Abbasi houses, hospitality is reflected in the stepped access that filters insiders from outsiders, ornate guest reception rooms, and discrete service areas. The open plan layout dissolves the host/guest thresholds, with differentiated finishes based on access. Conversely, Ekbatan apartments prioritise communal living, evident in television-oriented furniture arrangements and enmeshed family/entertaining zones. However, this communal focus often sacrifices privacy [11, 15].

2.3.2 Environment

The Abbasi House incorporates nature through internalised courtyard gardens, indoor water features, and trees. Traditional elements such as windcatchers and screens filter air and light, creating a protective, climate-moderated environment. In contrast, Ekbatan Apartments prioritise modern conveniences, utilising mechanical heating, cooling, and lighting systems. Terraces connect residents to distant views, but this design approach often results in severance from local climate contexts [22].

2.3.3 Concept of Home

The concept of home in these dwellings highlights contrasting ideologies. The traditional Abbasi House emphasises security, with an inward focus that prioritises a refuge for family life to unfold. Personalised rooms are designated based on generation and role, with a performative display outwards. In contrast, the contemporary Ekbatan Apartments showcase modern lifestyles, with spaces designed to signify visible attainment. This outward-focused approach reflects a shift in the concept of home, aligning with modern lifestyles and achievements [17, 60].

2.4 Housing and Cultural Identity

Housing and cultural identity of particular relevance is literature analysing housing semiotics related to cultural perspectives, traditions, and place identity. A home holds great meaning, becoming "a repository and mediator of cultural norms" [52]. Both internally and externally, components of house design from layouts to façade details carry cultural codes. Regional differences in domiciles related to climate, values, family structures, etc., generate vernacular architectural languages.

For example, traditional Iranian courtyard homes had inward-focused, introverted morphology representing cultural notions like privacy and hospitality [19]. Elements like screened balconies, decorative screens, and entry sequences allowed or denied access and visibility per cultural protocols. In Indonesia, raised, open Javanese joglo houses reflected beliefs in hierarchy and communal lifestyles [50]. Thus, housing contains a rich system of signs embodying what a cultural group finds meaningful in dwelling. The literature argues evolving standardised housing erases these critical localised identities.

2.5 *Cultural Identity Loss and Placelessness in Modern Housing*

A predominant critique across scholarship is the loss of cultural meaning occurring with modernist, globally-standardised housing [1, 39, 43, 47]. As traditional vernacular dwellings evolved using localised materials into unique physical forms fitting their settings, contemporary housing has largely abandoned this connection to context. Literature examines the resulting negative social and psychological implications of this placelessness and lack of localised cultural symbols.

Models developed for buildings in different climates and societies may not fully capture the nuances of privacy considerations in traditional Iranian architecture. Therefore, it is essential to approach the study of traditional buildings with a nuanced understanding of their cultural, social, and environmental contexts [2]. Architecture represents a culturally impacted response to climatic, cultural, and social conditions. Also, investigating the cultural dimensions of space and privacy can provide valuable insights into the importance of understanding traditional buildings within their specific cultural frameworks [20]

By incorporating relevant references and considering the unique cultural and environmental factors at play, researchers can gain a deeper understanding of how traditional buildings function and evolve in contemporary contexts, contributing to a more nuanced analysis of cultural continuity in architecture.

2.6 *Critiques of Global Homogeneity*

Several authors critically examine the global spread of repetitive, decontextualised housing models that ignore or underserve local cultural perspectives. Influential works like Relph's Place and Placelessness argue "placelessness is the casual eradication of distinctive places and the making of standardised landscapes" via replicable built environments lacking ties to community identity, history and environment [47]. Jiven and Larkham extend this concept, defining "placelessness" as "a weakening or loss of place identity" and the erosion of "the sense of place" via imported architectural styles [26].

Knox examined the rapid globalisation of a "world home culture" based on Western notions of housing, analysing superficial inclusion of token ornamental motifs without preserving cultural needs in layout and functionality [61]. These patterns occur even when architectural differences are somewhat maintained, as meaning still shifts. For example, Danby shows Malay houses changed cultural connections as concrete homes imitated wooden designs rather than renewing adaptations to settings [13]. Experts criticise abandonment of cultural perspectives in housing as not only diminishing diversity of built environments, but detrimentally impacting community bonds and localised systems [18]. As Kellet and Napier state, "as traditional dwellings are replaced by contemporary housing the social and cultural patterns of the community tend to break down" [30].

2.7 Psychological Dimension

The literature connects this cultural dilution in housing to negative psychological effects on dwellers. Rapaport and Eckhoff [46] introduced a conceptual "environmental meaning" paradigm linking culture, design, and psyche. He argued shared social values manifest physically in vernacular designs tuned to locale and lifestyle, supporting user interaction and fulfillment of cultural needs. In contrast, the erosion of cultural symbols diminishes inhabitants' "responsive capacity" and environmental support [46].

Expanding on environmental psychology ties, several studies revealed standardised, placeless housing reduces residential satisfaction and disengages dwellers. For example, Twigger-Ross and Uzzell [63] showed levels of place bonding and home identification were significantly higher for residents of culturally-connected houses than non-descript homes [63]. Inadequate privacy related to imported layouts also increased Iranian occupant dissatisfaction [53]. The existing literature emphasises that as traditional dwellings adapted over generations to reflect and enable meaningful practices, disruption of this cultural resonance impairs "at-homeness" [55]. Rethinking modern housing to restore cultural symbols thus has importance for inhabitant wellbeing.

2.8 Place Identity Preservation Approaches

Given the consequences of inadequate cultural perspective in contemporary housing, experts advocate for approaches maintaining the continuity of socio-physical resonance in built environments. Strategies at various scales exist to inject "place-connectedness" back while allowing necessary housing evolution.

Architecture must strike a balance between history and modernity, incorporating global influences while maintaining a distinctive cultural identity, as modernisation and globalisation impact architectural practices [4]. By exploring preservation strategies that emphasise the retention of place identity, researchers can mitigate the negative impacts of cultural homogenisation and preserve the distinctiveness of Iranian architectural traditions.

Cultural diversity and historical information about heritage buildings may encourage more pro-social behaviours [9]. By incorporating psychological dimensions into the preservation of place identity, researchers can delve into the emotional and cognitive connections that individuals have with their living spaces [35]. With this holistic approach to preservation tactics that prioritise the maintenance of place identity, experts might lessen the adverse effects of cultural homogenisation and maintain the unique characteristics of Iranian architectural traditions.

2.9 Public Policy and Planning Directions

Urban policies, plans, and regulations play a key role in guiding housing development to respect vernacular essence. Authors recommend localised policies steer construction using appropriate cultural configurations. For example, Knox advocated for increased governance to foster regionally attuned architecture through coded planning rules and consultations ensuring consensus on house styles suiting communal values versus importing arbitrary foreign trends [31]. Roaf highlights how traditional building techniques, specifically the use of windcatchers in the Iranian city of Yazd, are not merely relics of the past but rather hold vital lessons for modern urban planning, especially in the context of climate change and resource management [48]. By learning from these traditional methods, cities can work towards resilience against climate change, effectively addressing both water scarcity and adapting to changing environmental conditions. Therefore, understanding the implications of Iranian lessons is crucial for forming sustainable urban frameworks that utilise historical wisdom to tackle twenty-first century challenges.

Some cities have shifted zoning rules or ordinances to incentivise or enforce continuity of place character. For instance, new guidelines in Amsterdam mandated approval processes for housing proposals confirm designs match properties and streetscapes nearby [40]. As Gouverneur notes, legal frameworks enable shaping of housing from macro to micro: "from territorial coherence down to the physical definition of the collective house, its rooms and walls" [23].

The recommendations put forth by the authors emphasise the need for localised policies that guide construction based on appropriate cultural configurations, ensuring that architectural designs align with communal values and regional aesthetics. By promoting stronger governance and implementing structured planning regulations, cities can cultivate a unique sense of identity and protect against the imposition of random foreign influences [57]. The shift towards zoning rules and ordinances that incentivise or enforce continuity of place character, as seen in initiatives like those in Amsterdam, highlights the role of legal frameworks in shaping housing at both macro and micro levels, ultimately contributing to the preservation of cultural heritage.

2.10 Community Participation Tactics

Planning literature also widely advocates for the substantive participation of end users in housing design processes to transfer cultural priorities [54]. Informing architects of the meanings, rituals, preferences and problematics embedded in indigenous housing traditions gives continuity [30]. This empowers communities to manifest their identity physically while allowing for sustainable improvements.

Tactics like workshops, model-making, role-playing simulations, and demonstrative site visits effectively capture cultural insights from locals to integrate into evolving house configurations and plans [25]. For example, Tipple found Bangladeshis provided critical inputs on fundamentals like privacy gradients, gender separation customs, and multipurpose spatial needs applicable even in higher-density modern scenarios [62]. Such investigative techniques can reconnect people to place identity through housing.

2.11 Hybridised Vernacular Forms

Finally, emerging architectural strategies blend updated construction with embedded regional distinctions and infrastructures, renewing cultural resonances. These approaches celebrate and sustainably reinterpret native morphological housing languages. For instance, Adeokun explores Yoruba housing complexes maintaining identity through reinvented courtyards, inset balconies, and privacy/exposure patterns in contemporary materials like concrete [1]. Kütükçüoğlu reimagined angular wooden Black Sea houses using modular tidal construction logic for relocated Turkish fishing villages [32]. Such "critical regionalism" feasibly transfers cultural codes into current housing configurations through reworked vernacular dialogues with context [34].

This literature substantiates the need and methods to inject continuity of cultural perspective on dwelling into inevitable housing evolution across contexts, safeguarding meaning.

2.12 Western Models and the Systems in Iranian Architecture

While Western architectural models often prioritise aesthetic appeal and standardised functional design, they frequently overlook the intricate environmental systems embedded in traditional Persian architecture. For instance, wind towers, or badgirs, exemplify a sophisticated understanding of passive cooling techniques, utilising natural airflow to regulate indoor temperatures. This innovative use of local climatic conditions demonstrates a nuanced relationship between architecture and environmental context, which often remains unacknowledged in contemporary Western practices.

Roaf [49], underscores the significance of such traditional systems in efficiently managing natural resources. The qanat system, for example, exemplifies an ancient ingenuity that ensures sustainable water management in arid environments, while badgirs serve to harness natural wind patterns for cooling [49]. These technologies not only reflect cultural adaptation to environmental challenges but also emphasise a holistic understanding of resource quality that bridges architecture, culture, and sustainability. In contrast, the modern Ekbatan apartment complex in Iran may prioritise a uniform design that conforms more closely to global architectural trends, often neglecting these sophisticated environmental systems. By examining the implications of this design choice through a semiotic lens, we can better understand how the loss of these cultural symbols and practices contributes to a disconnection from traditional environmental wisdom.

3 Methodology

This study utilizes a qualitative, comparative case study method to analyse and contrast the cultural semiotics of traditional and contemporary Iranian housing. A case study approach enables an in-depth examination of architectural meaning within samples across history. The methodology incorporates the following:

3.1 Case Study Selection

As representative cases, the Abbasian traditional courtyard house and modern Ekbatan apartment complex were strategically sampled to compare the continuities and divergences of signs and symbols over time. These cases reflect key periods in Iranian housing, provide contrasting layouts and details, and have documentation available on configurations.

The intricately decorated Abbasian house in Kashan dates from the late eighteenth century Qajar era and contains emblematic Persian introverted geometry focused inwards on a central garden [56]. Ekbatan apartments built in 1977 represent early modernist high-rise dwellings in Iran explicitly referencing Western styles and standards in a car-centric, extroverted design [3]. Differences in form, style, and eras provide a rich basis for examining shifts in cultural encoding.

3.2 Semiotic Concepts for Housing

Analysis To decode and evaluate the embodied cultural signs within the traditional courtyard and contemporary apartment case studies, architectural elements are mapped using key semiotic concepts (see Fig. 1).

Spatial configurations, ornamentations, and functional zones are analysed for their communication of identity issues like introversion vs extroversion, privacy calibration, hospitality rituals, gender roles, communal lifestyles and environmental responses using these concepts.

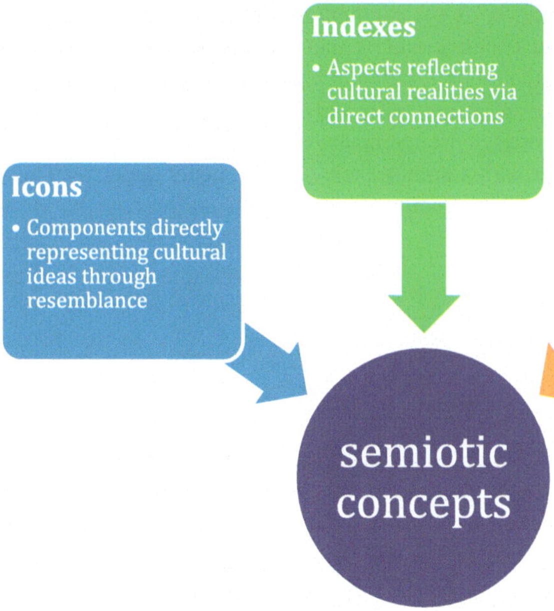

Fig. 1 Semiotic concepts

3.3 Comparative Analysis

Through cross-case examination, variations in the manifestation of cultural signs and symbols across traditional and contemporary case studies are identified and assessed. Instances of design continuity and place-rootedness as well as discontinuity and placelessness amidst housing modernisation are highlighted using the coded semiotic lens.

By applying this comparative method, the analysis elucidates the (re)production, erosion, and potential renewal of cultural housing meanings across the pivotal transition captured in the samples. Findings are important for balancing tradition and progress in Iranian housing issues. This revised methodology explains the strategic selection of representative case studies, outlines the semiotic conceptual tools applied, and provides more specifics on the cross-case comparative analysis process. Please let me know if you have any other suggestions for improving the research method description.

4 Results

4.1 *Comparative Semiotic Analysis*

Applying the semiotic conceptual framework across the Abbasian courtyard house and Ekbatan apartment case studies reveals continuities and divergences in the embodiment of cultural signs and symbols amidst Iran's housing evolution. The analysis maps and examines meanings related to introversion vs extroversion, hospitality, privacy, gender and family dynamics, environmental responses, ornament and craft, and conceptualisations of the home itself.

4.2 *Morphology: Introverted to Extroverted Forms*

One of the most salient transformations evident in the housing cases is the shift from courtyard-centered inward-looking homes to outward-facing, exposed apartments dissociated from context—dynamics directly signifying changes in cultural perspectives on interaction with community and nature.

4.2.1 Courtyard Introversion (Abbasian House)

The ornate Abbasian house deploys layered buffers and spatial progression to carve an inner sanctum sheltered from neighbors and the street, reflecting the fundamental cultural priority of privacy and family-focused living. The thick outer walls conceal inner activities under simple roofs, while internally the rooms organise around a central garden open to the sky but enclosed on four sides. This inverted focus withdraws attention inwards to a "secret orchard" for the family [24].

The blocked sightlines (see Fig. 1), indirect entry sequence forcing turns, screened balconies called Shabestan, muted lighting, and secluded rear service zones maintain introversion. As Bani Masoud explains, such traditional spatial organisation based on a "centrifugal scheme of family privacy represented a physical model of the Iranian mentality" [5]. The home shrinks exposure to the community, with select openings allowing monitored interaction. The courtyard morphology operates iconically, indexically, and symbolically to signify cultural emphases on familial bonds and female modesty over outward engagement.

4.2.2 Extroverted Opening (Ekbatan Apartments)

In contrast, Ekbatan apartments feature fully exposed, protruding balconies and windows continuously open to the surrounding neighborhood (see Fig. 2). The dissolution of buffers between interior and exterior spheres represents a transition in relative valuations of communal versus private realms. Sons explain this outward posture connects to Western-influenced aspirations for nuclear families, individualisation, and social mingling [16].

The street-facing windows, absence of screening devices, and positioning of spaces intended for guests and gatherings along the perimeter signify shifting cultural perspectives. Sons argue whereas the courtyard house allows the family to "withdraw into itself," the displayed apartment facilitates "communication with the outside

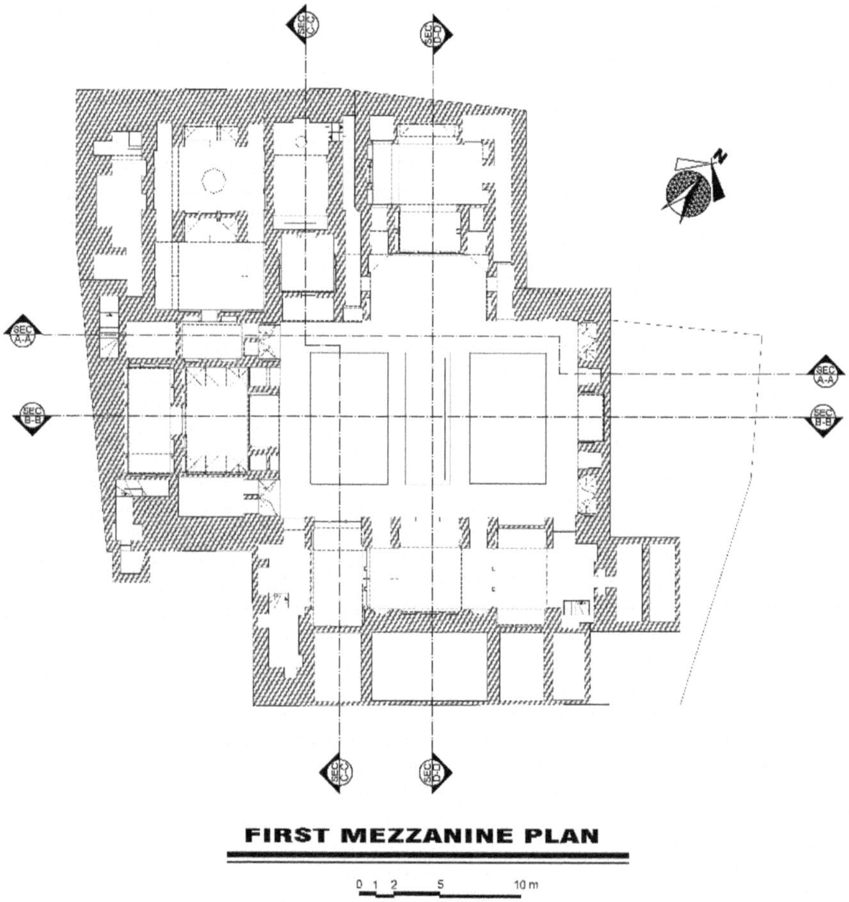

Fig. 2 Sight line analysis of Abbasian introverted house. *Source* Authors

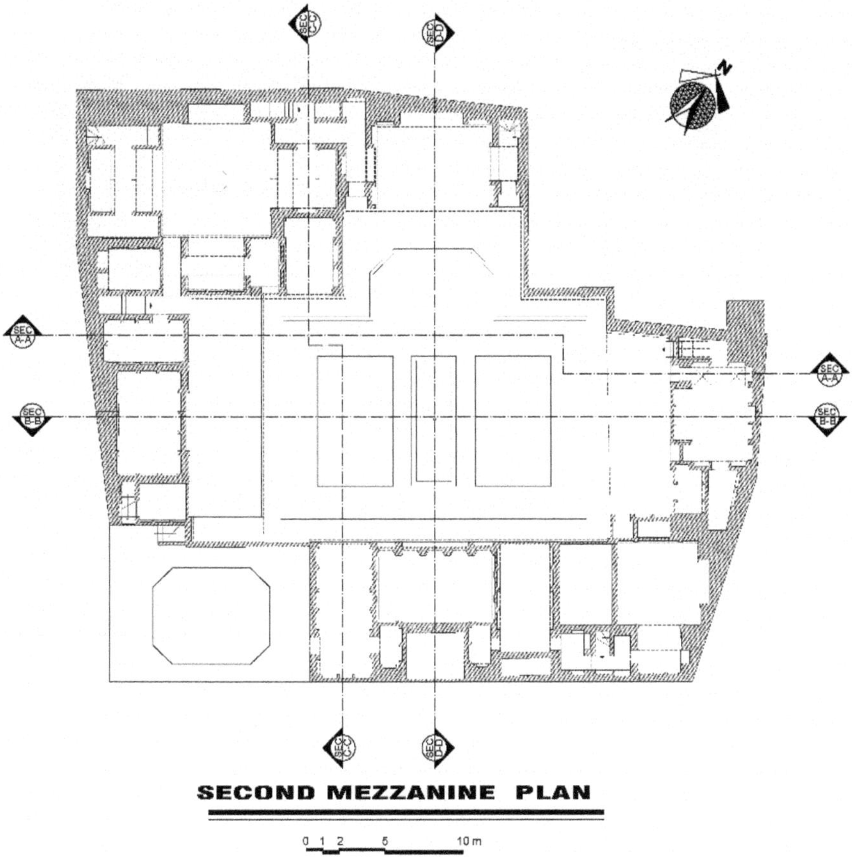

Fig. 2 (continued)

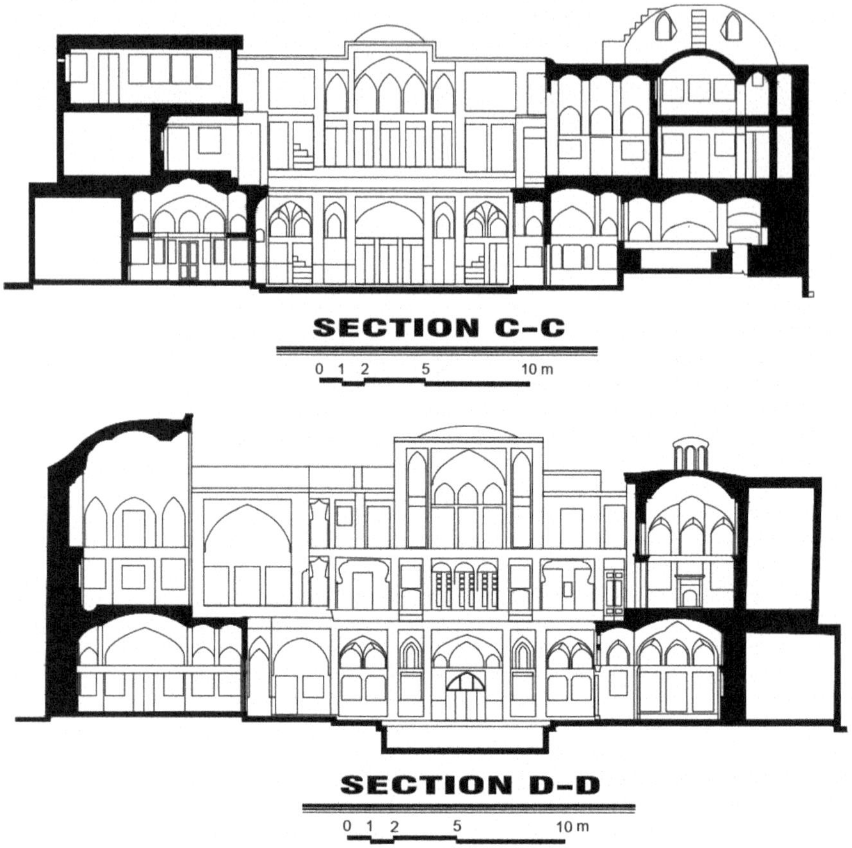

Fig. 2 (continued)

world" [64]. The extroverted volumes indexed and symbolised a cultural realignment from familial collectivism towards augmented external interaction.

4.3 Host/Guest Rituals and Regulations

Related to introverted versus extroverted postures, the case studies encode very different signs regarding hospitality, access, and insider/outsider rituals—realms that carry deep cultural resonance in Persian residences.

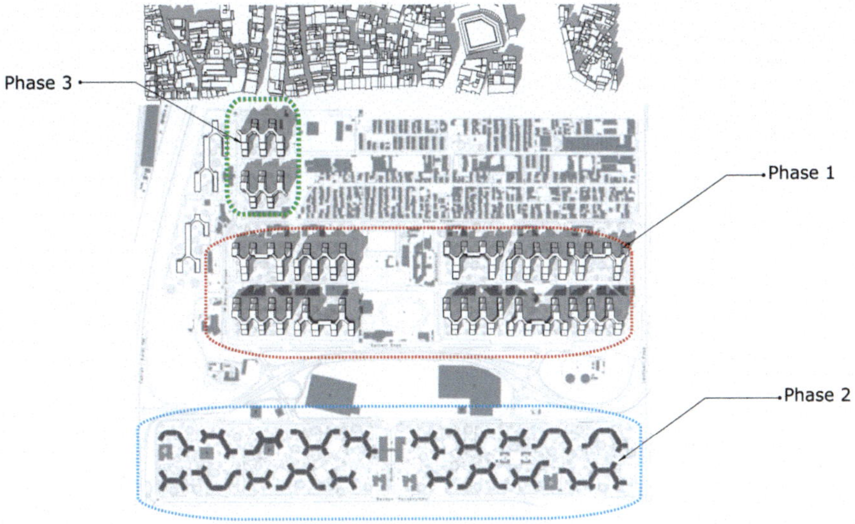

Fig. 3 Extroverted form of Ekbatan apartment. *Source* Authors

4.3.1 Invitation and Progression (Abbasian House)

To enter the Abbasian house is to traverse a sequence of thresholds each filtering access and visibility between zones based on degrees of familiarity. The outer rooms house guests and events involving non-relatives, the inner wing admits friends and close relations, and the private basement spaces and self-contained mother's suite restrict entry to the immediate family [21]. Guest reception areas feature ornamental focal points (see Fig. 3), while service rooms hide discreetly.

This stepped invitation calibrated by proximity differentiates insider and outsider through the legible façade signs, transitional corridors, selective visibility, and differentiated finishes apparent in the floorplan (see Fig. 4). The coding regulates social mixing and humility via allowing gradual penetration aligned with cultural etiquettes. Sons argues the "basic architectural principle was to provide every house with a series of spaces for receiving guests" [59], ordering contact.

4.3.2 Open Integration (Ekbatan Apartments)

In contrast, the Ekbatan dwellings largely dissolve barriers between hosts and guests within a singular exposed volume. Shin et al. [58] argue the adoption of European-style free floorplans created more fluid, casual interfaces between residents and visitors centered on a living room (see Fig. 5).

Fig. 4 Elaborate entry vestibule for greeting guests. *Source* Authors

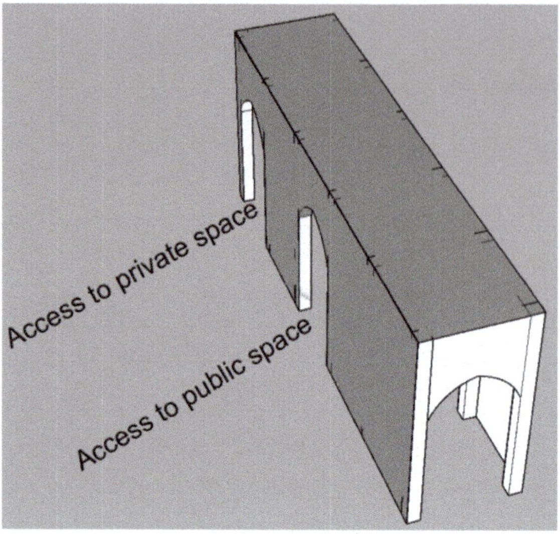

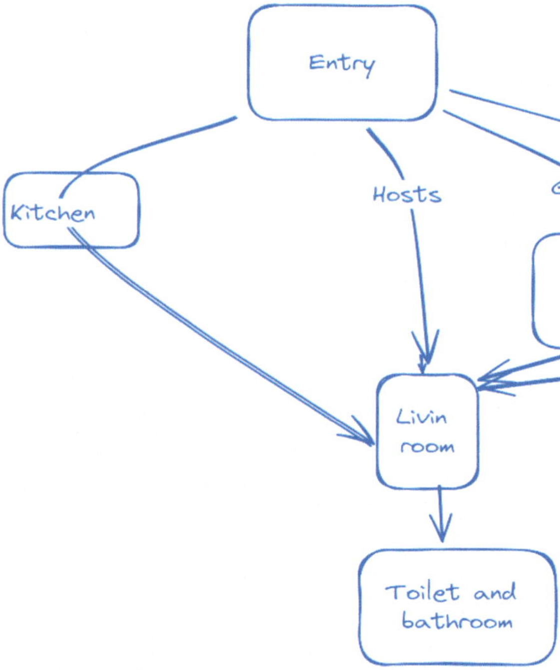

Fig. 5 Open plan living room merges hosts and guests. *Source* Authors

They state this "*demonstrates a relaxation, in which the encounter between public life, family life and individual life is played out openly*" compared to segmented gradations dictating interactions [58]. The dissolving thresholds Symbolise shifting priorities from hierarchical guest reception rituals towards integrated entertaining, indexed by television-oriented furniture arrangements. While more taxation on residents' privacy, the enmeshed organisation carries aspirational cultural meanings associated with modernity.

4.4 Embedded Signals: Ornament, Materials and Components

Beyond volumetric morphologies, the intricate array of architectural components, embellishments, finishes, and materialities coding the housing cases transmit additional cultural signs at the micro-scale regarding aesthetics, environmental responses, and meaning infusion into the dwelling.

4.4.1 Ornate Embellishments (Abbasian Houses)

Heavy ornamentation and artisanal details suffuse surfaces throughout Abbasian interiors and outer doors, contrasting the largely unadorned, earthen exterior walls. These intricate finishes manifest a desire for visual richness and decorative beauty within the domains of family life. Floral Molahkat tiles, massive mirror mosaics (see Fig. 6), and plaster muqarnas fill walls and ceilings, elevating interiors beyond bare necessities towards craft and adornment.

Nazari and Hashemi Nik [42] argue decorative patterning of Persian residences historically related to political power and aristocratic identity, with richer families displaying intricacy [42]. But on a cultural level, such applied embellishments also stem from cherishing visual pleasure embedded in surfaces.

Across history, ornamentation techniques have been employed to enrich the ambiance of spaces, incorporating elements such as murals, carvings, textiles, and mosaics. Components like brick patterning, stained glass, and wooden door inlays saturate and celebrate the sensory materiality within. These elements not only serve as symbols of social status and historical lineage but also reflect a deeper connection to the sensory materiality and visual aesthetics that have long been cherished in Iranian architectural traditions.

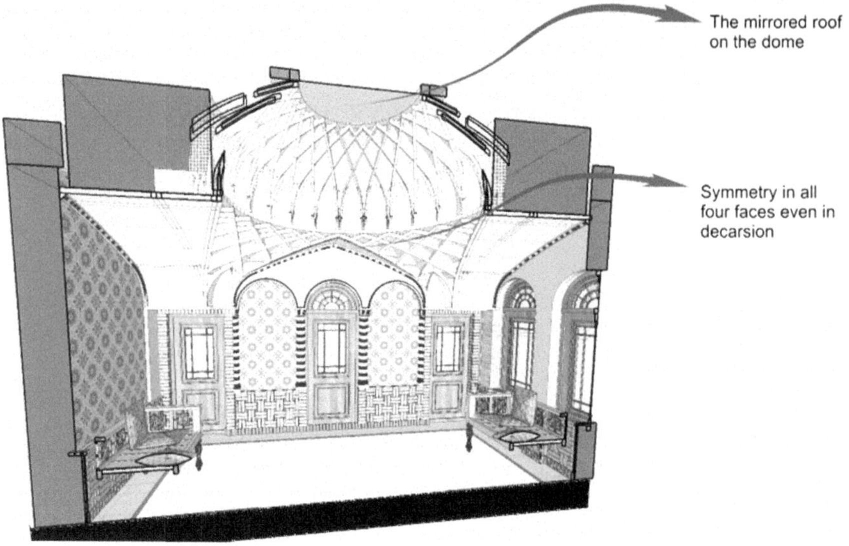

Fig. 6 Elaborately mirrored ceiling in *Shahneshin* guest room. *Source* Authors

4.4.2 Modular Components (Ekbatan Apartments)

While Abbasian houses nest distinct figures within ornamental tableaus, the Ekbatan design utilises serial, factory-made elements easily combined and substituted. Standard windows, doors, cabinetry, and tiles replicate across floors and units, snapping together via flexible grids (see Fig. 7) divorced from hand-craft. This industrial modularity introduces continuity and accessibility through enabled mass production, but sacrifices bespoke detail.

The interchangeable nature of parts allows resident control over finishes to personalise, but comprised integrated artisanship in favor of scalable construction reflecting value shifts. Machine-made modular materials also enable easier maintenance and replacement than customised joinery, prioritising convenience. The use of mirrors or applied wallpapers attempts the rehabilitation of warmth. Serial panels and framing signify cultural alignments with technological reproducibility over handwork devotion.

4.5 Environmental Responses: Light, Air and Nature

A third realm encoded divergently across cases is the role of environmental elements like light, wind, plants, and water features holding cultural import. The way these elements are depicted across different cases showcases diverse cultural reactions and interpretations. In traditional courtyard houses, these elements are seamlessly

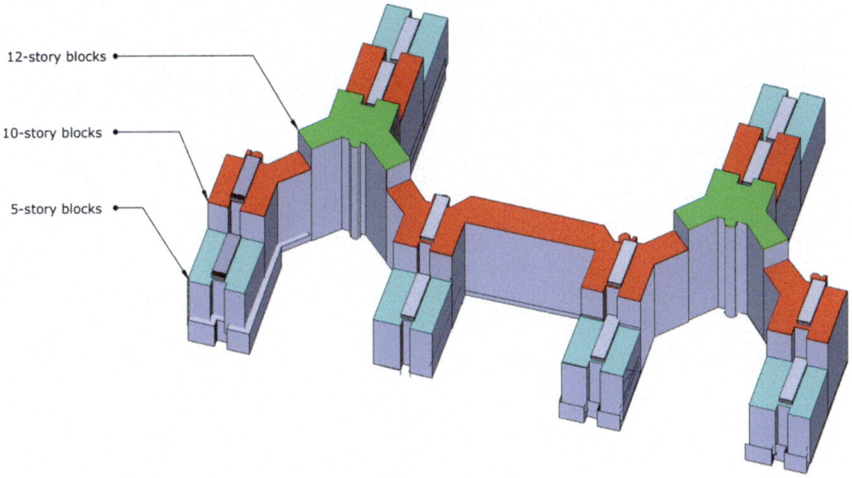

Fig. 7 Extensive modular repetition. *Source* Authors

integrated into the design to establish a harmonious connection between the built environment and nature. Light is utilised for interior illumination, wind is harnessed for natural ventilation, plants are included to introduce greenery indoors, and water features serve both aesthetic and practical purposes. Noteworthy aspects that set the Abbasi house apart in this regard include the following:

- The building has summer and winter parts, and in each season of the year, certain parts of the building are used according to the weather conditions of that season.
- The courtyards are like a pit where the cool air settles at night and is used during the hot day. In addition to regulating the air, these yards block the wind and create shade.
- The intricate decorative elements, such as stained glass windows and wooden door inlays, interact with natural light to create a visually stunning and dynamic atmosphere within the living spaces.

Traditional design principles can offer valuable insights into designing modern buildings by emphasising aspects such as natural ventilation, daylighting, and thermal comfort. The use of wind towers for better comfort conditions at different times of the day, and traditional design elements can inform modern building practices. For example, the concept of passive cooling techniques, like wind towers or shading devices, can be integrated into contemporary building designs to enhance energy efficiency and occupant comfort [37]. Additionally, traditional layouts that prioritise natural light and airflow can inspire modern architects to create sustainable and user-friendly spaces. By studying how traditional design principles address climatic challenges and user needs, architects and designers can adapt and incorporate these strategies into modern building designs.

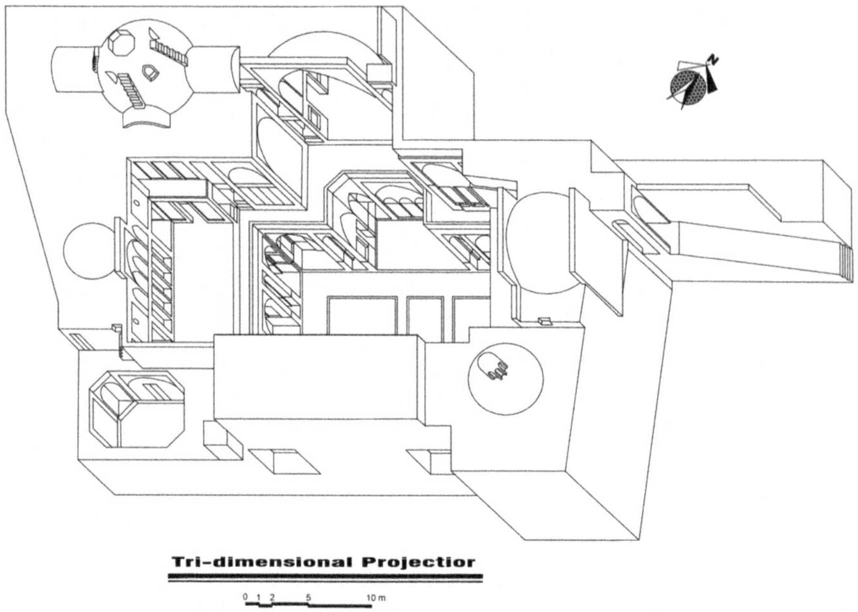

Fig. 8 Section cut showing enclosure of courtyard gardens. *Source* Authors

4.5.1 Internalised Nature (Abbasian House)

Courtyards, miniature gardens, fountains, and water channels bring elements of nature internally into sheltered residential spheres rather than exposing interiors fully to external weather (see Fig. 8). This internalisation allows both environmental connection and protection. The surrounding rooms open via balconies and shaded loggias towards the courtyard "outside" at the center, providing distal garden views and airflow without compromising isolation.

As Mazumdar [36] notes, vernacular arrangements demonstrate integrated eco-consciousness within protective settings, regulating light and air. Strategically placed small openings, screens, stained glass, and shutters filter illumination. Windcatchers ventilate internally, stealthing airflow through ducts. At the micro scale, this environmental responsiveness concentrated within introverted sanctums maintains cultural emphases on enclosed family settings with select climate moderation.

4.5.2 Exposure and Control (Ekbatan Apartments)

While the Abbasian house concentrates nature inward, Ekbatan apartments open fully outward to weather, climate, and urban surroundings through continuous walls

Fig. 9 Extensive glazing opens apartments fully to exterior climate. *Source* Authors

of glass (see Fig. 9). Sons connects this to Western notions of dominating environments, compared to working within conditions [38]. The modules allow control over mechanical heating, cooling, and embroidery to compensate for glazed exposure.

Terraces also project residents out towards vistas rather than internalising scenes. While increased environmental intervention supports individuals, the severing from geo-climate contexts and rehabilitation via technology signifies changed ideological environmental relationships. Opening interiors utterly sacrifices grounded aliases towards independence—an important cultural dialectic.

4.6 Conceptualisations of Home Itself

Finally, the spaces encode differing cultural conceptualisations of the home itself—whether refuge, forum, or performative sphere. The Abbasian house serves as a secure, family-focused shelter in contrast to the entertaining-centered Ekbatan apartment displaying modern lifestyles.

4.6.1 Domestic Sanctuary (Abbasian House)

As Bani Masoud summarises, introverted traditional homes provide "a place of repose at the heart of which their private life could unfold" centering inhabitants' psychological being [5]. Multiple chambers give family members personalised space (see Fig. 10), contrasting integrated Western living rooms. The home offers refuge inward rather than connecting residents outwards. Sons argues spatial layering contributes to a "sense of security and individuality," sustaining family structure [5]. Domesticity revolves around multi-generational bonds.

4.6.2 Performative Display (Ekbatan Apartments)

In the contemporary Ekbatan apartment, the continuous visibility from street-facing windows makes a home a stage set performing modern living. Sons explains "*life no longer unfolds sheltered inside the house but is moved forward, put on show,*

Fig. 10 Personalised rooms for various family members. *Source* Authors

Fig. 11 Windows openly exhibit family life. *Source* Authors

presented to the outside world" via transparent views [64]. Rooms are arranged to display family relationships and class aspirations rather than retreat inwards (see Fig. 11).

Exposure sacrifices privacy in favour of broadcasting the attainment of cosmopolitan lifestyles. Thus, housing shifts from haven to performative declaration through architectural coding.

5 Discussion

The application of semiotic theory unpacking cultural signs embodied across the traditional courtyard and contemporary apartment residences in Iran unveils conserved and transformed perspectives coded into contrasting housing forms. Regarding longstanding identifier themes like introversion, hospitality punctuation, environmental responses, and conceptualisations of dwelling functions, the symbolic meanings physically manifested diverged substantially despite attempts to sustain identity. The inward-focused morphology, access negotiation sequences, ornamental interiority, and protective stances of Abbasian dwellings indexed and reinforced traditionally fundamental cultural tropes like family centrality, female modesty, and gradations between insiders and outsiders. In contrast, the extroverted volumes, visibility, modularity, and performative exposure of Ekbatan apartments signified migrations towards Western-associated markers of progress like social fluidity, technological control, and nuclear families—altering symbolic resonances.

While offering modern amenities and options, the discontinued encoding seen in window transparency, dissolution of screens and thresholds, overall outward focus, and transferal of craft to industry signifies a migration of fundamental cultural perspectives now requiring rehabilitation. As Rapaport argued, the loss of established socio-environmental affiliations risks "responsive capacity". The social needs and practices upheld via the architectural coding of traditional housing forms maintain relevance even as technologies upgrade. The shifts away from climatically-attuned introverted quarters suited to communal lifestyles fail to transfer critical place-connected codes sustaining inhabitants. As Duncan emphasises, designed environments not only serve functions but communicate their supporting "*structure of social interests*" [17], which signifies a drastic change.

6 Conclusions

This cultural semiotic analysis uncovers how traditional worldviews and contemporary transitions are encrypted within contrasting residential spaces in Iran. Introverted morphologies, intricate ornamentation, and infrastructures supporting family-centered living have shifted toward extroverted volumes, modular designs, and technologically regulated units tailored to modern lifestyles. The divergence in environmental, social, and aesthetic signs embedded in these spaces reflects the profound impact of imported architectural styles, reconfiguring cultural meanings and societal affiliations. While traditional courtyard houses embodied communal living, environmental harmony, and a deep connection to cultural heritage, contemporary apartments prioritise individualism, functionality, and a departure from established social dynamics. This transformation is not just architectural but penetrates the very fabric of Iranian identity, as residential spaces shape social interactions and the sense of belonging.

The application of conceptual semiotic tools provides a framework for decoding how cultural perspectives are physically manifested across different eras. As Gouverneur aptly states, "*the history of the home speaks volumes about the history of society*" [23], tracing these architectural encodings helps bridge heritage and progress, ensuring a symbiosis between place and its inhabitants. The analysis highlights the need to balance symbolic preservation with the evolution of housing to maintain cultural continuity.

The study's findings underscore the critical role architectural design plays in both reflecting and shaping cultural values and societal norms. By contrasting traditional courtyard houses with modern apartments, this research reveals how cultural symbols are expressed through architectural elements, influencing the lived experiences of inhabitants. Ultimately, this research contributes to the broader discourse on the intersection of culture and architecture, calling for thoughtful consideration of cultural signs and symbols in the design and development of residential spaces.

References

1. Adeokun CO (2013) The Orowa house: a typology of traditional Yoruba architecture in Ileife, Nigeria. In: Conference: West Africa bulit enivironment research (WABER)
2. Alnaim MM (2024) Cultural imprints on physical forms: an exploration of architectural heritage and identity. Adv Appl Sociol 14(3):141–160
3. Alitajer S, Molavi Nojoumi G (2016) Privacy at home: analysis of behavioral patterns in the spatial configuration of traditional and modern houses in the city of Hamedan based on the notion of space syntax. Front Archit Res 5(3):341–352. https://doi.org/10.1016/J.FOAR.2016.02.003
4. Al-Zrigat ZM (2020) The impact of globalization on neoliberal architecture: how modern architecture has become a tool for control and compliance. J Civil Environ Eng 10(4). https://doi.org/10.37421/jcde.2020.10.350
5. Bani-Masoud A (2021) Contemporary architecture in Iran: from 1925 to the present. Independently published
6. Broadbent G (1994) Recent developments in architectural semiotics. Semiotica 101(1–2):73–102. https://doi.org/10.1515/SEMI.1994.101.1-2.73/HTML
7. Broadbent G, Ward A (1969) Design methods in architecture (Issue 6). Lund Humphries
8. Chandler D (2007) Semiotics: the basics. https://doi.org/10.4324/9780203014936/SEMIOTICS-BASICS-DANIEL-CHANDLER
9. Chng KS, Narayanan S (2017) Culture and social identity in preserving cultural heritage: an experimental study. Int J Soc Econ 44(8). https://doi.org/10.1108/IJSE-10-2015-0271
10. Choay F (2019) Urbanism and semiology. In: The city and the sign. https://doi.org/10.7312/gott93206-009
11. Cross C (1996) Urban design: ornament and decoration. Urban Des Int 1(2). https://doi.org/10.1057/udi.1996.26
12. Csikszentmihalyi M, Halton E (1981) The meaning of things: domestic symbols and the self. https://books.google.com/books?hl=en&lr=&id=5sWxHTayN0wC&oi=fnd&pg=PR9&dq=The+meaning+of+things:+Domestic+symbols+and+the+self&ots=G2W6kJXRRG&sig=DtDRD00AZ12V_NMFmPuRPbmKYwQ
13. Danby M (2013) Architecture in the Islamic World. In: Islam in the modern world (RLE politics of Islam). Routledge, pp 132–141
14. Diba D (2012) Contemporary architecture of Iran. Archit Des 82(3):70–79. https://doi.org/10.1002/AD.1406
15. Diba D, Dehbashi M (1977) Trends in modern Iranian architecture. Researchgate.Net. https://www.researchgate.net/profile/Mozayan-Sharif/publication/265219982_Trends_in_Modern_Iranian_Architecture/links/60f404cafb568a7098b9cc62/Trends-in-Modern-Iranian-Architecture.pdf
16. Dienel HL, Shirazi MR, Schröder S, Schmithals J (2017) Citizens' participation in urban planning and development in Iran. https://doi.org/10.4324/9781315572154
17. Duncan JS (1973) Landscape taste as a symbol of group identity: a Westchester County village. Geogr Rev 63(3). https://doi.org/10.2307/213943
18. Falamarzi R, Bahrambeigy B, Ahmadi M, Rajabzadeh A (2018) High-performance multi/many-core architectures with shared and private queues: network processing approaches. J High Speed Netw 24(2). https://doi.org/10.3233/JHS-180583

19. Farahani LM, Motamed B, Jamei E (2016) Persian gardens: meanings, symbolism, and design. Landscape Online 46(1). https://doi.org/10.3097/LO.201646
20. Fish R, Church A, Winter M (2016) Conceptualising cultural ecosystem services: a novel framework for research and critical engagement. Ecosyst Serv 21:208–217
21. Ghobadian V (1998) Climatic analysis of the traditional Iranian buildings, Tehran, Iran
22. Gottdiener M (2019) Culture, ideology, and the sign of the city. In: The City and the sign. https://doi.org/10.7312/gott93206-011
23. Gouverneur D (2014) Planning and design for future informal settlements: shaping the self-constructed city. Routledge. https://doi.org/10.4324/9781315765938
24. Hajian M, Alitajer S, Mahdavinejad M (n.d.) The influence of courtyard on the formation of Iranian traditional houses configuration in Kashan. Armanshahrjournal.Com. Retrieved from 28 Aug 2022. From https://www.armanshahrjournal.com/article_108573.html?lang=en
25. Jenkins J (2012) Building the revolution: soviet art and architecture, 1915–1935. Des Issues 28(4):86–92. https://repository.falmouth.ac.uk/id/eprint/1484
26. Jivén G, Larkham PJ (2003) Sense of place, authenticity and character: a commentary. J Urban Des 8(1). https://doi.org/10.1080/1357480032000064773
27. Kamali M (2024) The interaction of history and modern thought in the creation of Iran's architecture by investigating the approaches of past-oriented architecture. Front Archit Res 13(3). https://doi.org/10.1016/j.foar.2024.01.003
28. Karimi H, Adibhesami MA, Ghasemi M, Sepehri B (2023) Traditional dwellings in four middle eastern cities: adaptation strategies to harsh climate in privacy. https://doi.org/10.1007/978-981-99-3675-5_8
29. Karimi H, Adibhesami MA, Hoseinzadeh S, Movafagh S, Estalkhsari BM, Garcia DA (2024) Solar energy integration in heritage buildings: a case study of St. Nicholas Church. Energy Rep 11:4177–4191
30. Kellett P, Mark N (1995) Squatter architecture? A critical examination of vernacular theory and spontaneous settlement with reference to South America and South Africa. Tradit Dwellings Settle Rev 7–24
31. Knox PL, Castells M (1995) The Informational city. Information technology, economic restructuring, and the urban-regional process. Geogr J 161(1). https://doi.org/10.2307/3059942
32. Kütükçüoğlu B (2023) The emergence of Mediterraneanism discourse in modern Turkish architecture and the special position of Cengiz Bektaş. A/Z ITU J Fac Archit 20(2). https://doi.org/10.58278/0.2023.6
33. Lazutina TV, Pupysheva IN, Shcherbinin MN, Baksheev VN, Patrakova GV (2016) Semiotics of art: language of architecture as a complex system of signs. Int J Environ Sci Educ 11(17)
34. Lefaivre L, Tzonis A (1981) The grid and the pathway. In: Architecture in Greece, vol 15, pp 167–178
35. Manzo LC, Perkins DD (2006) Finding common ground: the importance of place attachment to community participation and planning. J Plann Lit 20(4). https://doi.org/10.1177/0885412205286160
36. Mazumdar S (2024) How does design affect culture? In: Theorizing built form and culture. Routledge, pp 280–289
37. McCabe C, Roaf S (2013) The wind towers of Bastakiya: assessing the role of the towers in the whole house ventilation system using dynamic thermal modelling. Archit Sci Rev 56(2):183–194
38. Mirzaei V, Sajadzadeh H, Khanlu N (2021) Physical and spatial modeling of Iranian house in hot and dry climate (case study: Kashan Abbasian house). Q J Human Geogr 13(1):136–153. https://www.magiran.com/p2225776
39. Mugerauer R (1987) The concept of dwelling: on the way to figurative architecture. Landscape J 6(2). https://doi.org/10.3368/lj.6.2.157
40. Municipality of Amsterdam (2020) woningbouwplan_2018_2025_web. https://www.amsterdam.nl/bestuur-organisatie/volg-beleid/stedelijke-ontwikkeling/documenten/?PagClsIdt=14631290#PagCls_14631290

41. Nasser N (2022) Beyond the veil of form: developing a transformative approach toward Islamic sacred architecture through designing a contemporary Sufi centre. Religions 13(3). https://doi.org/10.3390/rel13030190
42. Nazari S, Hashemi Nik R (2019) Documentation of the Yazdi-Bandi vault at the plaza of Sadr-e-A'zam, Tehran. Int J Archit Heritage 14(4):541–567. https://doi.org/10.1080/15583058.2018.1553076
43. Norberg-Schultz C (1987) On the way to figurative architecture. Places 4(1)
44. Parsaee M, Parva M, Karimi B (2015) Space and place concepts analysis based on semiology approach in residential architecture: the case study of traditional city of Bushehr, Iran. HBRC J 11(3):368–383. Retrieved from 2 Dec 2022. From https://www.sciencedirect.com/science/article/pii/S1687404814000571
45. Peirce C, Buchler J (1902) Logic as semiotic: the theory of signs. https://books.google.com/books?hl=en&lr=&id=Q6txDwAAQBAJ&oi=fnd&pg=PA100&dq=Logic+as+semiotic:+The+theory+of+signs&ots=XUgX-SV5Va&sig=LGxUf6Qb1YawQsJZHdleR4CuL7o
46. Rapaport RA, Eckhoff WS (1990) Monitoring linear alkyl benzene sulfonate in the environment: 1973–1986. Environ Toxicol Chem 9(10). https://doi.org/10.1002/etc.5620091003
47. Relph E (1976) Place and placelessness, vol 67. Pion London
48. Roaf S (2008) The traditional technology trap (2): more lessons from the windcatchers of Yazd. In: Proceedings of the conference on passive and low energy architecture, pp 19–21
49. Roaf S (2016) Resource quality cascades in traditional low exergy technologies. https://www.irbnet.de/daten/iconda/CIB14196.pdf
50. Robertson S (2012) Significant pavilions: the traditional Javanese house as a symbolic terrain. https://doi.org/10.26190/unsworks/15443
51. Sadalla EK, Vershure B, Burroughs J (1987) Identity symbolism in housing. Environ Behav 19(5):569–587. https://doi.org/10.1177/0013916587195002
52. Saile DG (1985) The ritual establishment of home. In: Home environments. https://doi.org/10.1007/978-1-4899-2266-3_4
53. Salehi RD (2019) A study on architectural characteristics of traditional Iranian ordinary houses in Mashhad, Iran. https://doi.org/10.5281/ZENODO.3299483
54. Sanoff H (2011) Multiple views of participatory design. Focus 8(1). https://doi.org/10.15368/focus.2011v8n1.1
55. Seamon D (2013) Place attachment and phenomenology: the synergistic dynamism of place. In: Place attachment: advances in theory, methods and applications. https://doi.org/10.4324/9780203757765
56. Sedighi M (2020) Rethinking the architecture of Shushtar-Nou: a forgotten episode of architectural regionalism in 1970s Iran. Int J Islamic Archit 9(1). https://doi.org/10.1386/ijia_00007_1
57. Sepe M (2013) Planning and place in the city: mapping place identity, pp 1–333. https://doi.org/10.4324/9780203076149/PLANNING-PLACE-CITY-MARICHELA-SEPE
58. Shin WS et al (2023) Analyzing competency level of architectural designers and Its improvement measures. J Archit Inst Korea 39(9):229–236. https://doi.org/10.5659/JAIK.2023.39.9.229
59. Siadtan S, Pourjafar M (2014) Iranian-Islamic architecture justification graph test (case examples: Rasoulian House in Yazd, Khans in Masuleh). Theor Stud New Technol Archit Urban Plann 4(3):27–39. https://bsnt.modares.ac.ir/article-2-4527-fa.html
60. Sonara H, U Eco (1979) A theory of semiotics. Pdcnet.Org. https://www.pdcnet.org/semioticscene/content/semioticscene_1979_0003_0001_0075_0075
61. Sumners C, Knox A (2005) Archaeoastronomical research and public education in a digistar computer-generated planetarium. In: Fountain JW, Sinclair RM (eds) Current studies in archaeoastronomy: conversations across time and space, pp 555–569
62. Tipple AG (2000) Extending themselves : user-initiated transformations of government-built housing in developing countries, p 358

63. Twigger-Ross CL, Uzzell DL (1996) Place and identity processes. J Environ Psychol 16(3). https://doi.org/10.1006/jevp.1996.0017
64. Zolfaghari SM, Pons O, Nikolic J (2023) Sustainability assessment model for mass housing's interior rehabilitation and its validation to Ekbatan, Iran. J Build Eng 65. https://doi.org/10.1016/j.jobe.2022.105685

Deteriorating Urban Walkability: Configurational Lessons from the Transformation of the Historic Core of Kerman

Aminreza Iranmanesh

Abstract The incremental development of organic urban tissues is influenced by a multitude of interconnected factors. This study delves into the intrinsic complexities of these forms, highlighting human-scale qualities often overlooked in contemporary urban design and planning. Specifically, it examines historic urban tissues in the hot and dry regions of Iran, exemplified by the city of Kerman. These urban areas were historically characterised by walkability, accessibility, and passive thermal comfort—attributes that contribute to sustainable urban forms. This paper explores the patterns of access networks in Kerman, focusing on how primary public spaces like bazaars and marketplaces interconnect with secondary residential areas. It aims to analyse the urban typo-morphological patterns that link various city parts to the main public realms, along with the micro-climatic adjustments facilitating these connections. The research highlights how recent alterations such as top-down urban development, infrastructural changes, and the introduction of vehicular roads, have significantly transformed the configurational characteristics of Kerman's once-walkable urban tissue. These transformations have obscured many micro-climatic solutions born from centuries of incremental urbanism, especially in the historic core. Utilising temporal mapping of the city's access network from the first aerial image in 1957 to the present, the study employs Space Syntax and statistical classification to analyse each temporal stage. The analysis classifies access network elements based on the to-movement and through-movement measures and provides comparisons between different temporal stages of core transformations. The temporal stages are selected based on the introduction of significant new elements into the grid. The findings reveal the impact of new roads on traditional mobility patterns, showing a shift favouring vehicular over pedestrian movement. This disruption is particularly evident in transitional spaces, which are crucial for pedestrian flow. While significant historic buildings are preserved, the pedestrian access network has deteriorated. The outcome shows that the decline in walkability is a result of the formalisation of the grid, which prioritises the creation of longer, more connected vehicular pathways through the

A. Iranmanesh (✉)
Faculty of Architecture and Fine Arts, Final International University, Via Mersin 10, TRNC, Girne, Türkiye
e-mail: aminreza.iranmanesh@final.edu.tr

A. Cheshmehzangi and S. Roaf (eds.), *Persian Vernacular Architecture*, Urban Sustainability, https://doi.org/10.1007/978-981-96-1116-4_9

city, often at the expense of pedestrian-friendly routes. The patterns identified in this study offer valuable insights for future urban planning, especially in preserving and revitalising the remaining parts of Kerman's historic core.

Keywords Urban morphology · Sustainable urban design · Space syntax · Kerman · Iran · Historical urban tissue · Walkability · Temporal mapping · Urban planning

1 Introduction

Historic urban forms are the outcome of incremental transformation processes that occur over long timespans [8]. The historic urban tissue in this regard can be considered the outcome of the collective socio-spatial wisdom of generations of people adapting urban forms to local conditions including climate, morphology and social and economic developments. The architectonic and urban solutions developed through these incremental changes are often the outcome of generations of trial and error resulting in innovations in space making that are more efficient, comfortable, and manageable. What is more, social and spatial urban structures are continually in the process of becoming [10]; they can be considered as evolving landscapes being continually refined to provide new answers to emerging issues. While scholarly discourse on vernacular architecture focuses more on issues such as comfort at the building scale [60, 62], the embedded vernacular genius in the urban grid as a whole requires further investigation, offering insights into designing more resilient and sustainable future cities (see Susan Roaf and McGill [64]). While buildings are designed to foster thermal comfort for indoor activities, a significant part of everyday life unfolds outside these structures. The outdoor life of the city is interwoven into connective tissues that have historically evolved to facilitate social interactions while providing reasonable comfort in external environments [61]. With climate change bringing these issues to the forefront once again, more walkable and passive urban solutions, embedded within the configurational characteristics of urban forms, can significantly contribute to reducing energy consumption and pollution [63].

Looking through this lens, lessons that can be learned from these processes can provide valuable insight into creating more efficient and resilient urban forms, particularly considering the changing climate and growing population (see Susan Roaf et al. [62]; Sharifi and Murayama [67]; UN-Habitat 75]). The social, economic, cultural, and morphological fabrics that interweave to form the urban tapestry are challenging to comprehend all at once [3]. Yet, it can be argued that remnants of these complexities manifest as reflections in the physical built environment. Therefore, exploring the morphological forms and configurational characteristics of urban spaces can provide insight into human experience and ingenuity [24–26]. By studying how streets intersect, and public spaces emerge, it is possible to gain valuable insights into the underlying dynamics that have shaped cities over time. This deeper understanding not only enriches our appreciation of urban landscapes but also provides practical knowledge that could be used to design more liveable, sustainable, and inclusive environments.

This paper, focuses specifically on the configurational characteristics of urban access networks. These networks, comprised of streets, pathways, and external arteries, form the lifelines of cities, facilitating movement, connectivity, and accessibility. By analysing their layout, connectivity, and spatial organisation, the study aims to uncover the underlying principles that govern urban form and function that are designed by people over time. This exploratory study hopes to shed light on the intricate relationships between urban morphology, configurational patterns of historic urban fabric, and the enduring vernacular wisdom that shapes our cities. The paper provides a temporal reading into the urban form of Kerman.

The historic urban fabric of Kerman has seen many radical transformations. The majority of these transformations were alterations aimed at fitting urban fabric for better vehicular movements. The initial state that became the basis for this research is the first aerial imageries that were taken from the city in 1956. Even in this snapshot, these alterations have already begun. Two major roads cut through the historic fabric during the first Pahlavi dynasty. Before this date, the Pahlavi dynasty started the modernisation of major urban centres. This modernisation attempt came with the introduction of new streets often in contrast with the dense pedestrian-oriented urban tissue. As a result, many parts of the historic fabric were lost to make way for new rods.

2 A Brief Overview of Top-Down Urbanisation in Iran from Pahlavi Era to Present

The Pahlavi dynasty in Iran, spanning from the early twentieth century until the late 1970s, saw significant urban alterations and transformations under the umbrella of the modernisation agenda. Under the leadership of Reza Shah Pahlavi and later his son Mohammad Reza Shah Pahlavi, Iran underwent rapid urbanisation and industrialisation, leading to profound changes in its urban fabric [46, 47]. Although Pahlavi monarchs put immense effort into the preservation and restoration of historical monuments, the modernisation of Iran, sometimes, came at the expense of its less significant historic fabric [1, 71]. These transformations mainly emerged as creating more auto-centric urban networks which was often in contrast with the dense historic tissue [48]. This period witnessed the demolition of some urban tissue to make way for large-scale infrastructure projects, such as highways, railways, and modern buildings (see Javanmardi [33]). Additionally, urban planning initiatives sought to impose European-inspired gridiron layouts and architectural styles, displacing traditional Iranian urban forms and architectural typologies. While these changes brought about improvements in transportation, sanitation, and housing, they also led to the loss of cultural identity and heritage. Historic neighbourhoods were fragmented, and existing communities were disrupted, contributing to social tensions and cultural

Fig. 1 Examples of vehicular arteries cutting through historic fabric in Iran. *Source* Maps generated from publicly open source maps on Google Earth 2023

dislocation. Moreover, the emphasis on modernisation neglected the preservation and maintenance of historic sites, leading to the deterioration and loss of many irreplaceable heritage assets.

After the Islamic revolution of 1979, the same top-down car-centric and infrastructure-based patterns in urban planning continued. This was, in part, facilitated by the rapid population growth in a relatively short period that increased the demand for new urban infrastructure [12, 17, 65, 72]. The complexity of changes in demographic profile during this period is complex and would be outside the limited scope of this study (for more information, see [12]). Initially, the new administration put little value in the preservation of historic tissue and priority was given to making cities toward a more auto-centric image. During this time, even more arteries were cut through historic tissue around the country. Although in recent years the importance of preserving the urban fabric and urban perspectives has found its way to the pallet of planners and local municipalities, many of the radical interventions in the past have left many historic cores with scares. Overall, it can be argued that in the past century, although a lot of emphasis was put on monumental buildings, the urban tissue comprising of residential and less-known buildings has been ignored, this led to the decline of integration, local walkability, and many vernacular qualities that made the core of the cities liveable [32, 67] (see Fig. 1).

3 Case Study: Kerman

Kerman, a city located in the south-eastern part of Iran, is notable for its dense historic fabric, deeply influenced by its geographical setting and historical developments. The city's urban morphology reveals a compact, traditional urban layout mainly constructed by mudbrick, adobe, and brick. The urban tissue is characterised by narrow, winding paths and a dense fabric of residential and commercial spaces, reflecting a response to both social practices and the arid desert climate of the region. According to the Köppen climate region classification, Kerman is considered a cold desert climate (BWK) with cold dry winters and semi-hot dry summers [42]. Accordingly, the urban layout has evolved to maximise shade and minimise wind exposure, enhancing comfort in both hot summers and cold winters.

The historic fabric of Kerman showcases a typology of architecture unique to the region, including bazaars, caravanserais, mosques, and traditional houses, all interconnected through a network of public spaces and streets that embody the city's socio-economic history and cultural identity [50]. Central to the urban grid, the bazaar in Kerman is characterised by its linear configuration, diverging from the typical concentric layouts observed in other bazaars across Iran. Within the city's heart lie caravanserais, surrounded by a network of other significant structures, including mosques, baths, public amenities, and educational institutions, all of which are integral to the bazaar's framework [18, 43]. These elements are arranged in a hierarchical structure that emphasises community interaction and the importance of trade and religion in the city's development. Typo morphological aspects of Kerman reveal the evolution of urban forms over time, influenced by factors such as defensive needs, trade routes, and the availability of water through Qantas (traditional underground irrigation channels) [5, 45], which have shaped the city's layout and architecture [30]. The population of the city has grown significantly after the Islamic Revolution, and top-down urban planning approaches have left their mark on the shape of the city [65]. The case of Kerman serves as an example of the complex interplay between traditional urban forms and modern development strategies in Iran. The rapid expansion of a once compact, interconnected network of movement into a large auto-centric city has impacted the socio-spatial qualities of vernacular urban forms. In this case, the temporal introduction of vehicular arteries cutting through the historic fabric can be used to understand the impact of radical interventions on transformations of movement patterns in vernacular urban forms. By examining the configurational shifts within Kerman's historic core, the case of Kerman highlights the broader impacts of auto-centric development on socio-spatial dynamics and its capacity for change. Thus, this study aims to uncover lessons learned from past urban planning practices, providing insights that can guide future efforts to balance modernisation with the preservation of cultural heritage and social vibrancy in Kerman.

4 Materials and Methods

4.1 Space Syntax: Exploring Configurational Patterns in Urban Forms

Space syntax is a set of theoretical and evidence-based frameworks that explores and evaluates the spatial configuration of spatial networks [22, 23, 26]. It operates on two fundamental premises, first, that the layout of spaces affects human behaviour and vice versa and none are passive aspects of the other. Second, spaces can be understood in their relation with the whole network and not in isolation [39, 77, 78].

Traditionally, space syntax analysis is traditionally performed on axial maps which represent the most direct lines of sight (where movement is possible) within a spatial network using the principle of convex space [4, 22, 26, 73]. Although other mapping techniques such as road centreline have been proposed [74], this study performs the analysis based on the axial map as road centreline cannot be accurately defined based on historic satellite imageries. Axial maps can provide a tangible perspective in integrated historic urban tissue where the paths evolved based on the natural movement of people interacting with the city [20]. On the basis of the axial map, space syntax provides many types of analysis such as depth, integration, and choice. This paper uses integration and choice for the analysis as they have been shown to be significant predictors of pedestrian movement [27, 68]. Integration measures how connected (central) each space is to the network as a whole, indicating potential for social interaction and movement. Choice, on the other hand, assesses the number of times a path is used within the shortest routes across the network, highlighting the paths most likely to be used and their strategic importance within the spatial configuration.

The evidence-based literature makes a strong case for the predictability of natural urban movements (pedestrians in particular) through urban access networks [27, 34, 44, 57, 68], thus, the predictability of movement, in this case, is used to explore the potential patterns of movement in the past and how it changed the socio-spatial patterns in the case of Kerman. Space syntax has been utilised to investigate potential past patterns of movement in urban access networks [13, 77, 81].

Space syntax is particularly useful in exploring urban temporal patterns and incremental changes due to its ability to quantitatively assess the spatial structures that influence human interaction and movement patterns over time [19, 29, 79]. By applying space syntax analysis, urban planners and researchers can identify how changes in spatial configuration affect the functionality and sociability of urban areas [29, 56]. This insight enables the prediction of how modifications to the urban fabric might influence future patterns of use, social cohesion, and accessibility [16, 39]. Furthermore, understanding the configurational transformation of urban fabric can provide insight into understanding the resiliency of urban forms facing alteration and change. This understanding, in turn, can help guide the design of more integrated, inclusive, and adaptable urban environments. Through the lens of space syntax, the dynamic and evolving nature of urban spaces can be explored, revealing the intricate relationship between spatial design and societal behaviours [2, 39].

In the context of Kerman, space syntax analysis has been used to explore housing typologies and patterns [35, 41, 80] and exploration of archaeological sites [70]. The configuration of the historic core has been thematically explored by Karimi [36], Karimi [37], Nejad [52], Ghoraba and Tabibian [14] and Kermani and Luiten [40] to the best of the author's knowledge.

4.2 Temporal Mapping

The study uses some temporal mapping of access networks based on the major alteration imposed on the historic grid. The initial mapping of the access network was conducted on the 1957 aerial image of the city. This image was superimposed on the current satellite imageries using QGIS. Accordingly, the extent of the study remains the available area visible in the 1957 image although the city has grown significantly in size. The temporal mapping consists of six states of the urban access network; the pre-1957 stage was created based on the continuity of urban paths as there is no prior formal documentation of the city [36, 50], this map was cross-referenced with the analysis of Karimi [36] exploring historic layers of the city dating back to the 1800s (also see Karimi [37]). Moreover, four stages after 1957 were selected based on the major alteration that changed the urban grid in a significant way (Fig. 2). The axial map was generated by identifying the longest lines of sight where movement appeared feasible. Given that many of the city's primary historic public thoroughfares are encapsulated within covered bazaars, an overlay process was employed. The detailed mapping of these bazaars was superimposed upon both aerial and satellite imagery. This methodological approach was aimed at producing an axial map that comprehensively represents the full spectrum of potential movements throughout the city.

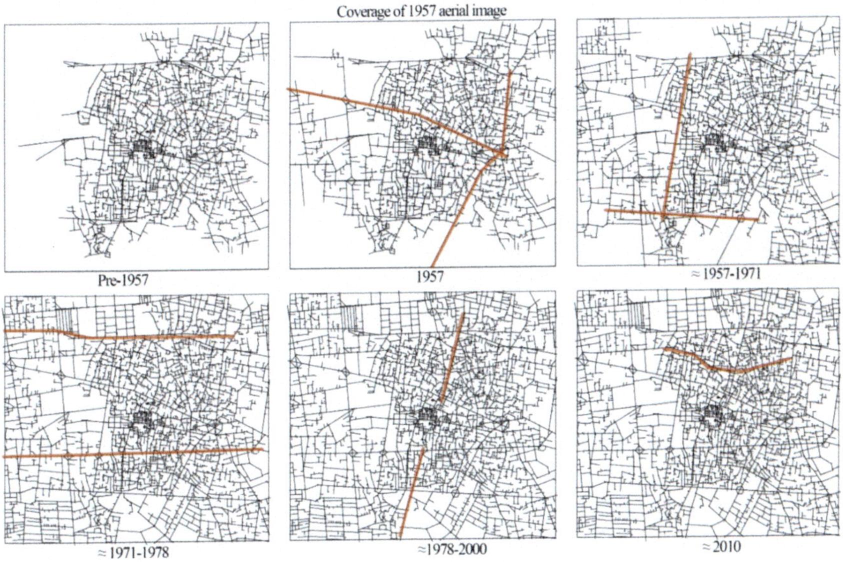

Fig. 2 Temporal mapping of urban access network, estimations made by the author based on available satellite and aerial imageries. *Source* The Author

5 Analysis and Discussion

The space syntax analyses were conducted using the DepthMap X application developed at UCL [9]. The integration and choice analyses were performed for each temporal stage in both local and global measures. The global measure includes all axes into account in their spatial relation with one another on the dual graph while the local measures are performed with a topological restriction. The local measures in this case were performed at $n = 3$ steps which is often used for axial analysis (see Salheen and Forsyth [66]; van Nes [76]; Van Nes and Yamu [77, 78]). All analyses are represented in Fig. 3. It is evident from the progression of integration and choice pattern that the straight lines cut into the urban fabric immediately become the most integral and in-between part of the urban access network. In the initial stage of the analysis (pre-1957) bazaar is the central axis of the city, it is both a destination (to movement) and a passage (through movement). The bazaar in this stage is both locally and globally integrated with the network indicating a high degree of intelligibility; through this perspective, while walking in the city one will eventually end up in the bazaar (similar findings have been reported by Karimi [36]). This represents the foreground spaces of the city which are the vessels of micro economy and social interaction [21, 24, 49]. Considering that the flow of information through a network is facilitated by weak social ties [15], these spaces can be considered the beating heart of the social structure of the city.

The organic development of the Kerman urban network underscores the role of foreground spaces in facilitating not only economic transactions but also in fostering social connections and communal identity [37, 40, 52]. Different axes of the bazaar seem to have formed based on the ways of trade and how caravans were entering the city to facilitate trade [52]. This is in line with the space syntax discourse arguing for the simultaneous and interactive development of land use and urban networks over time [27, 51, 58]. Foreground spaces like the Ghale Mahmoud Bazaar have historically evolved as the economic and social parts of the access network of Kerman. The linear configuration of the main arteries made them central to the city's trade and social life (also see Hakimian and Lak [18]; Lak and Hakimian 43]). This linear form may have contributed to their inability to remain central to the network when the urban layout was altered. What is more, the entrance gate to the city which leads to the starting point of this bazaar was removed and replaced by a growing urban fabric that did not consider this path in the development process. This shift underscores the critical role that spatial configuration plays in maintaining the socio-economic vitality of urban areas.

The analyses show configurational characteristics of the bazaar show a sudden and rapid decline with the introduction of new alternative auto-centric paths. As Fig. 3 represents, by stage 5, the entire bazaar (red in pre-1957), configurationally, becomes similar to the less integrated parts of the city resembling background residential spaces. Such transformations can relegate once vibrant bazaars to the background of urban life, aligning them more with residential spaces in terms of integration within the city's spatial network.

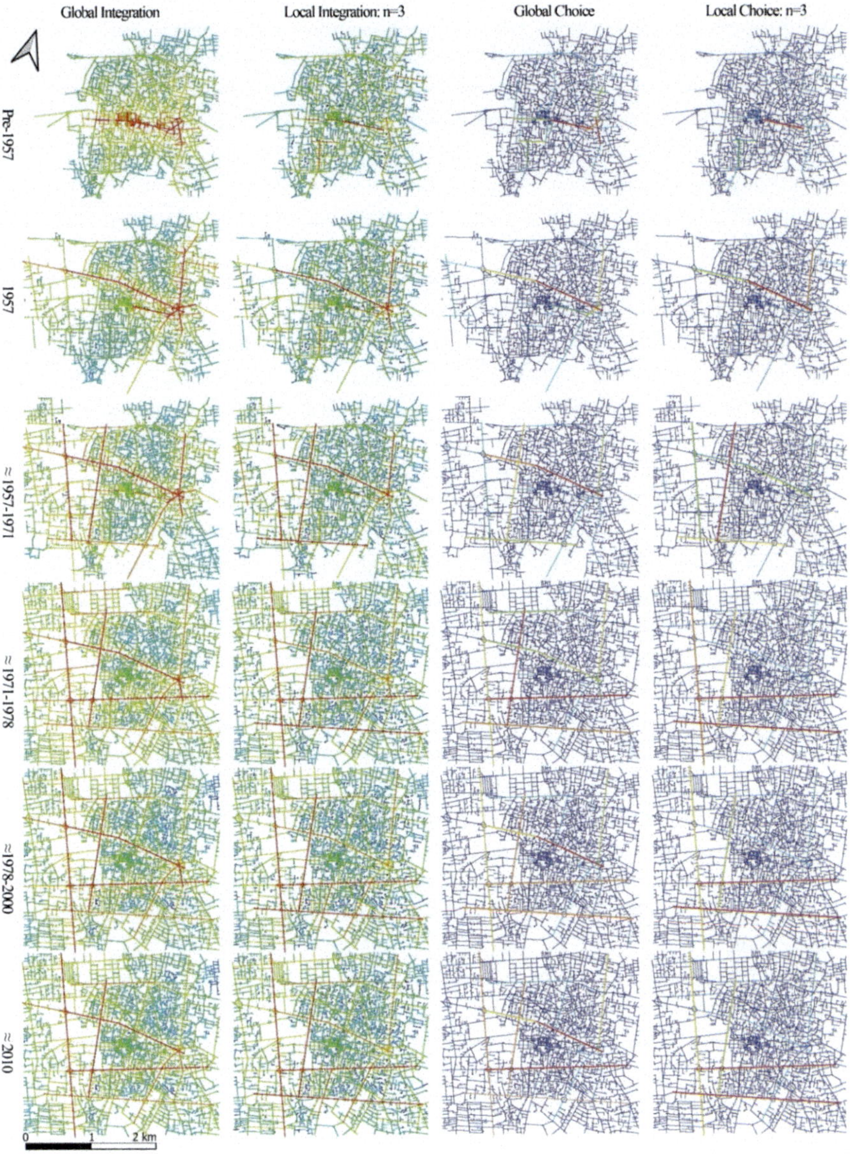

Fig. 3 Space syntax analyses of the historic core of Kerman for each temporal synapse. *Source* The Author

Looking through the holistic lens of configurational theory into the urban access network, the decline of foreground spaces due to auto-centric development poses a challenge to the city's ability to sustain its socio-cultural heritage and economic diversity [28]. In this scenario, the remaining foreground spaces of the bazaar might slowly transform into mere touristic showcases rather than being an integral part of the everyday life of the urban core. The imposition of formal paths in many areas, incrementally designed for facilitating pedestrian use, underscores the pivotal role urban configuration plays in preserving historical landscapes. As evident from the result of this study, despite these covered spaces offering aesthetically pleasing, thermally comfortable alternatives for movement, shielded from rain and wind, there's a noticeable trend towards their underutilisation. This observation points to a broader issue: the configurational qualities of traditional urban access networks are often overlooked in scholarly discourse. While the architectural aspects of vernacular architecture receive considerable attention, the collective ingenuity behind the creation of these urban access networks over time is less explored.

Formalisation of the grid reduces the natural topological steps required for common urban commuting [77], this, however, seems to favour vehicular travel over pedestrian movement. This shift not only affects the economic vitality of these areas but also impacts the social and cultural fabric of the city, as the opportunities for spontaneous social interactions and the traditional microeconomics supported by these spaces diminish. This is visible in parts of the bazaar that have been disconnected due to the introduction of new paths (also see Kermani and Luiten [40]). While these transformations are vividly present in the visual representation of space syntax measures, merely looking at the numbers presents a different reality (Table 1). The mean values for both integration and choice (average of all measures for each item) show a gradual increase with each temporal iteration. This seems to be a natural side effect of imposed straight axes on an existing deformed grid. These straight lines, in a dense organic grid, increase the overall accessibility of the access network, both at local and global levels. Yet, they diminish how the urban grid is being experienced by the people.

This phenomenon is perhaps most starkly observed along the Ghale Mahmoud Bazaar, which runs perpendicular to the newly introduced vehicular paths (also see Orszt and Savary [54]). This bazaar, once a major pedestrian artery of the historic city, has seen a decline over time, with the majority of its shops shuttered and the inns closely associated with it rapidly falling into disrepair. Contrasting this, sections of the bazaar that run parallel to the new paths have not experienced such rapid deterioration and continue to function, albeit with a noticeable decline in pedestrian traffic as one moves away from the main core of the bazaar, as illustrated in Fig. 4. These covered areas were originally designed to offer pedestrians—then the primary mode of inner-city transit—protection against the extremes of weather, including hot summers and cold, dry winters, yet even these qualities have failed to retain sufficient foot traffic following the introduction of vehicular pathways, leading to a socio-spatial decay. Within the framework of space syntax, spaces are not considered background to human activities [38, 39], accordingly, these decays in socio-spatial structures can be considered the outcome of the rapid transformations in the configurational

Table 1 Mean values for integration, choice and line length for temporal stages of access network

Year		Integration [HH]	Integration [HH] R3	Choice	Choice R3	Line length
Pre-1957	Mean	0.55	1.43	39,984.0	33.0	55.3
	Std. deviation	0.16	0.53	128,473.9	83.3	58.5
1957	Mean	0.79	1.48	29,478.6	44.4	59.2
	Std. deviation	0.24	0.60	119,219.5	225.3	77.9
1957–1971	Mean	0.96	1.55	26,776.1	61.6	62.3
	Std. deviation	0.30	0.66	132,511.4	444.9	98.3
1971–1978	Mean	1.30	1.73	23,629.0	117.4	67.8
	Std. deviation	0.31	0.75	132,445.5	1240.4	125.2
1978–2000	Mean	1.37	1.78	21,515.2	125.2	70.6
	Std. deviation	0.30	0.75	124,716.9	1268.8	128.5
2010	Mean	1.38	1.78	21,136.2	127.3	71.6
	Std. deviation	0.30	0.76	121,681.6	1280.2	129.4

characteristics of the urban grid. Hillier's argument for the movement economic [21] indicates that land use and movement develop in co-dependency, thus, in the long run, one is not superior to the other. This is also in line with the theory of natural movement that is well documented in space syntax literature as co-presence is facilitated by the centrality characteristics of the space and space is being shaped and reshaped by the co-presence of social structures [27].

Another phenomenon that can be seen during the transformations of this case is the increasing size of urban axes. The mean length of axes increases from 55 m in pre-1957 to 71 m in 2010s. The size of the urban block is an indicator of a healthy and

Fig. 4 Right: the main core of the bazaar, middle: the main perpendicular bazaar branching out of the core and left: abandoned Ghale Mahmoud Bazaar. *Source* The Author

walkable city, offering alternative paths for reaching destinations [31], smaller paths and blocks tend to provide a more even distribution of pedestrian movement. In this case, although the imposition of a formal grid over an organic core has increased the overall integration, it has also reduced the number of available alternative paths. This seems to concentrate the pedestrian traffic into the main straight paths and reduce the usage of the less dominant organic part of the grid.

Tracking the transformation of the choice (through-movement) shows a similar shift (see Fig. 3). In the earlier stages of the urban grid, the bazaar is also the in-between space of the city. Space with high choice value are spaces that are more frequently travelled through considering all origin-destinations in a grid. This indicates that moving from one neighbourhood to another required travelling through the central bazaar, not necessarily for shopping, but as the most comfortable, and accessible part of the grid. As argued by Niazi and Ozdil [53] these qualities have made the bazaar of Kerman, historically, a pedestrian-friendly artery of variety, diversity, mixed-use, and core of urban vitality. Yet, the betweenness characteristic of the bazaar quickly fades away being replaced by the formal grid surrounding it. The same issue can be explored from the perspective of human experience within the urban grid. The natural complexity and human scale of the pedestrian paths created places attached to the collective memory of the inheritance facilitating a sense of identity and belonging [6, 7]. On the other hand, long urban axes without proper human scale provide less stimuli capable of expediting the formation of urban memory or identity (Fig. 5).

These changes have not remained significant only in foreground spaces (bazaar), the background residential spaces have also been impacted. Neighbourhoods were the essential social structure of the historic core of Kerman [37, 69], highlighting privacy as a key characteristic of their morphology. Many of the new changes have reduced the degree of privacy in some parts of Kerman [80] while disconnecting many neighbourhoods from the central public arteries of the city [11]. These changes, coupled with reoccurring earthquakes have slowly led to the displacement of the local families causing more deterioration in the process.

Fig. 5 Examples of the new formal grid cutting through the historic urban tissue. *Source* The Author

6 Conclusions

This paper explored the configurational transformation of the access network of Kerman from a temporal space syntax analysis perspective. The importance of the study lies in the lesser-known impact of radical top-down approaches imposed on traditional urban tissue in the region. The modernisation process came with the need for more auto-centric design facilitating the transport demands of the growing population of the city. Within the span of approximately 70 years, new arteries were cut through the compact and complex structure of the urban network of Kerman. While the monumental buildings were given priority for preservation, less attention has been paid to the qualities of urban fabric. These changes led to the gradual transformation of the configurational characteristics of the city. Hence, patterns of movement were changed within the city rendering many of the primary pedestrian accesses less integrated into the day-to-day functioning of the city, and thus less useful.

One of these configuration patterns is the intricate balance between the foreground and background spaces of the city. Traditionally, in the city of Kerman, the covered bazaar played the major role of foreground spaces, connecting multiple neighbourhoods and city gated within a comfortable covered pedestrianised path with multiple degrees of hierarchy. However, the introduction of new vehicular pathways has shifted the balance, drawing the integration and choice core—the key areas of connectivity and movement based on the Space Syntax analysis—toward these new roads, reducing the bazaar's centrality in the urban network. This shift, became stronger with the addition of each new pathway cut through the market, rendering many parts of the bazaar into background spaces. The opposite can be observed in many residential neighbourhoods. Because the new streets were not supposed to demolish the main monumental buildings, hence the majority of these paths were carved directly through residential urban tissue. This brought many of the previous background spaces to the foreground of the city creating graduation changes in the socio-spatial structure of the city.

Moreover, this case study of Kerman reveals the challenges that dense, organically developed urban networks face in adapting to the rapid pace of modern life. The shift towards top-down auto-centric urban development strategies, contrasting with the bottom-up processes that historically shaped urban spaces, has led to a gradual reconfiguration of movement patterns. The historic core of the city, particularly the bazaar, remains vibrant and functional. However, many secondary axes within the urban grid have lost their original configurational utility. Initially, the city's center served both as a destination and a transitional space. Radical alterations to the urban grid have significantly diminished these dual roles, impacting the area's navigability and its integral function within the larger urban context.

The shift towards auto-centric development highlights a significant transformation in urban planning, favouring vehicular traffic over pedestrian mobility and social interaction. This transition challenges urban sustainability, resilience, and social equity, necessitating a balanced approach to modernisation that preserves the socio-cultural and economic significance of traditional urban spaces. As shown in the analysis, perhaps is too late to save large parts of deteriorated urban tissues, yet, lessons that can be learned from these configuration characteristics can provide insight into designing better and more integrated urban spaces that are human-centric and more sustainable.

To address these challenges, urban planners and policymakers must consider strategies that enhance the vibrancy and integration of historical urban areas. Initiatives could include promoting pedestrianisation, and through bottom-up analytical and qualitative studies before decision-making processes. Looking through the lenses of configurational morphology, mere rehabilitation of the introduction of interesting land use might not be sufficient in the long run. Efforts need to be made to improve the integration of the historic fabric with the newly developed parts. Effective pedestrian arteries that lead to the junctions connecting to the bazaar can mitigate the deterioration of the historic tissue by increasing foot traffic and micro-economy.

In the end, and in reflecting on the challenges posed by climate change, it is essential to recognise that many Western-rooted urban design solutions may not universally apply across all contexts. This study highlights the importance of drawing from the rich configurational patterns of historic Iranian cities, which evolved not only through the efforts of individuals but as the outcome of the collective genius of a resilient and adaptive society. These patterns were designed to provide comfort while simultaneously facilitating spaces for social, economic, and cultural interactions. By integrating these traditional patterns, it is possible to create sustainable environments that honour privacy, comfort, and environmental resilience, demonstrating that the wisdom of our past holds the key to addressing the challenges of our future.

6.1 Limitations of the Study

A one-dimensional look at the city and its intricate multifaceted complexities do not often render a comprehensive understanding of urban forms. This study focused on the configurational dimensions of urban from a temporal standpoint, yet, other dimensions must be simultaneously explored both from the pentalogies of urban tissue deterioration and for planning for more resilient future urban forms [67]. The development of the temporal axial map in this study was limited to the available aerial and satellite imageries as well as the observations of the author and might be subject to some errors. Future studies could include a comprehensive analysis of the entire city in its current form, with a closer examination of specific spaces in more detail. Additionally, exploring how enhanced pedestrian pathways and cycling-focused routes could support tourism and introduce economically beneficial functions to historic spaces would be valuable, particularly in the context of Kerman's hot and dry climate. The study is also limited to its contextual characteristics of Kerman and overgeneralisation of the outcome is better avoided. In the end, it must be noted that space syntax as a method has certain limitations and is subject its criticism in its approach [55, 59], yet the consistent evidence-based outcome of the analysis (such as presented in this paper) indicates merits in its approach to reading urban tissue based on the fundamentals of affordance, visibility, and movement.

References

1. Abdollah-Khan-Gorji B (1997) Urban form transformations: the experience of Tehran before and after the 1979 Islamic Revolution. PhD Dissertation, University of Southern California
2. Ahmed B, Hasan R, Maniruzzaman KM (2014) Urban morphological change analysis of Dhaka City, Bangladesh, using space syntax. ISPRS Int J Geo-Inf 3(4):1412–1444. https://www.mdpi.com/2220-9964/3/4/1412
3. Alexander C (2002) The nature of order: the phenomenon of life. The Centre for Environmental Structure, California
4. Batty M, Rana S (2004) The automatic definition and generation of axial lines and axial maps. Environ Plann B: Plann Des 31(4):615–640. https://doi.org/10.1068/b2985
5. Beaumont P (1971) Qanat systems in Iran. Hydrol Sci J 16(1):39–50
6. Cheshmehzangi A (2020) Identity of cities and city of identities. Springer, Singapore
7. Cheshmehzangi A (2021) Remembering through naming and palimpsests: case study example of the city of Shiraz, Iran. In: Cheshmehzangi A (ed) Urban memory in city transitions: the significance of place in mind. Springer Singapore, Singapore, pp 243–282. https://doi.org/10.1007/978-981-16-1003-5_7
8. Conzen MRG (1960) Alnwick, Northumberland: a study in town-plan analysis. Trans Pap (Inst Br Geogr) (27):iii–122. https://doi.org/10.2307/621094
9. DepthmapX Development Team (2017) DepthmapX (Version 0.6.0) [Computer software]. Retrieved from 20 Feb 2023
10. Dovey K (2009) Becoming places: urbanism/architecture/identity/power. Routledge, London. https://doi.org/10.4324/9780203875001
11. Fani Molki F (2017) A design research towards developing an appropriate domestic architecture in Iran: a case study of Kerman. University of Sheffield
12. Fanni Z (2006) Cities and urbanization in Iran after the Islamic revolution. Cities 23(6):407–411. https://doi.org/10.1016/j.cities.2006.08.003
13. Fladd SG (2017) Social syntax: an approach to spatial modification through the reworking of space syntax for archaeological applications. J Anthropol Archaeol 47:127–138. https://doi.org/10.1016/j.jaa.2017.05.002
14. Ghoraba N, Tabibian M (2017) Developing an applied model for explaining the mental structure of cognitive maps of people through spatial-morphological analysis of existing urban textures, case study: historical texture of Kerman. Bagh-E Nazar 14(54)
15. Granovetter MS (1973) The strength of weak ties. Am J Sociol 78(6):1360–1380
16. Griffiths S (2012) The use of space syntax in historical research: current practice and future possibilities. In: Greene M, Reyes J, Castro A (eds) Proceedings of the 8th international space syntax symposium
17. Habibi M (2009) Az shar ta shahr. University of Tehran
18. Hakimian P, Lak A (2023) A conceptual framework for auditing the physical integrity of historical districts: case studies of Kerman and Shiraz Bazaars in Iran. J Cult Heritage Manage Sustain Dev 13(4):642–661. https://doi.org/10.1108/JCHMSD-10-2020-0154
19. Hallowell GD, Baran PK (2013) Suburban change: a time series approach to measuring form and spatial configuration. J Space Syntax 4(1)
20. Hegazi YS, Fouda M (2019) Re-imaging Rosetta historic core through space syntax. Archnet-IJAR: Int J Archit Res 13(3):645–669. https://doi.org/10.1108/ARCH-05-2019-0109
21. Hillier B (1996) Cities as movement economies. Urban Des Int 1(1):41–60. https://doi.org/10.1057/udi.1996.5

22. Hillier B (1996) Space is the machine: a configurational theory of architecture. Press Syndicate, Cambridge
23. Hillier B (1999) Centrality as a process: accounting for attraction inequalities in deformed grids. Urban Des Int 4(3):107–127. https://doi.org/10.1057/udi.1999.19
24. Hillier B (2012) Studying cities to learn about minds: some possible implications of space syntax for spatial cognition. Environ Plann B: Plann Des 39(1):12–32. https://doi.org/10.1068/b34047t
25. Hillier B, Burdett R, Peponis J, Penn A (1986) Creating life: or, does architecture determine anything? Archit Comportement/Archit Behav 3(3):233–250
26. Hillier B, Hanson J (1984) The social logic of space. Cambridge University Press, Cambridge. https://doi.org/10.1017/CBO9780511597237
27. Hillier B, Penn A, Hanson J, Grajewski T, Xu J (1993) Natural movement: or, configuration and attraction in urban pedestrian movement. Environ Plann B: Plann Des 20(1):29–66
28. Inglês AC, Cannas L, Heitor T (2023) Urban design, architecture and space syntax in the conception of public spaces—a look at Luanda's revitalization. In: Mora PL, Viana DL, Morais F, Vieira Vaz J (eds) Formal methods in architecture. Springer, Singapore
29. Iranmanesh A, Kamalipour H (2023) A configurational morphogenesis of incremental urbanism: a comparative study of the access network transformations in informal settlements. Cities 140:104444. https://doi.org/10.1016/j.cities.2023.104444
30. Iranmanesh L (2011) Architectural background of Kerman. J Lang Culture 2(3):34–38
31. Jacobs J (1961) The death and life of great American cities. Random House, New York
32. Jamei E, Ahmadi K, Chau HW, Seyedmahmoudian M, Horan B, Stojcevski A (2021) Urban design and walkability: lessons learnt from Iranian traditional cities. Sustainability 13(10)
33. Javanmardi L (2020) Urbanism and dictatorship. A study on urban planning in contemporary history of Iran, Second Pahlavi: 1941–1979. Dissertation to the conferral of the academic degree, Bauhaus-Universita t Weimar
34. Jiang B (2009) Ranking spaces for predicting human movement in an urban environment. Int J Geogr Inf Sci 23(7):823–837. https://doi.org/10.1080/13658810802022822
35. Karbasi A (2023) The developmental history of space configuration in the last hundred year houses of Najafabad. Monthly Sci J Bagh-e Nazar 20(118):5–22
36. Karimi K (1997) The spatial logic of organic cities in Iran and the United Kingdom. In: 1st International space syntax symposium
37. Karimi K (2000) Urban conservation and spatial transformation: preserving the fragments or maintaining the 'spatial spirit.' Urban Des Int 5(3):221–231. https://doi.org/10.1057/palgrave.udi.9000012
38. Karimi K (2012) A configurational approach to analytical urban design: 'space syntax' methodology. Urban Des Int 17(4):297–318. https://doi.org/10.1057/udi.2012.19
39. Karimi K (2023) The configurational structures of social spaces: space syntax and urban morphology in the context of analytical, evidence-based design. Land 12(11)
40. Kermani AA, Luiten E (2010) Preservation and transformation of historic urban cores in Iran, the case of Kerman. WSEAS Trans Environ Dev 6(1):53–62
41. Khozaei Ravari F, Hassan AS, Abdul Nasir MH, Mohammad Taheri M (2022) The development of residential spatial configuration for visual privacy in Iranian dwellings, a space syntax approach. Int J Build Pathol Adapt (ahead-of-print). https://doi.org/10.1108/IJBPA-05-2021-0080
42. Kottek M, Grieser J, Beck C, Rudolf B, Rubel F (2006) World map of the Köppen-Geiger climate classification updated
43. Lak A, Hakimian P (2018) A new morphological approach to Iranian bazaar: the application of urban spatial design theories to Shiraz and Kerman bazaars. J Archit Conserv 24(3):207–223. https://doi.org/10.1080/13556207.2018.1543821
44. Lerman Y, Rofè Y, Omer I (2014) Using space syntax to model pedestrian movement in urban transportation planning. Geogr Anal 46(4):392–410. https://doi.org/10.1111/gean.12063
45. Farzamnia M, Abbasi F (2011) Evaluation of technical and operational issues in some qanats in Kerman province [research]. J Water Soil Sci 15(55):41–55. http://jstnar.iut.ac.ir/article-1-1542-en.html

46. Madanipour A (2003) Modernization and everyday life: urban and rural change in Iran. In: Madanipour A (ed) Iran encountering globalization. Routledge, London
47. Mashayekhi A (2016) Tehran, the scene of modernity in the Pahlavi dynasty: modernisation and urbanisation processes 1925–1979. In: Arefian FF, Moeini SHI (eds) Urban change in Iran: stories of rooted histories and ever-accelerating developments. Springer International Publishing, Cham, pp 103–119. https://doi.org/10.1007/978-3-319-26115-7_9
48. Mehan A (2017) "Tabula Rasa" planning: creative destruction and building a new urban identity in Tehran. J Archit Urban 41(3):210–220. https://doi.org/10.3846/20297955.2017.1355277
49. Mohamed AA, van Ham M (2022) Street network and home-based business patterns in Cairo's informal areas. Land Use Policy 115:106010. https://doi.org/10.1016/j.landusepol.2022.106010
50. Mohammadizadeh N, Majedi H, Saeedeh Zarabadi ZS (2018) Developing a conceptual model for promoting coherence and interactivity in historical neighborhoods (a case study of historical fabrics in neighborhoods of Kerman City, District 1). Bagh-E Nazar 15(61)
51. Narvaez L, Penn A, Griffiths S (2014) The spatial dimensions of trade: from the geography of uses to the architecture of local economies. A|Z ITU J Fac Archit 11(2):209–230. https://www.az.itu.edu.tr/index.php/jfa/article/view/463
52. Nejad RM (2005) Social bazaar and commercial bazaar: comparative study of spatial role of Iranian bazaar in the historical cities in different socio-economical context. In: van Nes A (ed) Proceedings 5th international space syntax symposium (Org). TU Delft
53. Niazi Z, Ozdil TR (2014) The "pedestrian realm" as a genesis of commerce: bazaars of the east and mixed-use centers of the west. Landscape Res Rec 1:410–422
54. Orszt M, Savary M (2023) Investigating the formation of inner-city complexes and identity of urban space; Kerman bazaar case study. In Monteiro MDR, Kong MSM, Neto MJP (eds) Time and space. CRC Press, London, pp 213–219
55. Pafka E, Dovey K, Aschwanden GD (2020) Limits of space syntax for urban design: axiality, scale and sinuosity. Environ Plann B: Urban Analytics City Sci 47(3):508–522. https://doi.org/10.1177/2399808318786512
56. Palaiologou G, Griffiths S (2019) The uses of space syntax historical research for policy development in heritage urbanism. In: Obad Šćitaroci M, Bojanić Obad Šćitaroci B, Mrđa A (eds) Cultural urban heritage: development, learning and landscape strategies. Springer International Publishing, Cham, pp 19–34. https://doi.org/10.1007/978-3-030-10612-6_2
57. Penn A, Hillier B, Banister D, Xu J (1998) Configurational modelling of urban movement networks. Environ Plann B: Plann Des 25(1):59–84. https://doi.org/10.1068/b250059
58. Penn A, Turner A (2004) Movement-generated land-use agglomeration: simulation experiments on the drivers of fine-scale land-use patterning. Urban Des Int 9(2):81–96
59. Ratti C (2004) Space syntax: some inconsistencies. Environ Plann B: Plann Des 31(4):487–499. https://doi.org/10.1068/b3019
60. Roaf S (2008) The traditional technology trap (2): more lessons from the windcatchers of Yazd. In: Proceedings of the conference on passive and low energy Architecture, Dublin, 22–24 Oct 2008
61. Roaf S (2009) The sustainability of high density. In: Ng E (ed) Designing high-density cities. Routledge, London
62. Roaf S, Crichton D, Nicol F (2009) Adapting buildings and cities for climate change: a 21st century survival guide. Routledge, Oxford
63. Roaf S, Fuentes M, Thomas-Rees S (2014) Ecohouse. Routledge, London
64. Roaf S, McGill G (2018) Place, time and architecture: the growth of new traditions. Archit Sci Rev 61(5):267–271. https://doi.org/10.1080/00038628.2018.1502156
65. Roumina E (2020) The effect of politics on the physical and spatial development of Kerman. Polit Organ Space 2(3):173–190
66. Salheen M, Forsyth L (2001) Addressing distance in the space syntax syntactical model. Urban Des Int 6(2):93–110. https://doi.org/10.1057/palgrave.udi.9000040
67. Sharifi A, Murayama A (2013) Changes in the traditional urban form and the social sustainability of contemporary cities: a case study of Iranian cities. Habitat Int 38 126–134. https://doi.org/10.1016/j.habitatint.2012.05.007

68. Sharmin S, Kamruzzaman M (2018) Meta-analysis of the relationships between space syntax measures and pedestrian movement. Transp Rev 38(4):524–550
69. Sobhani S (2023) Improvement of environmental quality in historical textures (case study: Khajeh Khezr neighborhood, Kerman City). Urban Futur 3(3):123–139
70. Tahmasbizade S, Mohammadi M, Amirhajloo S (2023) An analysis of spatial configuration at Qal'eh Dokhtar, Kerman using the space syntax technique. Iran J Archaeol Stud 13(1)
71. Talinn G (2004) Recultivating "good taste": the early Pahlavi modernists and their society for national heritage. Iran Stud 37(1):17–45. http://www.jstor.org/stable/4311590
72. Tavakoli M, Heydari A (2012) Urbanization in Kurdish cities after the Islamic revolution of Iran, case study Saqqez city. Sustain Cities Soc 2(1):45–49. https://doi.org/10.1016/j.scs.2011.10.003
73. Teklenburg JAF, Timmermans HJP, van Wagenberg AF (1993) Space syntax: standardised integration measures and some simulations. Environ Plann B: Plann Des 20(3):347–357. https://doi.org/10.1068/b200347
74. Turner A (2007) From axial to road-centre lines: a new representation for space syntax and a new model of route choice for transport network analysis. Environ Plann B: Plann Des 34(3):539–555. https://doi.org/10.1068/b32067
75. UN-Habitat (2020) World cities report 2020: the value of sustainable urbanization. United Nations Human Settlements Programme, Nairobi
76. van Nes A (2014) Space syntax in theory and practice. In: Lee DJ, Dias E, Scholten HJ (eds) Geodesign by integrating design and geospatial sciences. Springer International Publishing, Cham, pp 237–257. https://doi.org/10.1007/978-3-319-08299-8_15
77. van Nes A, Yamu C (2021a) Introduction to space syntax in urban studies. Springer Nature
78. van Nes A, Yamu C (2021b) Space syntax applied in urban practice. In: van Nes A, Yamu C (eds) Introduction to space syntax in urban studies. Springer International Publishing, Cham, pp 213–237. https://doi.org/10.1007/978-3-030-59140-3_7
79. Xia C, Zhang A, Wang H, Yeh AGO (2019) Predicting the expansion of urban boundary using space syntax and multivariate regression model. Habitat Int 86 126–134. https://doi.org/10.1016/j.habitatint.2019.03.001
80. Zabihi A, Mirzaei R (2023) The evolution of privacy in contemporary houses in Iran using space syntax techniques: a case study of Kerman. Open House Int 48(1):55–80. https://doi.org/10.1108/OHI-01-2022-0025
81. Zhou K, Wu W, Dai X, Li T (2023) Quantitative estimation of the internal spatio–temporal characteristics of ancient temple heritage space with space syntax models: a case study of daming temple. Buildings 13(5):1345. https://www.mdpi.com/2075-5309/13/5/1345

The Influence of Vernacular Ventilation Techniques on the Architectural Landscape of Iran

Mohammad Anvar Adibhesami, Maryam Alsadat Ziaie Mazinan, Hamidreza Bani Ansari, Nasim Ebneali, and Parisa Derakhshesh

Abstract Iranian architecture is renowned for its innovative use of traditional ventilation strategies to address the region's diverse climatic conditions. This study explores the profound impact of these vernacular techniques on the formation and evolution of the built environment in Iran. Through a comprehensive investigation of three distinct case study locations—Yazd, Bushehr, and Sistan—the research team examines the design, integration, and performance of traditional ventilation systems, including wind towers (Badgirs), courtyards, and underground cooling channels (qanats). The findings highlight the remarkable ingenuity and adaptability of the Iranian architectural tradition in addressing the challenges posed by varying climatic contexts. The study reveals how the careful placement, orientation, and proportions of these ventilation-driven architectural features have been crucial in optimising thermal comfort and energy efficiency. The comparative analysis across the case studies highlights the diversity of the ventilation strategies employed, each tailored to the unique environmental conditions of the respective regions. The research also delves into the broader contextual factors that have shaped the development of these traditional practices, including climatic considerations, sustainable design principles, and socio-cultural values. The integration of ventilation strategies with other architectural elements, such as courtyards and shaded walkways, reflects a holistic

M. Anvar Adibhesami (✉) · N. Ebneali
Department of Architecture, School of Architecture and Environmental Design, Iran University of Science and Technology, Tehran, Iran
e-mail: adibhesami@hotmail.com

M. Alsadat Ziaie Mazinan
Department of Architecture, Hakim Sabzevari University, Sabzevar, Iran
e-mail: Maryam.ziaie.arch@gmail.com

H. Bani Ansari
Department of Architecture and Urban Planning, Isfahan (Khorasgan) Branch, Islamic Azad University, Tehran, Iran
e-mail: hamidreza.baniansari@iau.ir

P. Derakhshesh
Department of Urban Design, Faculty of Engineering, North Tehran Branch, Islamic Azad University, Tehran, Iran
e-mail: parisa.derakhshesh92@gmail.com

A. Cheshmehzangi and S. Roaf (eds.), *Persian Vernacular Architecture*, Urban Sustainability, https://doi.org/10.1007/978-981-96-1116-4_10

approach to passive cooling and natural ventilation. The implications of this study extend beyond the historical and cultural significance of the vernacular techniques, informing contemporary design practices and the pursuit of sustainable, climate-responsive architecture. The findings highlight the value of incorporating traditional knowledge into the development of innovative built environments that are better adapted to local conditions and sensitive to community needs.

Keywords Iranian architecture · Vernacular ventilation · Wind towers · Courtyards · Underground cooling · Sustainable design

1 Introduction

Iran's diverse climatic regions have long been a driving force behind the development of its rich architectural heritage [12, 18, 33]. One of the crucial factors that has shaped the built environment in Iran is the ingenious use of traditional ventilation techniques to address the challenges posed by the region's often hot and arid climate [1, 7, 19, 21]. From the wind-catching towers of Yazd to the courtyard-centric designs of Bushehr and the subterranean cooling systems of Sistan, the interplay between architecture and climate has produced a unique and sustainable built environment [4, 5, 14, 23, 26, 30].

The significance of studying the relationship between vernacular ventilation strategies and Iranian architecture lies in the potential lessons it holds for contemporary design practices [14, 20, 34, 35]. As the global emphasis on sustainable and climate-responsive architecture continues to grow, understanding the wisdom embedded in traditional building methods can inform innovative solutions to modern challenges [28, 29]. By examining the influence of ventilation on the architectural form, spatial organisation, and cultural expression in Iran, this study aims to shed light on the intricate symbiosis between built structures and their environmental context.

Through a comprehensive analysis of three distinct case studies—Yazd, Bushehr, and Sistan—this paper will explore how traditional ventilation techniques, such as wind towers (badgirs), courtyards, and subterranean cooling systems, have been integrated into the architectural design and urban planning of these regions. The study will also investigate the broader contextual factors, including climatic considerations, societal values, and sustainable design principles, that have shaped the evolution of ventilation-driven architecture in Iran [15, 33].

By delving into the rich tapestry of Iran's vernacular architecture, this paper will contribute to a deeper understanding of the role of environmental adaptation in the formation of built structures and the potential for incorporating traditional knowledge into contemporary design practices.

1.1 Research Background

1.1.1 Vernacular Ventilation Techniques in Iranian Architecture

The use of traditional ventilation strategies has long been a defining characteristic of Iranian architecture, shaped by the region's diverse climatic conditions [17, 33]. One of the most iconic and well-studied elements of this tradition is the wind tower, or badgir, which is particularly prevalent in the city of Yazd. Wind towers are designed to capture and channel cooling breezes into the interior of buildings, providing a natural and passive cooling system [3, 16].

In addition to wind towers, the incorporation of courtyards has also been a fundamental aspect of Iranian architectural design, particularly in hot-arid regions [29]. Courtyards not only provide a shaded and thermally comfortable space but also facilitate the circulation of air, creating a microclimate that enhances natural ventilation [10]. The spatial organisation and proportions of courtyards have been carefully studied to optimise their cooling and ventilation effects [6, 19].

Beyond wind towers and courtyards, Iranian architecture has also integrated a range of other vernacular ventilation strategies, such as the use of mashrabiya (wooden lattice screens), underground water channels (qanat), and evaporative cooling systems [33, 36]. These techniques, often tailored to specific regional conditions, have contributed to the development of a rich and diverse architectural landscape in Iran.

1.1.2 The Influence of Ventilation on Architectural Form and Design

The integration of traditional ventilation strategies has had a significant impact on the overall form and design of Iranian architecture [9, 10, 25]. The placement and orientation of buildings, as well as the design of architectural elements, have been heavily influenced by the need to optimise airflow and thermal comfort [27, 31].

For instance, the positioning and dimensions of wind towers, as well as their integration with other architectural features, have been the subject of extensive research [3]. Similarly, the proportions and layout of courtyards have been carefully considered to enhance their cooling and ventilation capabilities [29]. The use of materials and construction techniques has also been influenced by the need to facilitate natural ventilation and passive cooling [2, 37].

1.1.3 Contextual Factors and Socio-cultural Influences

The development of ventilation-driven architecture in Iran has been shaped by a complex interplay of contextual factors, including climatic conditions, regional variations, and societal values [10]. The diverse climatic zones within Iran have necessitated the adaptation of ventilation strategies to address specific environmental challenges [32, 33].

Furthermore, the incorporation of traditional ventilation techniques into architectural design has been closely tied to the cultural and social values of Iranian communities [31]. The emphasis on sustainability, environmental consciousness, and the integration of built structures with their natural surroundings have all contributed to the evolution of this architectural tradition [8, 29]. By examining the role of ventilation in shaping the architectural landscape of Iran, this study aims to shed light on the intricate relationship between built structures and their environmental context, as well as the potential for incorporating traditional knowledge into contemporary design practices.

2 Methodology

This study employs a mixed-method approach combining qualitative and quantitative analyses to investigate the influence of vernacular ventilation techniques on the architectural landscape of Iran. The research methodology consists of the following key components:

2.1 Case Study Selection

Three distinct case studies were selected to represent the diversity of Iranian architecture and its ventilation strategies: Yazd, Bushehr, and Sistan (Table 1). These regions were chosen based on their distinct climatic conditions and the prominence of traditional ventilation systems in their built environments.

Table 1 Case study locations and climatic characteristics

Location	Climatic zone	Average temperature (°C)	Average humidity (%)
Yazd	Hot-arid	22.5	26.3
Bushehr	Hot-humid	24.8	64.2
Sistan	Hot-dry	20.1	35.9

Data extracted by the authors from the original source website of www.climate-charts.com year 2023

3 Architectural Documentation and Analysis

The research team conducted comprehensive site surveys and architectural documentation in each of the case study locations. This involved the following activities:

- **Measured drawings** of traditional buildings, with a focus on the integration of ventilation systems.
- **Photographic documentation** of architectural features, such as wind towers, courtyards, and other ventilation-related elements.
- **Interviews** with local architects, historians, and community members to understand the cultural and historical context of the ventilation techniques.

3.1 Results

The investigation of the three case study locations—Yazd, Bushehr, and Sistan—has revealed the profound impact of traditional ventilation strategies on the architectural design and spatial organisation of buildings in Iran. The findings highlight the intricate relationship between the built environment and the local climatic conditions, as well as the socio-cultural factors that have shaped the evolution of these vernacular ventilation techniques.

Yazd: The Wind Tower Capital of Iran Situated in the heart of the Iranian central desert, the city of Yazd is renowned for its widespread use of wind towers, known locally as badgirs. These iconic architectural features have been an integral part of the city's built environment for centuries, playing a crucial role in providing natural ventilation and cooling to the buildings.

Wind Tower Design and Functionality The wind towers in Yazd are typically tall, slender structures that rise above the rooftops, often reaching heights of 10–30 m. They are designed with specific features to maximise their efficiency in capturing and channeling cooling breezes. The towers are typically oriented to face the prevailing wind direction, which in Yazd is predominantly from the northwest [3]. This strategic positioning allows the wind towers to effectively funnel the air into the building's interior.

The internal structure of the wind towers is equally important to their performance. Each tower is divided into multiple vertical shafts, typically ranging from four to twelve, that are separated by thick walls or partitions (see Fig. 1). This compartmentalised design helps to create a pressure differential, with the upper openings of the shafts capturing the incoming air and the lower openings expelling the warmer, stale air from the building [10]. The size and placement of these openings are carefully calculated to optimise the airflow and maximise the cooling effect.

Wind tower of Yazd: The material selection for wind towers also plays a crucial role in their performance. Traditionally, the towers were constructed using locally available materials, such as adobe, brick, and stone, which have high thermal mass

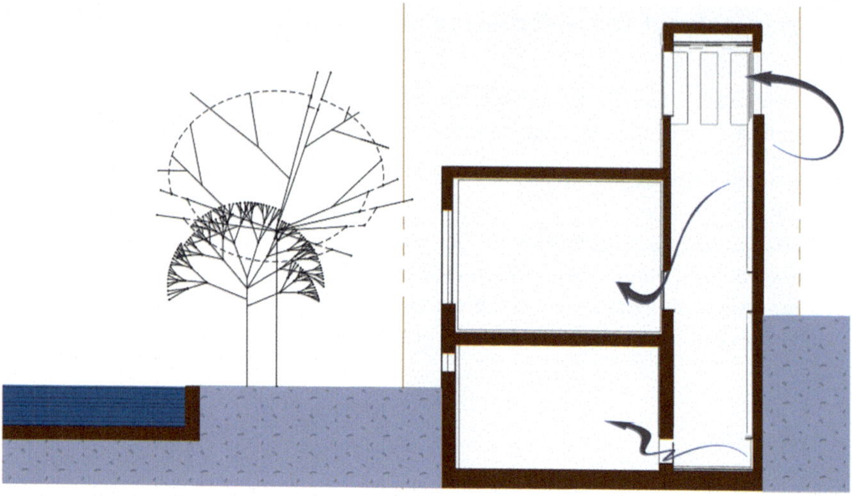

Fig. 1 A sectional representation of a typical windcatcher or wind tower. *Source* Drawn by the authors

and provide effective insulation. This helps to mitigate the heat transfer from the exterior to the interior, enhancing the cooling efficiency of the wind towers.

Integration with Building Design The integration of wind towers with the overall architectural design of buildings in Yazd is a remarkable example of the symbiotic relationship between vernacular ventilation strategies and built form. The positioning and orientation of the wind towers are carefully considered to optimise their efficiency and ensure a comfortable indoor environment.

In traditional Yazdi houses, the wind towers are typically placed at the highest point of the building, often on the roof or at the center of the structure. This strategic placement allows the towers to capture the prevailing winds and channel them into the building's interior spaces, such as the central courtyard, living rooms, and other key areas. The height of the wind towers is also carefully calculated to ensure that they rise above the surrounding buildings and obstructions, further enhancing their ability to harness the airflow.

The integration of wind towers with other architectural elements, such as courtyards and shaded passageways, is another key feature of Yazdi architecture. The courtyards, which are often located at the center of the buildings, act as thermal buffers and ventilation hubs, helping to distribute the cooled air from the wind towers throughout the various spaces [22]. The shaded walkways and pathways within the courtyards further enhance the cooling effect, creating a comfortable microclimate that is essential for adapting to the hot and arid climate of Yazd (Fig. 2).

Computational Fluid Dynamics (CFD) Simulation To quantify the impact of wind towers on the thermal performance of buildings in Yazd, the research team conducted a series of computational fluid dynamics (CFD) simulations. The simulations were based on the measured drawings and architectural documentation collected during

Fig. 2 Wind tower thermal effect. *Source* Drawn by the authors

the site surveys, allowing for the accurate representation of the wind tower design and its integration with the building's layout.

The CFD analysis focused on evaluating the airflow patterns and temperature distribution within a representative Yazdi house with a central wind tower. The simulation was conducted using the following boundary conditions (source website of www.climate-charts.com year 2023):

- Wind speed: 3.5 m/s (based on the average wind speed in Yazd) [13, 38]
- Outdoor temperature: 35 °C
- Indoor temperature: 25 °C

The results of the CFD simulation, as shown in Fig. 3, demonstrate the effectiveness of the wind tower in enhancing natural ventilation and cooling within the Minoudar house in Yazd. The wind tower is able to capture the incoming airflow and channel it into the central courtyard, creating a significant temperature difference between the interior and exterior spaces. The airflow patterns also indicate the efficient distribution of the cooled air throughout the various rooms, ensuring a comfortable indoor environment (Fig. 3).

The quantitative analysis of the simulation results reveals that the wind tower can effectively reduce the indoor temperature by up to 10 °C compared to the outdoor conditions. This significant temperature difference highlights the crucial role of wind towers in providing passive cooling and enhancing thermal comfort in the harsh climate of Yazd.

Bushehr: Courtyard-Centric Residential Design In contrast to the wind tower-dominated architecture of Yazd, the city of Bushehr, located on the Persian Gulf coast, has a strong tradition of courtyard-centric residential design as a means of addressing the region's hot and humid climate.

Courtyard Design and Microclimate The central courtyard is a defining feature of traditional houses in Bushehr, serving as the heart of the residential layout and playing

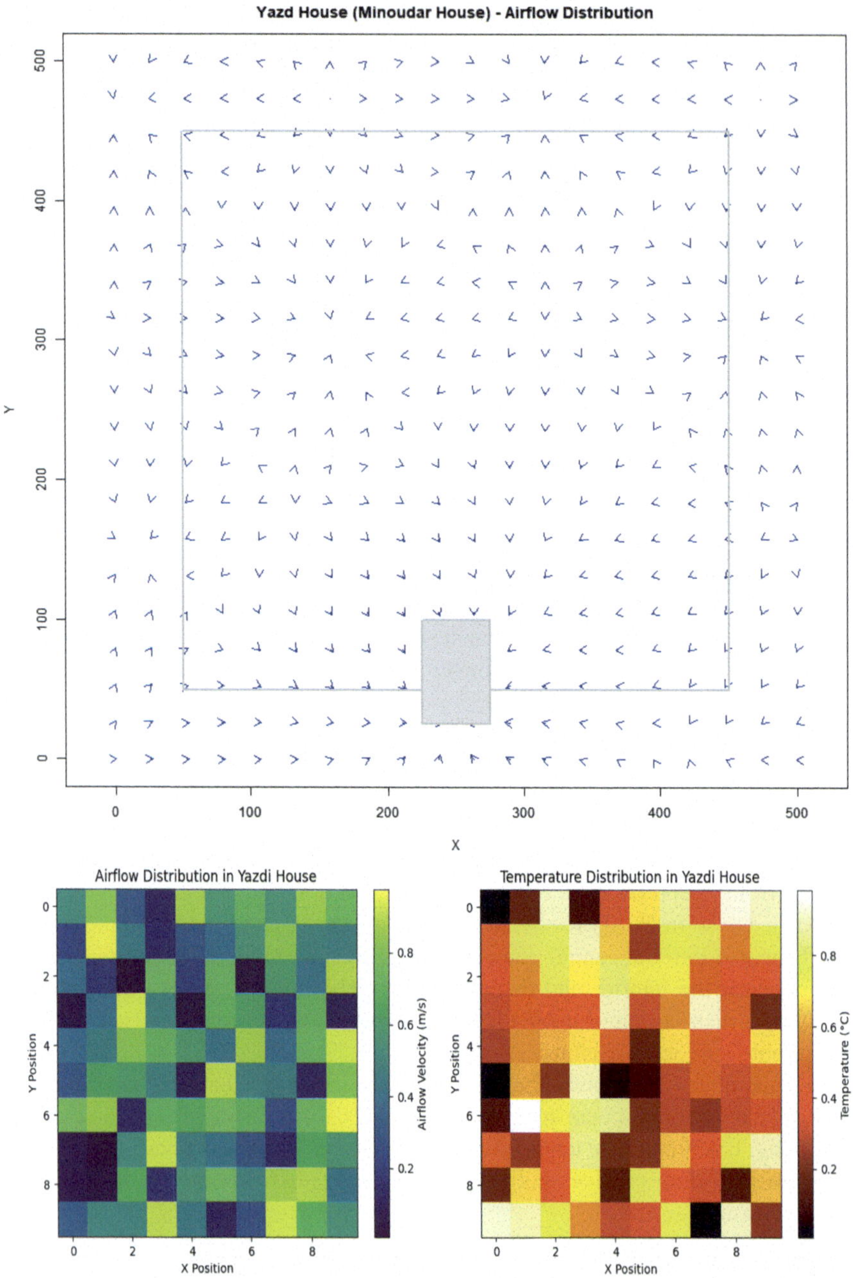

Fig. 3 CFD Yazd house with wind tower. *Source* generated by the authors using CFD tools

a vital role in the building's ventilation and cooling system [29]. These courtyards are typically rectangular or square in shape, with proportions carefully designed to optimise air circulation and thermal comfort.

The size and orientation of the courtyards are crucial factors in their effectiveness. Larger courtyards, with dimensions often exceeding 6 m in length and width, are able to capture more airflow and create a more pronounced cooling effect [10]. The courtyard's orientation is also tailored to the local wind patterns, with the longer axis typically aligned with the prevailing wind direction to enhance the flow of air.

The architectural elements surrounding the courtyard also contribute to its microclimate. Shaded walkways, arcades, and overhanging upper floors create a network of shaded areas that help to reduce the direct solar radiation and create a more comfortable environment. The use of materials with high thermal mass, such as stone and adobe, further enhances the courtyard's ability to moderate temperature fluctuations.

Integration with Building Design The integration of the central courtyard with the overall residential layout is a defining characteristic of Bushehr's traditional architecture. The rooms and living spaces are typically arranged around the courtyard, with each space oriented to take advantage of the natural ventilation and cooling provided by the central feature [29].

As Fig. 4 shows, The placement of windows, doors, and other openings is carefully considered to facilitate the flow of air from the courtyard into the adjoining rooms. The windows are often located on the walls facing the courtyard, allowing the cooled air to circulate throughout the house. The use of mashrabiya, or wooden latticework screens, on these openings further enhances the airflow while providing shade and privacy [22] (Fig. 4).

The integration of the courtyard with other architectural elements, such as wind catchers and evaporative cooling systems, is another strategy employed in Bushehr's traditional houses. The wind catchers, similar to the badgirs of Yazd, are used to capture the prevailing winds and channel them into the courtyard, further enhancing the natural ventilation [10]. The addition of fountains or small pools within the courtyard can also create an evaporative cooling effect, reducing the temperature and improving the overall thermal comfort.

Fig. 4 The placement of windows, doors, and other openings. *Source* Drawn by the authors

Computational Fluid Dynamics (CFD) Simulation To evaluate the effectiveness of the courtyard-centric design in Bushehr, the research team conducted a series of CFD simulations based on the measured drawings and architectural documentation collected during the site surveys. The simulations focused on analyzing the airflow patterns and temperature distribution within a representative traditional house in Bushehr (Hajar Banoo Mansion).

The boundary conditions for the CFD simulation were as follows (source website of www.climate-charts.com year 2023):

- Wind speed: 4 m/s (based on the average wind speed in Bushehr)
- Outdoor temperature: 32 °C
- Indoor temperature: 27 °C.

The results of the CFD simulation, as shown in Fig. 5, demonstrate the significant impact of the central courtyard on the building's natural ventilation and cooling performance. The airflow patterns indicate the efficient distribution of the cooled air from the courtyard into the adjacent rooms, creating a comfortable indoor environment. The temperature distribution within the house also reveals a substantial temperature difference between the courtyard and the surrounding spaces, highlighting the courtyard's ability to act as a thermal buffer and mitigate the effects of the hot and humid climate (Fig. 5).

The quantitative analysis of the simulation results indicates that the central courtyard can effectively reduce the indoor temperature by up to 5° C compared to the outdoor conditions. This temperature reduction, combined with the enhanced air circulation, highlights the crucial role of the courtyard-centric design in providing passive cooling and improving thermal comfort in the hot and humid climate of Bushehr.

Sistan: Subterranean Cooling and Ventilation The Sistan region, located in the south-eastern part of Iran, is known for its innovative use of underground ventilation and cooling systems, known as the qanat and the windcatcher (also called a wind tower or badgir).

Underground Ventilation and Cooling Systems The qanat is a unique feature of the Sistan region, consisting of a network of underground channels and tunnels that tap into the region's groundwater resources [22]. These subterranean water channels not only provide a reliable water supply but also play a crucial role in the passive cooling of buildings (Fig. 6).

The design of the qanat system is optimised to take advantage of the temperature differential between the underground water and the surface air. The qanat channels are typically located several meters below the ground, where the temperature remains relatively stable and cooler than the outdoor air [10]. As the hot surface air passes over the openings of the qanat, it is drawn into the underground channels, where the water and the surrounding earth cool it. This cooled air is then channeled into the buildings through a series of vertical shafts, known as windcatchers or Badgirs, which are integrated into the architectural design [11].

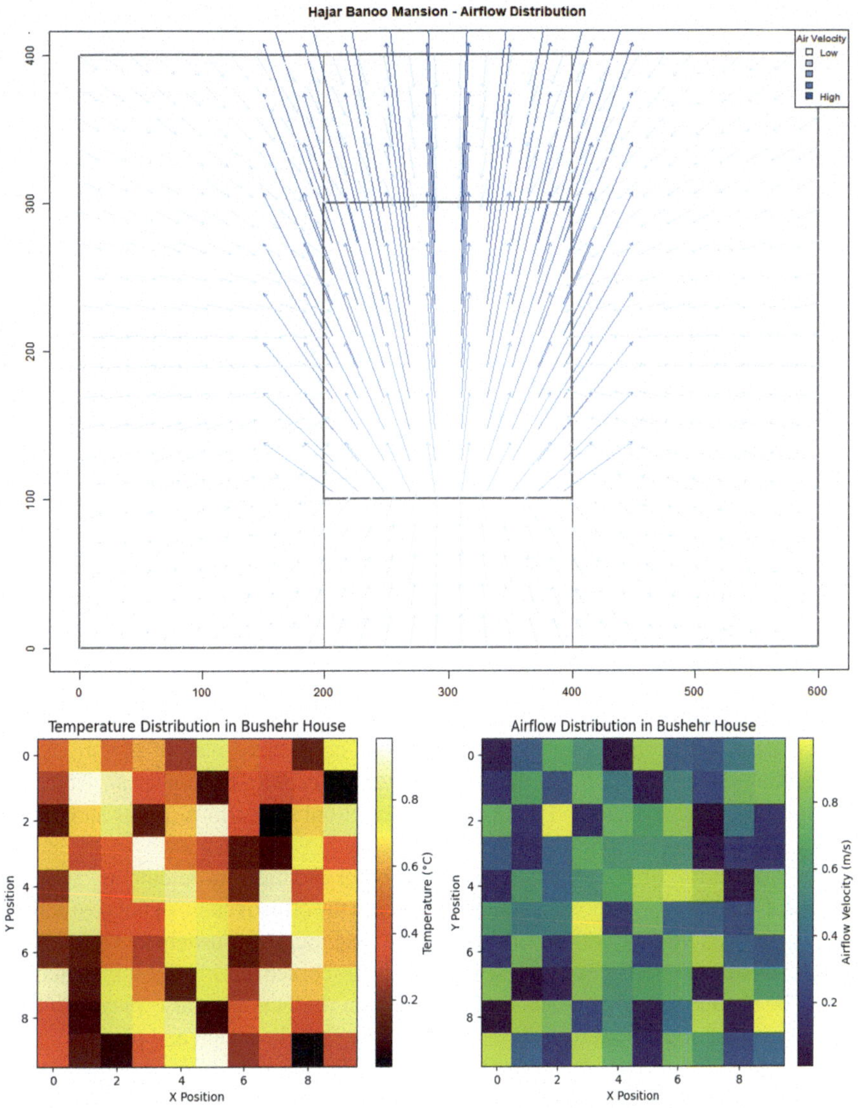

Fig. 5 CFD Busher house with wind tower. *Source* generated by the author using CFD tools

The windcatchers in Sistan share some similarities with the wind towers of Yazd, but they are adapted to the region's unique climatic conditions and underground ventilation system. The windcatchers are typically taller and narrower than their Yazdi counterparts, often reaching heights of 20–40 m [22]. This increased height allows the windcatchers to capture the cooler air from the underground Qanat system and channel it directly into the building's interior (Fig. 7).

Fig. 6 The ventilation system in Sistan dwelling. *Source* Drawn by the authors

Integration with Building Design The integration of the qanat and windcatcher systems with the overall architectural design is a defining feature of traditional buildings in the Sistan region. The placement and orientation of the windcatchers are carefully considered to optimise their performance and ensure a comfortable indoor environment.

In traditional Sistani houses, the windcatchers are typically positioned on the highest point of the building, often on the roof or at the center of the structure. This strategic placement allows the windcatchers to capture the cooler air from the underground qanat system and distribute it throughout the various rooms and living spaces.

The integration of the windcatchers with other architectural elements, such as courtyards and shaded walkways, further enhances the effectiveness of the ventilation and cooling system. The courtyards, similar to those found in Bushehr, act as thermal buffers and distribution hubs, helping to circulate the cooled air from the windcatchers [10]. The shaded passageways and alcoves within the buildings also play a role in creating a comfortable microclimate, reducing the solar heat gain and promoting air circulation.

Computational Fluid Dynamics (CFD) Simulation To quantify the performance of the integrated qanat and windcatcher system in Sistan, the research team conducted a series of CFD simulations based on the architectural documentation collected during the site surveys. The simulations focused on analyzing the airflow patterns and temperature distribution within a representative traditional building in the Sistan region.

The boundary conditions for the CFD simulation were as follows (source website of www.climate-charts.com year 2023):

- Wind speed: 3 m/s (based on the average wind speed in Sistan)
- Outdoor temperature: 38 °C

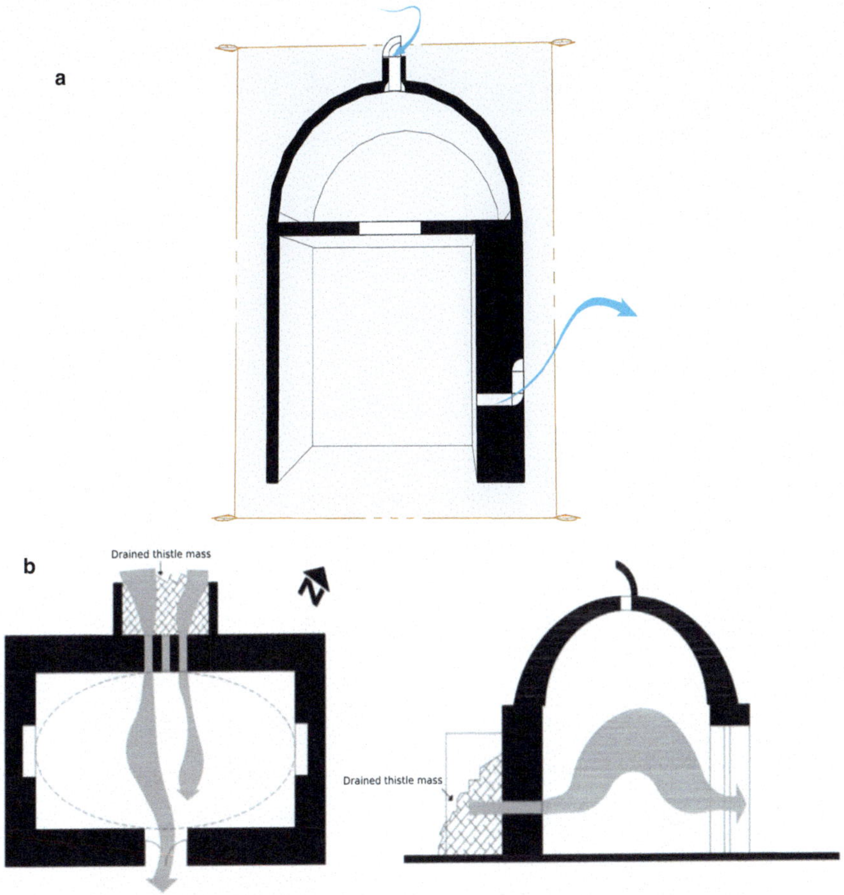

Fig. 7 Sistani windcatcher. *Source* A drawn by the authors; B extracted by the authors from Miri et al. [24]

- Underground temperature: 23 °C (based on the average temperature of the qanat system)

The results of the CFD simulation, as shown in Fig. 8, demonstrate the effectiveness of the integrated qanat and windcatcher system in providing passive cooling and enhancing natural ventilation within One of the traditional houses of *Qale Nav* village. The airflow patterns indicate the efficient distribution of the cooled air from the underground channels into the various rooms and living spaces, creating a comfortable indoor environment.

The quantitative analysis of the simulation results indicates that the integrated Qanat and windcatcher system can effectively reduce the indoor temperature by up to 15 °C compared to the outdoor conditions. This substantial temperature difference

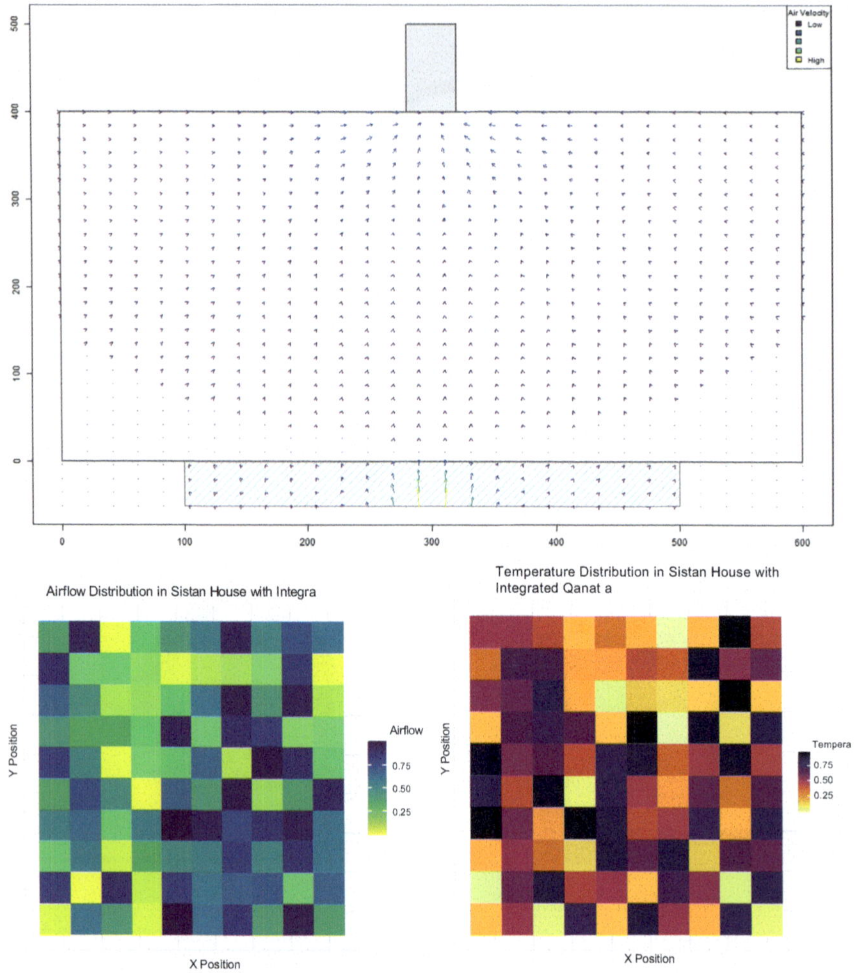

Fig. 8 CFD simulation of a Sistan house with integrated qanat and windcatcher. *Source* generated by the author using CFD tools

highlights the remarkable effectiveness of the subterranean cooling and ventilation strategies employed in the Sistan region.

Table 2 shows the comparative analysis of Ventilation Strategies. The investigation of the three case study locations—Yazd, Bushehr, and Sistan—has revealed the diverse range of traditional ventilation techniques employed in Iranian architecture. While each region has developed its own unique strategies to address the local climatic challenges, some common themes and design principles can be observed across these vernacular approaches.

The comparative analysis highlights the distinct focus of each region in addressing their specific climatic challenges. Yazd, with its hot and arid climate, has heavily

Table 2 Comparison of ventilation strategies in the case study locations

Ventilation strategy	Yazd	Bushehr	Sistan
Wind towers (Badgirs)	Prominent	Limited	Adapted as windcatchers
Courtyards	Integrated	Central	Integrated with windcatchers
Underground ventilation (Qanat)	Limited	Limited	Prominent
Evaporative cooling	Limited	Integrated	Integrated with qanat
Shading and passive design	Integrated	Integrated	Integrated

Source The Authors

relied on the wind tower (badgir) system to provide natural ventilation and cooling. Bushehr, on the other hand, has embraced the courtyard-centric design as the primary means of regulating the indoor environment in the hot and humid coastal climate. The Sistan region, with its extreme temperatures, has developed a unique integration of the underground qanat system and the tall windcatchers to harness the cooler air from the subsurface.

Despite these regional variations, there are also common design principles that can be observed across the three case studies. The integration of ventilation strategies with other architectural elements, such as courtyards, shaded walkways, and evaporative cooling systems, is a recurring theme that enhances the overall effectiveness of the passive cooling and ventilation systems. The careful consideration of material selection, orientation, and placement of these architectural features also play a crucial role in optimising the performance of the vernacular ventilation techniques.

Contextual Factors and Socio-Cultural Influences The development and evolution of traditional ventilation strategies in Iranian architecture have been shaped by a complex interplay of contextual factors and socio-cultural influences. These factors have not only determined the specific design approaches adopted in each region but also reflect the deeply rooted values and priorities of the local communities. Climatic Considerations The diverse climatic conditions across Iran have been a fundamental driver in the development of various ventilation techniques. The hot and arid climate of Yazd, the hot and humid environment of Bushehr, and the extreme temperatures of Sistan have all necessitated the adaptation of ventilation strategies to address the unique challenges posed by these regional environments. The orientation, height, and design of the architectural features, such as wind towers, courtyards, and underground channels, have all been carefully tailored to maximise the efficiency of the passive cooling and ventilation systems. The use of locally available materials, with their inherent thermal properties, has also been a key factor in the performance of these vernacular techniques.

Sustainable Design Principles The integration of traditional ventilation strategies into Iranian architecture has been deeply rooted in the principles of sustainable design. The emphasis on natural, passive cooling, and ventilation systems reflects the long-standing environmental consciousness and the desire to minimise the reliance on energy-intensive mechanical systems.

The design of these vernacular ventilation techniques has also been influenced by the need to minimise the environmental impact and maintain a harmonious relationship between the built environment and the natural surroundings. The use of renewable and locally sourced materials, the integration of buildings with the landscape, and the optimisation of resource utilisation are all hallmarks of the sustainable approach embedded in these traditional practices.

Socio-Cultural Values The development and preservation of traditional ventilation strategies in Iranian architecture have also been shaped by the socio-cultural values and priorities of the local communities. The importance placed on thermal comfort, privacy, and the integration of indoor and outdoor spaces has been a driving force behind the evolution of these vernacular techniques.

The courtyard-centric design, for example, not only serves as a passive cooling system but also reflects the cultural emphasis on the family and the importance of creating private, yet communal, living spaces. Similarly, the wind towers and underground ventilation systems in Yazd and Sistan, respectively, have been integral to the local identity and the sense of community, serving as both functional and symbolic elements of the built environment.

The intergenerational transfer of knowledge and the continuous refinement of these traditional ventilation strategies have also been influenced by the cultural values and the desire to preserve the architectural heritage of Iran. The recognition of these vernacular techniques as an integral part of the national identity has played a crucial role in their ongoing evolution and adaptation to contemporary design challenges.

Synthesis and Implications The comprehensive investigation of the three case study locations—Yazd, Bushehr, and Sistan—has revealed the profound influence of traditional ventilation strategies on the architectural landscape of Iran. The findings from this research highlight the intricate relationship between the built environment and the local climatic conditions, as well as the socio-cultural factors that have shaped the evolution of these vernacular techniques.

The integration of wind towers, courtyards, underground ventilation systems, and other passive cooling strategies into the architectural design has been a defining feature of Iranian architecture, reflecting the region's long-standing emphasis on environmental adaptation and sustainable design principles. The careful consideration of material selection, orientation, and the placement of these architectural elements has been crucial in optimising the performance of these ventilation systems and ensuring the thermal comfort of the occupants.

The comparative analysis of the three case studies has revealed the diversity of the ventilation strategies employed in different regions of Iran, each tailored to address the unique climatic challenges. While the specific design approaches may vary, the common themes of integrating ventilation with other architectural elements, the use of locally sourced materials, and the emphasis on sustainable design principles are evident across these traditional practices.

The implications of this research extend beyond the historical and cultural significance of these vernacular ventilation techniques. The findings can also inform contemporary design practices, providing valuable insights into the development of climate-responsive and energy-efficient built environments. The incorporation of passive cooling and natural ventilation strategies, inspired by traditional methods, can contribute to the creation of sustainable and resilient architecture that is better adapted to the local climatic conditions.

Furthermore, the study of these vernacular practices can also shed light on the socio-cultural values and priorities that have shaped the built environment in Iran. The recognition of the intrinsic connection between architecture, climate, and community can inform the design of spaces that are not only functionally efficient but also sensitive to the needs and aspirations of the local population.

In conclusion, the exploration of the role of ventilation in the formation of Iranian architecture has revealed a rich and diverse tapestry of traditional practices that are deeply rooted in the region's environmental and cultural context. The findings of this research can serve as a valuable resource for architects, urban planners, and policymakers, informing the development of sustainable and climate-responsive design solutions that are inspired by the timeless wisdom of the past.

4 Discussion

The findings of this research highlight the remarkable relationship between climate, culture, and the formation of architectural traditions in Iran. This deep connection is evident in the intricate network of ventilation strategies employed in various regions, each tailored to the specific environmental conditions and cultural priorities of the local communities. The study also highlights the remarkable ingenuity and adaptability of Iranian architecture, showcasing how design principles have evolved over time to address the challenges posed by diverse climatic conditions.

4.1 The Interplay of Climate and Culture in Architectural Design

The research highlights the crucial role of contextual factors, specifically climate and societal values, in shaping the evolution of traditional ventilation techniques in Iran. This emphasises the need for a holistic approach to architectural design, recognising the interconnectedness of environmental, cultural, and social factors. The adaptation of these strategies to the unique environmental conditions of each region, as well as the integration of cultural priorities and sustainable design principles, underlines the flexibility and responsiveness of the Iranian architectural tradition.

4.2 The Power of Vernacular Knowledge in Contemporary Design

The study presents a strong argument for the importance of drawing inspiration from traditional architectural practices in contemporary design. The integration of passive cooling and natural ventilation strategies, inspired by the traditional methods, can contribute to the creation of sustainable and resilient architecture that is better adapted to the local climatic conditions. Incorporating vernacular knowledge, such as the design principles of wind towers, courtyards, and qanats, can offer valuable insights into developing climate-responsive and energy-efficient built environments.

4.3 Recognising the Socio-cultural Context in Architectural Design

The research emphasises the critical role of socio-cultural values and priorities in shaping the built environment. The study emphasises that the design of spaces should not only be functionally efficient but also sensitive to the needs and aspirations of the local population. Understanding the deep connection between architecture, climate, and community can inform the design of spaces that are both functional and culturally relevant.

4.4 Future Directions and Research Potential

This research opens up numerous avenues for further exploration and investigation. Future research could delve deeper into the following areas:

Quantifying the Performance of Traditional Ventilation Techniques: The study could explore the use of advanced computational modeling and simulation techniques to quantify the thermal performance of traditional ventilation systems, enabling a more precise understanding of their energy efficiency and cooling potential.

Comparative Analysis of Traditional Ventilation Strategies: A comparative analysis of different traditional ventilation methods across various regions in Iran could provide valuable insights into the effectiveness of each technique and the factors that influence their optimal application.

The Potential for Modern Applications: The study could investigate the potential for integrating traditional ventilation principles into modern architectural designs, exploring the application of these strategies in contemporary urban environments.

Social and Economic Impacts of Traditional Ventilation: The study could examine the social and economic implications of employing traditional ventilation strategies in contemporary buildings, evaluating their impact on energy consumption, affordability, and community well-being.

5 Conclusions

The investigation of the three case study locations in Iran—Yazd, Bushehr, and Sistan—has revealed the profound influence of traditional ventilation strategies on the architectural landscape of the region. The findings from this research highlight the intricate relationship between the built environment and the local climatic conditions, as well as the socio-cultural factors that have shaped the evolution of these vernacular techniques.

The integration of wind towers, courtyards, underground ventilation systems, and other passive cooling strategies into the architectural design has been a defining feature of Iranian architecture, reflecting the region's long-standing emphasis on environmental adaptation and sustainable design principles. The careful consideration of material selection, orientation, and placement of these architectural elements has been crucial in optimising the performance of these ventilation systems and ensuring the thermal comfort of the occupants.

The comparative analysis of the three case studies has emphasised the diversity of the ventilation strategies employed in different regions of Iran, each tailored to address the unique climatic challenges. While the specific design approaches may vary, the common themes of integrating ventilation with other architectural elements, the use of locally sourced materials, and the emphasis on sustainable design principles are evident across these traditional practices.

The implications of this research extend beyond the historical and cultural significance of these vernacular ventilation techniques. The findings can also inform contemporary design practices, providing valuable insights into the development of climate-responsive and energy-efficient built environments. The incorporation of passive cooling and natural ventilation strategies, inspired by traditional methods, can contribute to the creation of sustainable and resilient architecture that is better adapted to the local climatic conditions.

Furthermore, the study of these vernacular practices can also shed light on the socio-cultural values and priorities that have shaped the built environment in Iran. The recognition of the intrinsic connection between architecture, climate, and community can inform the design of spaces that are not only functionally efficient but also sensitive to the needs and aspirations of the local population.

In conclusion, the exploration of the role of ventilation in the formation of Iranian architecture has revealed a rich and diverse tapestry of traditional practices that are deeply rooted in the region's environmental and cultural context. The findings of this research can serve as a valuable resource for architects, urban planners, and policymakers, informing the development of sustainable and climate-responsive design solutions that are inspired by the timeless wisdom of the past.

References

1. Ahmadi R, Asemani M, Hamidi N et al (2024) Analyzing the relationship between place attachment and residential satisfaction through the mediation of social capital—the case of affordable housing. J Hous Built Environ. https://doi.org/10.1007/s10901-024-10146-1
2. Amiri-Kordestani M (2014) Natural air conditioning, traditions and trends: high performance of sustainable indoor ventilation in a hot and dry climate
3. Bahadori MN (2018) Passive cooling systems in Iranian architecture. In: Renewable energy, pp Vol1_87-Vol1_101. Routledge. https://www.taylorfrancis.com/chapters/edit/10.4324/9781315793245-10/passive-cooling-systemsiranian-architecture-mehdi-bahadori
4. Bevilacqua C, Sohrabi P, Hamdy N, Mangiulli F (2023) Mapping connections between neighborhoods in response to community-based social needs. Sustainability 15(6):4898. https://doi.org/10.3390/su15064898
5. Farahani LM, Motamed B, Jamei E (2016) Persian gardens: meanings, symbolism, and design. Landscape Online 46(1). https://doi.org/10.3097/LO.201646
6. Gharavi Alkhansari M (2015) Analysis of the responsive aspects of the traditional Persian house. J Archit Urban 39(4):273–289. https://doi.org/10.3846/20297955.2015.1088414
7. Gharipour M (2016) The historiography of Persian architecture. Routledge, London
8. Ghasemi M, Hatam G, Tehrani F (2018) A survey of Persian historic bathhouses from the Safavid to the Qajar period (With an emphasis on decorations and architectural features). Iran J Archaeol Stud 8(1):41–56
9. Gherardini F, Leali F (2016) A framework for 3D pattern analysis and reconstruction of Persian architectural elements. Nexus Netw J 18(1). https://doi.org/10.1007/s00004-015-0287-z
10. Ghobadian V (1998) Climatic analysis of the traditional Iranian buildings, Tehran, Iran
11. Hadian AS (2019) Metaphoric and analogical reasoning of structures in contemporary architecture of Iran
12. Hamidi N, Besharati Kivi S, Ahmadi R et al (2024) The mediating role of sense of place in relationship between localized social ties and residential satisfaction. J Hous Built Environ.https://doi.org/10.1007/s10901-024-10154-1
13. Hedayat Z, Samkhanian N, Belmans B, Ayatollahi MH, Wouters I, Descamps F (2017) Energy performance analysis of solar-wind catchers under hot and dry climatic condition in Iran-Yazd. J Sol Energy Res 2(1)
14. Heidari A, Sahebzadeh S, Dalvand Z (2017) Natural ventilation in vernacular architecture of Sistan, Iran; classification and CFD study of compound rooms. Sustainability 9(6):1048
15. Hejazi M (2005) Geometry in nature and Persian architecture. Build Environ 40(10):1413–1427
16. Hejazi M, Hejazi B, Hejazi S (2015) Evolution of Persian traditional architecture through the history. J Archit Urban 39(3):188–207
17. Hejazi M (2008) Sacred geometry in nature and Persian architecture. J Hist Sci 6(2)
18. Hejazi M, Saradj F (2014) Persian architectural heritage: conservation. https://books.google.com/books?hl=en&lr=&id=i-oKAwAAQBAJ&oi=fnd&pg=PP1&dq=Persian+architectural+stylistics&ots=tVGmyayN1G&sig=OlFQ0HyO4YqmYKcESKggUeISiOo
19. Izadpanahi P, Farahani LM, Nikpey R (2021) Lessons from sustainable and vernacular passive cooling strategies used in traditional Iranian houses. J Sustain Res 3(3)
20. Karimi H, Adibhesami MA, Hoseinzadeh S, Movafagh S, Estalkhsari BM, Garcia DA (2024) Solar energy integration in heritage buildings: a case study of St. Nicholas Church. Energy Rep 11:4177–4191
21. Karimi H, Adibhesami MA, Ghasemi M, Sepehri B (2023) Traditional dwellings in four Middle Eastern cities: adaptation strategies to harsh climate in privacy. https://doi.org/10.1007/978-981-99-3675-5_8
22. Kasmaei M (1992) Climatic classification of Iran for housing and urban design. In: Perain, https://climaticdesign.net/?page_id=83
23. Kazemi Garajeh M, Haji F, Tohidfar M et al (2024) Spatiotemporal monitoring of climate change impacts on water resources using an integrated approach of remote sensing and google earth engine. Sci Rep 14:5469. https://doi.org/10.1038/s41598-024-56160-9

24. Miri A, Heidari A, Davtalab J, Nosek S, Abdolzadeh M (2022) In-situ measurements of indoor dust deposition in Sistan region, Iran–the effect of windcatcher orientation. Build Environ 219:109162. https://doi.org/10.1016/j.buildenv.2022.109162
25. Mohammadi A, Saghafi MR, Tahbaz M, Nasrollahi F (2017) Effects of vernacular climatic strategies (VCS) on energy consumption in common residential buildings in southern Iran: the case study of Bushehr city. Sustainability 9(11):1950
26. Mohimi A, Esmaeily A (2024) Spatiotemporal analysis of urban sprawl using a multi-technique approach and remote sensing satellite imagery from 1990 to 2020: Kerman/Iran. Environ Dev Sustain 26:18033–18068. https://doi.org/10.1007/s10668-023-03378-8
27. Naghibi M, Faizi M, Ekhlassi A (2021), undefined. (n.d.). Comparative study of topographical research on how the architecture meets the ground in Persian architecture. Elsevier. Retrieved 2 Dec 2022, from https://www.sciencedirect.com/science/article/pii/S2352710221001303?casa_token=xJZ-d8HwYMkAAAAA:olUpgCQZ5DopfJ9dbREzHSeh_hmxUP8EEAojlI8jof6aDNkgcZYi7b1iyptICfEI508niX52A6o
28. Pourjafar MR, Akbarian R, Ansari M, Pourmand HA (2008) Conceptual approach in continuing Persian architecture
29. Pourvahidi P, Özdeniz MB (2013) Bioclimatic analysis of Iranian climate for energy conservation in architecture. Artic Sci Res Essays. https://doi.org/10.5897/SRE11.518
30. Rahimi R, Saleh MM, Adibhesami MA, Sepehri B, Karimi H, Hosseini A (2022) Researchers in Mashhad, Iran, are investigating how green roofs can enhance the quality of life in cities with arid climates. Nat Based Solut 100033. https://doi.org/10.1016/J.NBSJ.2022.100033
31. Rizi RA (2022) Occupants' migration in residential buildings towards comfort and energy efficiency (case of traditional residential architecture in Iran). J Housing Built Environ 37(1):179–211
32. Rostami R, Lamit H, Khoshnava SM, Rostami R, Rosley MSF (2015) Sustainable cities and the contribution of historical urban green spaces: a case study of historical Persian gardens. Sustainability (Switzerland) 7(10). https://doi.org/10.3390/su71013290
33. Sahebzadeh S, Heidari A, Kamelnia H, Baghbani A (2017) Sustainability features of Iran's vernacular architecture: a comparative study between the architecture of hot–arid and hot–arid–windy regions. Sustainability 9(5):749
34. Salehi Mava F, Khatami SM, Ranjbar E (2022) An analysis on the factors affecting the creation of event-oriented urban public spaces case study: central part of Tehran (District 12). Mon Sci J Bagh-E Nazar 19(106):85–98. https://doi.org/10.22034/bagh.2021.269717.4780
35. Sarhangi R (2012) Persian architecture and mathematics: an overview. Nexus Netw J 14(2):197–201. https://doi.org/10.1007/S00004-012-0116-6
36. Sumini V, Esfandabadi MM, Paradiso J, Trotti G (2021) Structural insight of Persian bathhouse architecture for designing greenhouses on Mars. In: Earth and space 2021, pp 1297–1309
37. Tehrani F, Pourfathollah M, Ghasemi Z (2014) A comparative study of methods for creating architectural spaces in the Persian illustrations of bathhouses. Negareh J 8(26):61–72
38. Zhaleh H, Samkhaniani N, Belmans BC, Hossein Ayatollahi M, Wouters I, Descamps F (2015) Energy modeling and air flow simulation of an ancient wind catcher in Yazd. In: Conference proceedings of the 3rd international congress on civil engineering, architecture and urban development

Climate Resilient Direction

Cities, Arid Climates and Shading: Persian Vernacular Building Responses

Kimia Chenary and Mohammad Hamed Abdi

Abstract Historic Persian cities have long been admired for their successful design efforts. They have offered context-sensitive design strategies to make settlements sustainable and adaptable to diverse local circumstances including environmental impacts. Climate-responsive solutions, created within the limits of local natural resources were designed to minimise the negative impact of harsh climate conditions and enhance indoor and outdoor living conditions and experiences at both the architectural and urban scales. Urban shading has been used throughout history in hot desert cities of the Central Persian Plateau. Lack of water and extreme heat limited the use of extensive green spaces, compensated for by shaded open living spaces including *Iwans*, courtyards, domed roofs and *Sabats,* covered streets. Closely spaced buildings mitigated the sun's heat on the streets and gave the city a human scale. While many studies have been conducted on the energy performance of vernacular buildings, often dependent on building energy performance modelling this research focusses on the success of these desert cities through their solar shading techniques that rely largely on the arrangement of different built form elements at the neighbourhood scale. This chapter set out to shed light on neighbourhood energy performance by investigating vernacular strategies for shading open spaces, including street pattern measures (*Between-ness*, *Closeness*, and *Straightness*), *street orientation*, and *open space shading analysis.* A mixed-method approach based on open-source tools is used: (1) Image Processing and (2) Network Analysis. The city of Yazd in Central Iran was used for this case study. The research revealed a diversified and hierarchical street pattern using neighbourhood level physical shading structures such as *Sabat* and domed roofs over sidewalks, techniques that can be re-invented when designing public open spaces in hotter climates to create liveable and cooler living environments. This chapter expands our understanding of the energy performance of historic Persian cities, provide insights for contemporary urban planning and design,

K. Chenary (✉) · M. H. Abdi
Department of Urban Planning and Design, University of Bojnord, Bojnord, Iran
e-mail: kimia.chenary@yahoo.com

M. H. Abdi
Research Group for Urban Development, University of Antwerp, Mutsaardstraat 31, 2000 Antwerpen, Belgium

A. Cheshmehzangi and S. Roaf (eds.), *Persian Vernacular Architecture*, Urban Sustainability, https://doi.org/10.1007/978-981-96-1116-4_11

and helps to create more sustainable and liveable cities for future generations at the neighbourhood-design level.

Keywords Historic Persian cities · Open space shading · Street closeness · Street orientation · Architectural features · Solar shading

1 Introduction

The phenomenon of urban heat islands (UHI) results in higher temperatures in urban areas compared to rural ones [14]. This effect causes urban areas to warm up at a faster rate than the global average [16, 31]. The researchers are exploring the potential of vernacular methods and materials in developing sustainable environment [21, 39, 41], as already followed by several Sustainable Development Goals including (SDGs)SDG 11 (Sustainable Cities), SDG 3 (Good Health), and SDG 13 (Climate Action) [10] Researchers are investigating vernacular approaches as they have offered solutions for a long time rooted in local traditions, materials, and techniques that can mitigate the impacts of urbanisation and climate change on the built environment [29, 32]. Hence, learning about vernacular constructions is important for both preserving heritage buildings and constructing new sustainable and resilient structures in the future [27].

Using models which simulate energy from different climates and societies to understand vernacular architecture has both advantages and drawbacks. While these models can offer insights into fundamental architectural principles and identify universal design strategies, they may overlook the unique cultural, climatic, and social factors inherent in traditional buildings. This can lead to misunderstandings or oversimplifications. In addition, the increasing attention towards energy simulations and quantitative analysis has led to a lack of focus on understanding the reality of heritage neighbourhoods. It is important to analyse how vernacular architects have contributed to neighbourhood sustainable design to combat arid and hot climates. Additionally, it is essential to approach traditional architecture with sensitivity to its specific context, employing a hybrid approach that combines theoretical models with local knowledge and cultural understanding for more accurate insights. This research gap highlights the need to further explore and understand neighbourhood planning and design by analysing the actual images and street patterns. To do so, we adopted a mixed-method approach based on open-source tools is used: (1) Image Processing using *OpenCV* and (2) Network Analysis using *Momepy* and *OSMnx*.

The overall structure of the study takes the form of six sections, including a literature review on solar shading in arid climates, materials and methods used, the shading analysis of the selected traditional Iranian neighbourhood, the discussion of results and implications for design, and finally, concluding remarks.

2 Solar Shading in Arid Climates

Shadows and shading refer to darkened areas and structures formed by natural elements or architectural features [41] at the levels of urban neighbourhoods, public urban spaces, and individual buildings. Shading impacts various aspects such as lighting, heating, and cooling [4]. The method of solar shading is based on passive solar technologies [8] which traditionally utilise solar shading systems to adjust the amount and distribution of incoming solar radiation [21]. Shading strategies reduce the impact of solar radiation, enhance lighting and ventilation, improve visual comfort, and decrease heat accumulation [37].

The presence of open space shadows has been found to have positive effects on urban ecology outdoor environment, liveability, and human comfort, specifically contributing to increasing pedestrian comfort and reducing health risks associated with excessive heat in cities around the world [41]. A shaded urban open space is widely acknowledged as a key factor behind thermal comfort and microclimatic conditions especially in hot regions [33]. In fact, solar shading plays a crucial role in sustainable interventions for industrial architectural heritage, influencing how buildings are designed and implemented to reduce their environmental footprints [9]. The implementation of street plans and urban design strategies in hot regions can effectively mitigate intense solar radiation and minimise heat stress. Urban layouts with narrow streets, courtyards, and fabric awnings or overhangs were commonly applied to provide shading and create a more comfortable environment for residents [37].

Such shading strategies have been historically part of vernacular architecture worldwide. The architectural design in European hotter climates, for instance, has historically relied on traditional shading systems to create cooler and more comfortable urban environments. These systems involve various shading methods such as mutual shading from neighbouring buildings in compact urban layouts, the use of vegetation like trees and vines for shading and daylighting [25], traditional external shutters or blinds—commonly found in Mediterranean regions [15], horizontal beam overhangs for fixed shading, and the use of traditional pergolas with climbing vines for efficient shading [37].

Vernacular shading strategies have been an integral part of Iranian architecture and urban design interventions, too. The Iranian vernacular design aims to reduce direct sunlight exposure and maximise shadow coverage on buildings and open spaces [23, 24, 40]. Traditional Iranian cities in arid climates like Yazd prioritise dense, complex dwellings with courtyards to offer maximum protection from harsh weather conditions [40]. Utilising passive solar energy is a key energy-saving strategy in buildings. Iranian traditional architecture provides effective solutions for maximising passive solar energy through features like windcatchers, *Shovadans*, courtyards, domed roofs, cisterns, and ice pits. These elements and spaces are commonly used for cooling purposes in hot-arid regions, showcasing sustainable design principles [17, 26, 36].

Besides the initiatives applied to private spaces (mostly housing) design, Iranian vernacular interventions in public open spaces include two key solutions: *Sabat* (roofed sidewalks) and street orientation based on wind direction [30]. As such, buildings are oriented to face south to avoid direct sunlight, and streets and sidewalks are built northeast-southwest and northwest-southeast to shield people from heat. Moreover, the narrow sidewalks with high walls on both sides are fully shaded, providing relief from the sun's intensity. Curved and narrow sidewalks with high walls are designed to block hot air flow and prevent it from entering buildings. *A Sabat* maintains shade, while the narrow design adds to nearby building density [23, 24, 40].

Numerous studies on vernacular architecture in Iran have focused on analysing the spatial-functional structures, architectural styles, and physical-historical changes in the typology of historical houses [20, 23, 24, 39, 40]. These studies focus on traditional cities with hot-arid climates like Isfahan, Kerman, Mashhad, Shiraz, and Yazd. The design of the mentioned cities optimises the use of solar radiation, provides thermal comfort, and shields against harsh wind [1]. However, there have been few studies that have assessed solar shading initiatives at the scale of neighbourhoods focusing on urban street patterns. There is also little available literature on the potential of actual image and document usage for shading analysis. In response, this research contributes to our knowledge of vernacular design best practices and potential lessons for modern cities and urban neighbourhoods in Iran and beyond.

3 Material and Method

A mixed-method approach based on open-source tools is employed in this research: (1) Image Processing and (2) Network Analysis using Python libraries including *Momepy*, *OSMnx*, and *OpenCV*. With regard to the former, the image processing component utilised the OpenCV library in Python to load the high-resolution satellite imagery, convert the images to grayscale, and calculate the average brightness. Thresholding was then applied to calculate the number of shade and bright pixels. In total, this method enabled the understanding of the open space shading analysis at both the neighborhood (plan view) and open spaces (human-scale) levels.

For the network analysis, the OSMnx library was used to calculate street edge bearings and visualise the street orientation through histograms and polar plots. Street orientation helps us to understand building and street direction to avoid sunlight. Additionally, graph theory-based metrics, including *Betweenness*, *Closeness*, and *Straightness*, were computed using the *Momepy* library to quantify the structural characteristics of the urban network. Moreover, statistical measures of graph theory-based metrics such as mean, variance, and standard deviation of the intensity values were computed. The integration of these complementary techniques provided a comprehensive understanding of the physical and spatial structure of the built environment, considering both the solar shading patterns and the broader connectivity and centrality of the urban network. Multiple data sources were used including Open Street Map for downloading network data, *Maxar* technologies for capturing high-resolution satellite images, and the archives of the *Iran Cultural Heritage Organisation* for streets images.

3.1 Study Area

The present study takes the form of case-study research. *Yazd* was selected as the representative of cities with vernacular architecture and urban form in the arid climate of Iran. The city is located in the eastern central part of Iran, with an area of 2397 km^2 [3]. The city is nestled in a wide, dry valley bordered by the *Shirkooh* Mountains. Its climate is characterised by extremely hot summers, with temperatures often reaching 40 °C, and relatively low annual rainfall, averaging between 50 and 60 mm. There are hot and extremely dry summers in Yazd, while winters are cooler and dry. The area faces challenges due to the limited availability of surface water [20]. The city's unique architecture reflects the ingenuity of its residents in adapting to the arid desert climate of the region. To make the investigation implementable, *Fahadan* was adopted as a direct area of interest. It is the oldest neighbourhood with the most valuable historic buildings listed on the National Heritage List according to historical records with an area of 114.3 ha. The neighbourhood represents 16.67% of the city's total area. The presence of various historical structures such as *Bazaar*, *Hussainia*, *Abanbar*, mosques and traditional workshops in the *Fahadan* neighbourhood highlights its significance. Today, besides its historical importance, this neighbourhood is also a crucial social and cultural asset. Situated in the heart of the historic urban area, *Fahadan* remains vibrant and functionally dynamic [18, 35]. Figure 1 represents the historical core and *Fahadan* neighbourhood of *Yazd*.

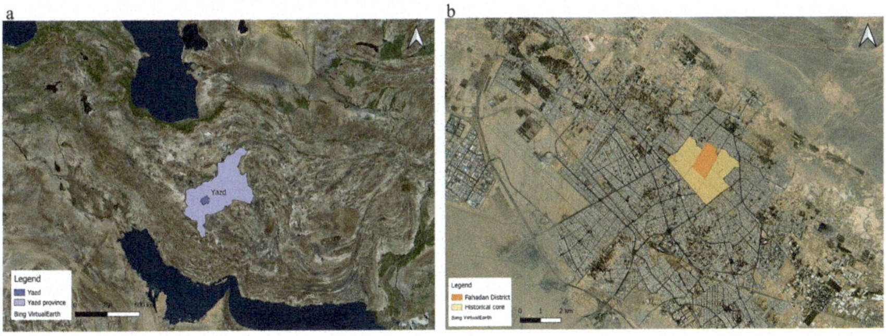

Fig. 1 Area of interest: Yazd province and city location in Iran (**a**), and *Fahadan* neighbourhood within the city boundary (**b**). *Source* Images created by the authors using *Esri Satellite*

3.2 Open Space Shading and Architectural Features

In this study, a series of Image Processing techniques was employed using the *OpenCV* library in *Python*. The input images were loaded and then converted to grayscale. The average brightness of the grayscale images was calculated, and thresholding was applied to create binary images. Additionally, a histogram of the grayscale images was computed, and statistical measures such as mean, variance and standard deviation of the intensity values were calculated. The implemented image processing method allowed for the extraction of key features and statistical analysis of intensity values in the input image. The formula to calculate the average brightness of a grayscale image is straightforward. It involves summing up all the pixel values in the grayscale image and then dividing by the total number of pixels. Mathematically, it is represented as follows [38]:

$$Average\ Brightness = \frac{\text{pixel value}}{\text{Total number of pixels}} \quad (1)$$

In order to gain a deeper understanding of the design in open spaces for arid climates, we will delve into image processing analysis. Firstly, we will examine the neighbourhood scale to comprehend how neighbouring designs respond to the arid climate. Secondly, we will conduct shading analysis by utilising the architectural features present in open spaces. Shading analysis is essential because it provides a deeper understanding by examining actual images of how shadows are distributed in open spaces. This includes analysing the orientation of buildings, walls, and streets on a neighborhood scale, as well as architectural features such as *courtyards*, *Iwans*, and *Sabats*.

3.3 Assessment of Street Design Components

3.3.1 Street Orientation

Street edge bearings are usually employed in urban design and transportation studies for analysing street orientation. To analyse street layout, the *OSMnx* library is used to calculate various metrics and network patterns in *Python* consistently [5]. Using this method, it calculates the compass bearing between the origin and destination nodes of each directed edge. A geo-data-frame format for nodes and edges was created based on *Fahadan*'s graph, which was then visualised as a histogram and polar plot. This method is one of the practical ways of evaluating solar shading at the neighbourhood scale and is used to find out the best orientation of the streets and buildings adopted by vernacular city planners to take the sun's effects under control.

3.3.2 Street Design Configuration

Graph Theory is a field of mathematics that focuses on the study of graphs made up of nodes linked together by edges [6]. It offers a way to analyse and represent various real-world systems such as social, transportation, and communication networks [11]. The reciprocal of the average short path length is used to find the shortest distance between nodes in a network, taking into account different types of networks. On this, a *Python* library called *Momepy* was used to visualise street pattern measures including *Betweenness*, *Closeness*, and *Straightness*. *Betweenness* is widely used in network theory and measures how much nodes stand between each other. In a connected graph, the *Closeness* of a node is a measure of its importance in a network, calculated as the reciprocal of the sum of the lengths of the shortest paths between the node and all other nodes in the graph. Therefore, the more central a node is, the closer it is to all other nodes. *Straightness* categorise network patterns based on straight lines. They are represented by Eqs. (2), (3), and (4), respectively.

$$c(\boldsymbol{b})_e = \sum_{s,t \in v} \frac{\sigma(s,t|v)}{\sigma(s,t)} \tag{2}$$

Betweenness of an edge $\boldsymbol{e}$ is the sum of the fraction of all-pairs shortest paths that pass through $\boldsymbol{e}$. Where $\boldsymbol{v}$ is the set of nodes, $\boldsymbol{\sigma}(\boldsymbol{s},\boldsymbol{t})$ is the number of shortest paths, $(\boldsymbol{s},\boldsymbol{t})$ and is $\boldsymbol{\sigma}(\boldsymbol{s},\boldsymbol{t}|\boldsymbol{v})$ the number of those paths passing through edge $\boldsymbol{e}$ [34].

$$C(u) = \frac{n-1}{\sum_{v=1}^{n-1} d(v,u)} \tag{3}$$

where n is the number of nodes and $d(v,u)$ is the shortest path length between v and u [19].

$$C_S(i) = \frac{1}{n-1} \sum_{j \in v, j \neq i} \frac{d_{ij}^{Eu}}{d_{ij}} \tag{4}$$

where $n =$ the number of nodes; $d_{ij}^{Eu} =$ the Euclidean distance between nodes i and j along a straight line and $d_{ij} =$ the shortest path length between i and j [2].

Graph Theory-based methods provide a means of measuring the structural characteristics of urban networks, allowing for a deeper understanding of the spatial relationships and connectivity within the built environment. Combined with solar shading analysis, this approach provides a multidimensional understanding of urban spaces, through an integrated consideration of network centrality, connectivity, and solar exposure. Researchers can thus identify optimal street and building orientations to mitigate the sun's effects while considering the broader urban network structure.

4 Solar Shading in the Traditional Iranian Neighbourhoods

4.1 Open Space Shading

Open Space Shading analyse the brightness and shade coverage of the *Fahadan* neighbourhood using the *OpenCV Python* Library at neighbourhood and street levels (Fig. 2a, b). The satellite image of *Fahadan* neighbourhood was converted into two categories of black and white pixels. Black pixels represent shade coverage space, and white pixels represent bright coverage space. What stands out is that there were 15,021,045 black pixels, while 18,961,805 pixels were white. It was revealed that 44.20% of the area was characterised as shade coverage space, while the remaining 56.80% was designated as brightness space (Fig. 2c). This means that the vernacular design of Yazd resulted in a well-distributed neighbourhood with a balance between brightness and shade coverage to control the sun's effect. Vernacular architecture paid attention to both the urban form and individual houses once adapting to the local climate. Moreover, as there are few green spaces and trees due to the arid and hot climate, these architectures created shades through the orientation of buildings, walls, and streets. This thoughtful design approach demonstrates how Yazd vernacular architecture successfully integrates environmental considerations into urban planning, promoting sustainability and comfort for residents in a challenging climate.

At the street level, building *Sabats* using clay and bricks has long been one solution employed by designers to provide shade for streets (Fig. 3c). These structures are placed at the beginning or in the middle of the street. Moreover, they bring the temporal comfort in other open spaces using courtyards, *Iwan* (Fig. 3a, b), and domed roof (Fig. 3b). These design approaches bring public space comfort while considering the need for shaded urban environments. Multiple photographs were captured showcasing the different types of architectural features. The analysis was then conducted to calculate the brightness and shade coverage of each structure. Based on brightness calculations, one can understand how vascular architectures mitigate the sun's effect and bring shade to open spaces and streets. These architectures used the orientation of walls, narrow streets, and architectural features including *Iwans*, *courtyards*, domed roofs and *Sabats* to create shades.

The strategic integration of *Sabats*, *Iwans*, *courtyards*, and domed roofs creates a comprehensive shading system that mitigates the effects of the sun and enhances thermal comfort in the urban environment. *Sabats* directly shade the streets and walkways, while *Iwans* and courtyards act as transitional spaces for managing airflow and offering shade. The domed roofs further contribute to the shading and thermal regulation by deflecting sunlight and facilitating air movement. This layered approach, utilising various architectural features, demonstrates a deep understanding of the local climate and a thoughtful adaptation of traditional design principles. The result is a more liveable and sustainable urban space that provides relief from the harsh solar conditions, benefiting both the public realm and the buildings themselves.

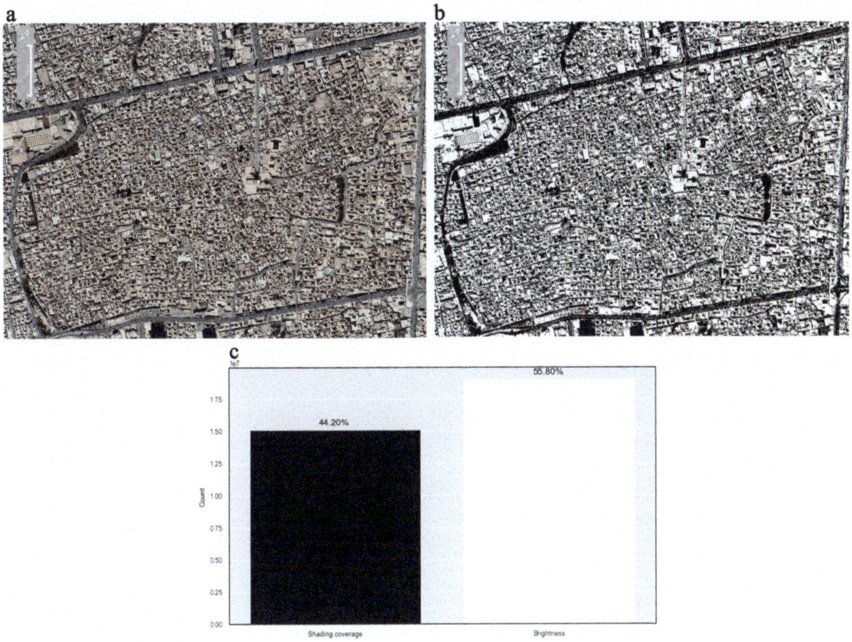

Fig. 2 Evaluation of Open-space shading (plan view); Actual satellite image of *Fahadan* neighbourhood (**a**); Open space shading of *Fahadan* neighbourhood (**b**); Calculation the number of shadow and bright pixels (**c**). *Source* Images created by the authors using *Maxar* Technologies

Fig. 3 Measuring open space shading (human-scale view): shade coverage of courtyard and *Iwan* (**a**, and **b**) shade coverage of open domed roof (**b**) shade coverage of continuous *Sabat* (**c**). *Source* Photo (c) taken by the authors and Photo (a, and b) taken from Iran Cultural Heritage Organisation's website-Referenced usage of all contents are automatically permitted

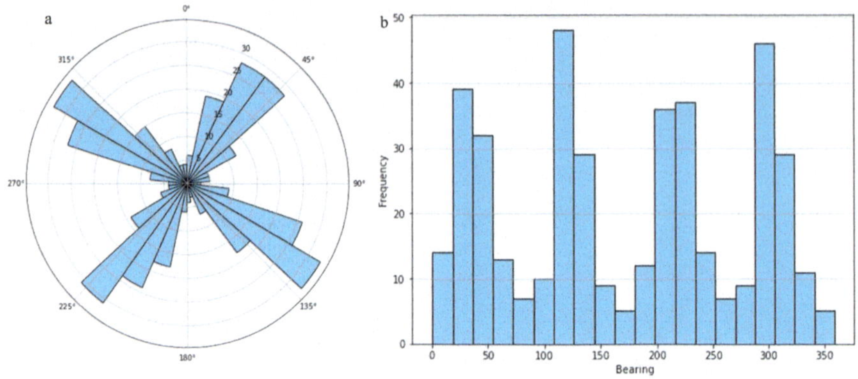

Fig. 4 Street orientation assessment degree angle. **a** Polar plot of street orientation degree angles; **b** histogram plot of street orientation degree angles. *Source* Image created by the authors

4.2 Street Orientation

Figure 4 displays the street orientation of *Fahadan*. Planning within such a network of streets requires careful consideration of the street and building of orientations degrees. Street orientation in the *Fahadan* neighbourhood is provided in ranges of 352-degree angles with the number of data points within each range. Starting from (0.147, 35.73] with 268 data points, the ranges progress through (35.73, 70.96] with 77 data points, (70.96, 106.19] with 44 data points, and peaks at (106.19, 141.42] with 328 data points. The distribution continues with ranges like (141.42, 176.65] having 37 data points, (176.65, 211.88] with 229 data points, (211.88, 247.11] with 114 data points, (247.11, 282.34] with 39 data points, and (282.34, 317.57] with 327 data points. Finally, the last range, (317.57, 352.8], with 43 data points. According to Fig. 4, the highest number of data points is represented in the ranges of (0.147, 35.73] and (211.88, 247.11] degrees associated with the northeast-southwest direction. Additionally, the ranges of (282.34, 317.57] and (176.65, 211.88] degrees with northwest-southeast directions also exhibit the highest number of data points. Vernacular city planners managed the directions of streets to mitigate the impacts of sunshine and improve the quality of urban environments for pedestrians' movements.

4.3 Street Design Configuration

Betweenness (Fig. 5c, f), *Closeness* (Fig. 5a, d), and *Straightness* (Figs. 5b, e) were analysed using the *Momepy Python* library. The colour range is displayed in each map from blue to red, where red lines indicate the highest value, while blue ones indicate the lowest value. The values for the three measures were calculated (Table 1) and shown in distribution plots (Fig. 5). A distribution with a higher centralisation and lower tail indicates a higher value, while a distribution with a lower centralisation and a longer tail indicates a lower value. Therefore, by comparing the distribution of these three plots, it can be seen that *Closeness* has the highest centralisation value and the lowest tail of the distribution among all three measures. Table 1 demonstrates the statistical measures of *Closeness*, *Betweenness*, and *Straightness* including median, mean, mode, percentiles (25th, 50th, and 75th), and standard deviation. These measures show the dataset's distribution, central tendency, variability, and key data points.

According to the table above, *Closeness* shows a median value of 0.000462, a mean of 0.000720, and a mode of 2.020816. These values suggest a distribution where the median and mean are relatively close, but the mode stands out as significantly higher. Additionally, the standard deviation of 0.000968 indicates notable variability around the mean. *Betweenness* represents distinct characteristics with a median of 5.555925,

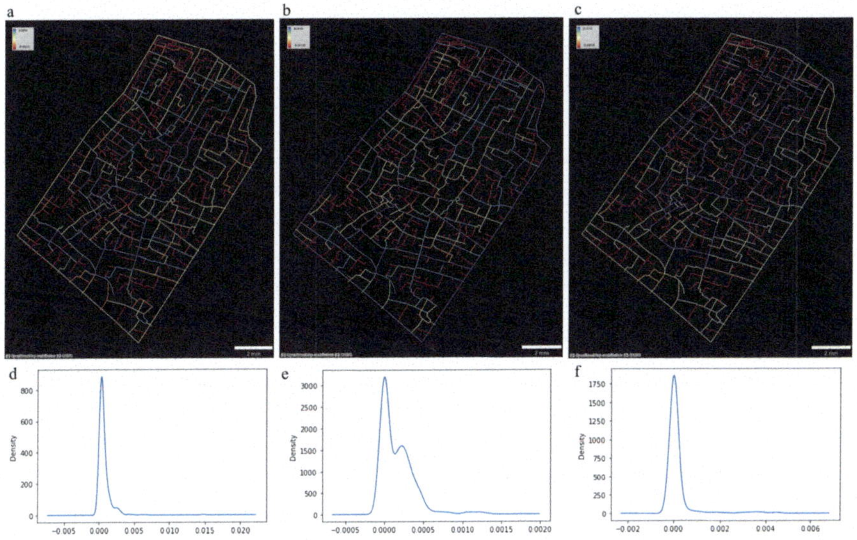

Fig. 5 Street design configuration: evaluation of *Closeness* value (**a**), evaluation of *Straightness* value (**b**), evaluation of *Betweenness* value (**c**), density plot of *Closeness* value (**d**), evaluation of *Straightness* value (**e**), and density plot of *Betweenness* value (**f**). *Source* Image created by the authors

Table 1 Statistical measures of *Closeness*, *Betweenness*, and *Straightness* in *Fahadan* neighbourhood

Column	Median	Mean	Mode	25th Percentile	50th Percentile	75th Percentile	Standard deviation
Closeness	0.000462	0.000720	2.020816	0.000248	0.000463	0.000815	0.000968
Betweenness	5.555925	0.000191	2.777962	0.000003	0.000006	0.000050	0.000666
Straightness	0.000117	0.000163	0.0	0.000000	0.000117	0.000275	0.000212

a mean of 0.000191, and a mode of 2.777962. The disparity between the median and mean values indicates potential outliers affecting the average. The mode shows more closely with the median compared to Closeness, while the standard deviation of 0.000666 suggests moderate variability around the mean. Straightness presents different patterns with a median of 0.000117, a mean of 0.000163, and a mode of 0.0. The lower median and mode values suggest a distribution skewed towards lower values. The standard deviation of 0.000212 indicates moderate variability around the mean. In summary, Closeness shows higher values across all statistical measures compared to Straightness. This indicates that the street pattern in the *Fahadan* neighbourhood represents an organic and close layout rather than consisting of straight streets. *Betweenness* also shows higher values in terms of median, mean, and mode compared to straightness. This means that the neighbourhood features more central streets.

5 Discussion and Implications

Yazd, an ancient Persian city with a 5000-year history is recognised for its arid desert climate. While studies focused more on modelling energy efficiency at the individual level, the city's success in solar shading lies in the cohesive arrangement of elements at the neighbourhood level. This research employed a mixed-method approach using open-source tools to analyse street patterns and actual images.

In line with [39], the results confirmed that the street orientation in Yazd is northeast-southwest and northwest-southeast by using *OSMnx* and the actual street network dataset. Furthermore, following the existing body of literature [23, 24, 30, 39] the present study's results put stress on compact and organic street design, using *Momepy* and street pattern measures including *Betweenness*, *Closeness*, and *Straightness*. What is surprising is that a balance of shade and bright areas in the plan view of the *Fahadan* neighbourhood was discovered, which was not already mentioned in previous studies. Vernacular architects effectively provide shade by utilising street orientation, a dense street pattern, and *Sabats* to keep the sun's effects under control.

Modern urban design can benefit from the historic Persian cities such as Yazd. The historic city of Yazd offers exemplars in traditional desert architecture, and its design principles can guide the creation of sustainable and culturally rich future urban environments. One of the foundational lessons of Yazd's traditional design is the importance of orientation and dense texture. The city's layout shows a thoughtful orientation that maximises natural ventilation and reduces heat gain, an approach critically relevant in designing modern buildings in hot climates. Furthermore, Yazd's dense urban fabric, consisting of narrow, shaded alleyways, promotes walkability and creates a human-scaled environment. This approach not only contributes to the creation of more walkable cities but also aids in reducing the urban heat island effect by minimising exposed surfaces and maximising shaded areas [7, 22].

Yazd's architecture also emphasises the value of inner courtyards. These spaces serve multiple functions; they provide privacy, act as a family's private oasis, and allow for solar shading. This insight can be applied to modern buildings and urban blocks, where integrated private and semi-private open spaces can enhance the liveability and sustainability of urban environments [28]. Additionally, traditional designs often prioritise natural ventilation through features like courtyards and strategically placed windows, which modern architects can incorporate to enhance airflow and maintain comfortable indoor climates with less energy consumption. The use of locally sourced materials in traditional architecture promotes sustainability and better thermal performance, a principle that modern buildings can adopt [13]. Moreover, harmonising with the landscape through gardens and natural features not only improves ecological footprints but also enhances occupant well-being [42].

Diverse traditional architectural features have been on the agenda in contemporary architecture, too. The wind towers mechanism, for instance, have been integrating into new designs including commercial buildings to improve natural ventilation efficiency, which tends to be crucial for energy conservation and occupant comfort. It can be illustrated by the Gold Souq (Central Market) in Sharjah, UAE, features an imposing wind catcher integrated into the market's design. Madinat Jumeirah in Dubai also showcases traditional wind catchers used in both the hotel and market areas. Likewise, the Tehran Museum of Contemporary Arts incorporated a modern modified wind catcher into its architecture [12, 26].

In addition, urban planners and designers can benefit from Persian vernacular architecture when planning neighborhoods and street designs. Designing street patterns can incorporate a moderate grid and curvilinear layout instead of a pure grid pattern. This allows individuals to enjoy narrow streets where the various shades from building walls protect them from the sun's heat. In fact, we can learn from vernacular architecture to prioritise human prosperity over speed. The historic city of Yazd offers vital lessons for future urban design and architecture embracing natural elements, fostering walkability, and creating multifunctional private and public spaces. Vernacular architects viewed neighbourhoods as a whole system. They focused on open spaces and life between buildings, emphasising connectivity to create a cohesive system. They aimed to incorporate ample shaded open spaces, a concept that can be applied to modern neighbourhood design interventions. By learning from Yazd's architecture and urban planning principles, modern cities can become more sustainable, liveable, and resilient against climate change. The blend of traditional wisdom with modern technology presents a promising avenue for designing future urban environments.

6 Concluding Remarks

The present study was designed to explore the vernacular shading strategies and approaches in the actual street patterns and architectural features of a hot, arid, desert city in Iran, Yazd. Examining the brightness and shade coverage of the case of the *Fahadan* neighbourhood showed that 44.20% of the area was shaded, while the remaining 56.80% was exposed to sunlight. Brightness calculations in *Fahadan* revealed how successfully vernacular architects provided shaded open spaces at the micro level. To neutralise the sunshine effects, *Fahadan* streets were oriented towards northeast-southwest and northwest-southeast. The higher levels of *Closeness* and *Betweenness* compared to *Straightness* were found by investigating the street pattern measures. *Fahdan*'s street pattern is characterised by a dense layout with lower straight streets. By analysing actual street patterns and images of *Fahadan*, it was found that vernacular architects create cooler environments by providing shade through open spaces.

By analysing actual street patterns and images of Fahadan, it was found that to what extend vernacular architects can contribute to cooler environments and open spaces through shading interventions in built environment. The strategic orientation of streets and the incorporation of architectural features like Iwans, courtyards, domed roofs, and Sabats (covered passageways) at the pedestrian scale are key to this approach. These architectural elements are interconnected, forming a cohesive network of shaded areas throughout the neighbourhood. The Iwans, with their deep recesses, provide shade at the building entrances, while the courtyards and domed roofs create shaded open spaces within the building domain. The Sabats, or covered passageways, link these shaded areas, allowing pedestrians to move through the neighbourhood in the comfort of the shade.

These vernacular design strategies, rooted in the local climate and cultural context, demonstrate a system of interlinked design strategies and interventions by which spaces users' comfort is targeted through passive cooling techniques. The findings of this chapter expanded our understanding of the energy performance of historic Persian cities provided insights for contemporary urban planning and design, and helped to create more sustainable and liveable cities for future generations at the neighbourhood-design level.

Data Availability Statement All Python codes used in this study, including shading coverage using OpenCV library, the street orientation using OSMnx library, and Street Design Components (*Betweenness*, *Closeness*, and *Straightness*) using Momepy library can be accessed via https://github.com/kimiachenari/kimiachenari.

References

1. Akbari H, Niazi Motlagh Joonaghani N (2022) Analysis of the geometric and natural properties of courtyards in historical houses of Isfahan (Iran). J Asian Arch Build Eng 21(5):1879–1890
2. Altaweel M, Hanson J, Squitieri A (2021) The structure, centrality, and scale of urban street networks: cases from pre-industrial Afro-Eurasia. PLoS ONE 16(11 November):e0259680–e0259680. https://doi.org/10.1371/journal.pone.0259680
3. Asadpour H, Taherian B, Asadpour A (2022) Application of green school site selection criteria to increase environmental sustainability; case study: Fahadan neighborhood of Yazd. Arman Arch Urban Dev 14(37):223–238
4. Bellia L, Marino C, Minichiello F, Pedace A (2014) An overview on solar shading systems for buildings. Energy Procedia 62:309–317. https://doi.org/10.1016/j.egypro.2014.12.392

5. Boeing G (2017) OSMnx: new methods for acquiring, constructing, analyzing, and visualizing complex street networks. Comput Environ Urban Syst 65:126–139. https://doi.org/10.1016/j.compenvurbsys.2017.05.004
6. Boeing G (2019) Urban spatial order: street network orientation, configuration, and entropy. Appl Netw Sci 4(1):1–19. https://doi.org/10.1007/s41109-019-0189-1
7. Bolouhari S, Barbera LV, Etessam I (2020) Learning traditional architecture for future energy-efficient architecture in the country; case study: Yazd city. Naqshejahan-Basic Stud New Technol Arch Plan 10(2). Retrieved from http://bsnt.modares.ac.ir/article-2-39832-en.html
8. Brito-Coimbra S, Aelenei D, Gloria Gomes M, Moret Rodrigues A (2021) Building façade retrofit with solar passive technologies: a literature review. Energies 14(6). https://doi.org/10.3390/en14061774
9. Castilla MV (2023) Analysis of performance strategies for the preservation of heritage buildings: the case of Puerto de Santa María. Buildings 13(5):1196
10. Chenary K, Pirian Kalat O, Sharifi A (2024) Forecasting sustainable development goals scores by 2030 using machine learning models. Sustain Dev n/a(n/a). https://doi.org/10.1002/sd.3037
11. Chenary K, Soltani A, Sharifi A (2023) Street network patterns for mitigating urban heat islands in arid climates. Int J Digit Earth 16(1):3145–3161. https://doi.org/10.1080/17538947.2023.2243901
12. Chohan AH, Awad J (2022) Wind catchers: an element of passive ventilation in hot, arid and humid regions, a comparative analysis of their design and function. Sustainability 14(17). https://doi.org/10.3390/su141711088
13. Chohan AH, Awad J, Elkahlout Y, Abuarkub M (2024) Evaluating windcatchers in UAE heritage architecture: a pathway to zero-energy cooling solutions. Ain Shams Eng J 102936. https://doi.org/10.1016/j.asej.2024.102936
14. de Speville CD, Seviour WJM, Lo YTE (2023) Predicting future UK nighttime urban heat islands using observed short-term variability and regional climate projections. Environ Res Lett 18(10):104044
15. Di Salvo S (2020) Façade solar control and shading strategies for buildings in the Mediterranean region. Environ Sci Sustain Dev 5(2):32–47
16. Du H, Zhan W, Voogt J, Bechtel B, Chakraborty TC, Liu Z, Hu L, Wang Z, Li J, Fu P (2023) Contrasting trends and drivers of global surface and canopy urban heat islands. Geophys Res Lett 50(15):e2023GL104661
17. Eskandari H, Saedvandi M, Mahdavinejad M (2017) The impact of Iwan as a traditional shading device on the building energy consumption. Buildings 8(1):3
18. Farashah MDP, Aslani E, Khademzade M (2018) Strategic planning of cultural tourism development in historic city of Yazd (case study: Fahadan neighbourhood). Almatourism: J Tour Cult Territ Dev 9(18):23–44
19. Fleischmann M, Feliciotti A, Kerr W (2022) Evolution of urban patterns: urban morphology as an open reproducible data science. Geogr Anal 54(3):536–558. https://doi.org/10.1111/gean.12302
20. Foruzanmehr A (2013) Residents' perception of earthen dwellings in Iran. Int J Urban Sustain Dev
21. Grynning S, Lolli N, Wågø SI, Risholt BD (2017) Solar shading in low energy office buildings-design strategy and user perception
22. Jamei E, Ahmadi K, Chau HW, Seyedmahmoudian M, Horan B, Stojcevski A (2021) Urban design and walkability: lessons learnt from Iranian traditional cities. Sustainability 13:(10). https://doi.org/10.3390/su13105731
23. Jasim IA, Farhan SL, Al-Maliki LA, AL-Mamoori SK (2021) Climatic treatments for housing in the traditional holy cities: a comparison between Najaf and Yazd cities. IOP Conf Ser: Earth Environ Sci 754(1):12017
24. Keshtkaran P (2011) Harmonization between climate and architecture in vernacular heritage: a case study in Yazd Iran. Procedia Eng 21:428–438
25. Liu S, Yin S, Hua J, Ren C (2023) 13—Climate-responsive architectural and urban design strategies for adapting to extreme hot events. In: Pacheco-Torgal F, Granqvist C-G (eds) Adapting

the built environment for climate change. Woodhead Publishing, pp 253–272. https://doi.org/10.1016/B978-0-323-95336-8.00002-0
26. McCabe C, Roaf S (2013) The wind towers of Bastakiya: assessing the role of the towers in the whole house ventilation system using dynamic thermal modelling. Archit Sci Rev 56(2):183–194
27. Moscoso-García P, Quesada-Molina F (2023) Analysis of passive strategies in traditional vernacular architecture. Buildings 13(8). https://doi.org/10.3390/buildings13081984
28. Nazemi P (2013) Necessity of urban transformation in introverted historic textures: the ancient Persian city of Yazd. J Plan Hist 13(1):50–67. https://doi.org/10.1177/1538513213507538
29. Nguyen AT, Truong NSH, Rockwood D, Le Tran AD (2019) Studies on sustainable features of vernacular architecture in different regions across the world: a comprehensive synthesis and evaluation. Front Arch Res 8(4):535–548. https://doi.org/10.1016/j.foar.2019.07.006
30. Özer DG, Yardımlı S, Shahriary A (2020) Sustainable street architecture and its effects on human comfort conditions: Yazd, Iran. A| Z ITU J Fac Arch 17(2):113–122
31. Pakzad E, Kalat OP, Pirzaman YK (2024). Comparative study of the state of sustainability between Iran and Turkey using the triangular sustainable development model. Environ Eng Manag J (EEMJ) 23(2)
32. Pardo JM (2023) Challenges and current research trends for vernacular architecture in a global world: a literature review. Buildings 13(1). https://doi.org/10.3390/buildings13010162
33. Peeters A, Shashua-Bar L, Meir S, Shmulevich RR, Caspi Y, Weyl M, Motzafi-Haller W, Angel N (2020) A decision support tool for calculating effective shading in urban streets. Urban Clim 34:100672. https://doi.org/10.1016/j.uclim.2020.100672
34. Porta S, Crucitti P, Latora V (2006) The network analysis of urban streets: a primal approach. Environ Plann B Plann Des 33(5):705–725. https://doi.org/10.1068/b32045
35. Rezaei N (2017) Resident perceptions toward tourism impacts in historic center of Yazd Iran. Tour Geogr 19(5):734–755
36. Roaf S (n.d.) 739*:* The traditional technology trap (2): more lessons from the windcatchers of Yazd
37. Stack A, Goulding J, Lewis JO (2000) Shading systems: solar shading for the European climates. European Commisson
38. Tan L Jiang J (2019) Image processing basics. In: Tan L, Jiang J (eds) Digital signal processing, 3rd edn, Chap 13. Academic Press, pp 649–726. https://doi.org/10.1016/B978-0-12-815071-9.00013-0
39. Tavassoli M (2016) Urban structure in hot arid environments. In: Strategies for sustainable development. This Springer Imprint Is Published by Springer Nature
40. Yazdi H, Mügge G, Movafagh S, Fallahi MH, Ludwig F, Moazen S (2023) The effect of improper interfering in the historical architecture on energy wasting (case study: Bibi-Roqayyeh house, Yazd, Iran). IOP Conf Ser: Earth Environ Sci 1196(1):12103
41. Yue N, Zhang Z, Jiang S, Chen S (2022) Deep feature migration for real-time mapping of urban street shading coverage index based on street-level panorama images. Remote Sens 14(8). https://doi.org/10.3390/rs14081796
42. Zhong W, Schröder T, Bekkering J (2022) Biophilic design in architecture and its contributions to health, well-being, and sustainability: a critical review. Front Arch Res 11(1):114–141. https://doi.org/10.1016/j.foar.2021.07.006

Human Sustainable Interaction Between Architecture and Climate in Shiraz Heritage Fabric for Designing Climate Resilience

Maryam Ekhtiari

Abstract Having accurate attitudes towards architecture along with proper actions can promote direct interactions between humans and their environment and help them have a better understanding of the climate. Clarifying the underlying beliefs and behaviors of building occupants that shape the interactions between them and their environment can enhance their understanding of the climatic impacts of the architecture around them. This study examines how the strength of the bonds between nature and climate within and outside traditional buildings have contributed to the preservation of the old architectural fabric of Iran. This study looks at the historic architecture of Shiraz on three levels: overall climatic integration of texture, climatic spatial configuration, and climatic components. Using mixed methods of research drawing upon logical argumentation. Analysis of the data shows that the Shiraz Heritage fabric follows the orders of four crucial climatic factors, such as light, air, soil, and water separately and together in unity. Bioclimatic moderation in Shiraz vernacular fabric is illustrated with the quality of climate-responsive architecture according to four climatic factors together. Moreover, the spatial configuration shows a sequence of linked spaces that constitutes interior spaces to the semi-open and open spaces and neighborhood spaces, which are organised as a matrix of interconnection spaces. This combination of linked organisations could support thermal-acoustic-visual diversity in architecture for diverse multisensory needs. Consequently, it could evoke climatic resilience due to human interaction with the place of living. This kind of organisation facilitates the correlation between climatic modulation. Therefore, building components and elements, such as passive ventilation and cooling, pools, and ponds act as interactive climatic elements. The results show that the architecture of the old fabric of Shiraz city is perfectly compatible with the environment. This architecture, with the accompaniment of human needs and desires as an inseparable part of a unified world, not only strongly supports human requirements with the climate but also helps them to raise and sublimate in the environment. In other words, the conclusions of this chapter in a simple comparison highlight the right attitude about the continuous life cycle on Earth. It represents that if a man-made space could

M. Ekhtiari (✉)
Department of Architecture, School of Art & Architecture, Shiraz University, Shiraz, Iran
e-mail: m_ekhtiari@shirazu.ac.ir

A. Cheshmehzangi and S. Roaf (eds.), *Persian Vernacular Architecture*, Urban Sustainability, https://doi.org/10.1007/978-981-96-1116-4_12

have a durable and smooth organisation like nature and ecology, this could retain its vitality and value. Everything considered, recognising the geographical and native criteria is necessary for architecture.

Keywords Climate-resilient architecture · Nature-adapted architecture · Old fabric of Shiraz · Passive strategies

1 Introduction

A climatic study is the basis of any logistic programming in a region and can strengthen the innate relationship between humans and their environment [31]. Nowadays, people face difficulties in the natural environment, because of the undesirable climatological effects of human influences. The continuation of measurable climate changes, such as deforestation and global warming, causes critical temperature fluctuations and extreme precipitation, which results in changes in extreme events, such as local rain and hail storms, unusual winters, vast and frequent summer droughts, and other climatic threats [56]. These events harm the environment, the climate, humans, and other creatures. For instance, expansions and contractions caused by temperature changes result in tensile or compressive stress called thermal stress, which decreases physiological activity in humans, plants, and animals [43].

To have a healthy life, humankind needs a desirable environment, which consists of natural and man-made ones. It is significant to care about the God-given natural environment to adapt to the climate. Human efforts in the natural environment are used to produce man-made environments, which are studied in architectural science. A desirable architecture helps people to intuit and accomplish a harmonious relationship with the environment [46]. Vernacular houses in Iran have established a significant relationship with the climatic variables of the context [57]. Climatic consideration is a fundamental view in designing vernacular houses in Iran that optimises energy consumption and sublimated comfort balance [41]. This anthropogenic quality re-orders human and architectural culture. Unfortunately, today, the construction method is applied without consideration for the traditional culture. This contributes to endangering the natural environment, to which we should also add human life. To address these environmental problems, it seems reasonable to offer solutions for preventing abnormal climate changes in man-made environments, especially architectural ones. For arranging this matter, there seems to be very little effort to make except for accurate identification of climate factors and precise human interaction with these factors. In this paper, the emphasis is on the mutual affection of Shiraz architecture in a historical fabric with climate factors to recognise the quality of interaction between the Iranian ancient context and the natural environment.

2 Background of the Study

The term resilience derives from the present participle of the Latin verb 'resilire', meaning "to jump back" or "to recoil", "spring back", or "bounce Back" [34], Which refers to the ability of systems to absorb changes of state variables, driving variables, and parameters, and persist [22]. Resilience is not only about resistance to change and conservation of existing structures but also, the capacity of a system to absorb disturbance and re-organise to still retain fundamentally the same function, structure, identity, and behaviors [17, p. 259]. Although there is no commonly accepted definition of resilience [7: 2], most definitions spotlight a capacity for successful adaptation in the face of disturbance, stress, or adversity [42: 129, 52]. According to social-ecological systems complexity as the integrated systems in which humans are part of nature [4], following instrumentally rational solutions for recognising building resilience coordinated by context, is an intelligible need [5].

Researchers identified some features that are significant for achieving resilience. Several properties contribute to system resilience, including integrative factors (coordination and involvement of different actors), fractal self-structure, self-regulation, adaptive capabilities, diverse categories, interconnectivity, redundancy, transformative capacity, buffer capacity (Threshold capacity), the existence of equilibrium in systems, nested hierarchical in the adaptive cycle [9, 21, 16, 23, 30, 58]. Urban spatial resilience has characteristics including form (spatial patterns), interaction (relationships, linkage, flows), and structure (rules) of urban places [6], robustness, stability, flexibility, resourcefulness, coordination capacity, redundancy, diversity, foresight capacity, independence, connectivity and interdependence, collaboration capacity, agility, adaptability, self-organisation, creativity and innovation, efficiency, and equity [55]. Therefore, spatial resilience addresses energy efficiency, reduction usage of non-renewable energies, thermal resilience, reducing environmental pollution, and using natural resource recycling technology. According to the importance of inner strengths and resources to demonstrate remarkable ecosystem resilience [18], this chapter investigates the human-sustainable interaction between historical architecture and the climate of Shiraz on three levels of overall climatic integration of texture, climatic spatial configuration, and climatic components.

3 First Level: Understanding Overall Climatic Integration of Texture

3.1 Order of the Earth (Soil) and Its Effect on the Old Fabric of the City

Shiraz's oval-shaped plain is encircled by relatively high mountain ranges (Bemo including Sabzposhan, Chehlmaqam, and Babakohi mountains, or Mortaza Ali, Saadi, Baramdelak and Ahmadi, Qalat and Mount Derak and other mountains) in the form of a boundary fence or defensive wall with limited ways through narrows (e.g. Tang Allah Akbar and Tang Ab Khan). The location of Shiraz is strategic for its preservation and safety. The slopes of the plain lead to eastern and southeastern directions [33, 35: 149–150, 45]. According to Mosalanejad's studies [37], the Shiraz plain is divided into three parts, one of which has strong and stable soil for construction and the other has fertile soil suitable for agriculture. Therefore, Shiraz has three parts, including two plain parts and one mountain part [37], with three types of soil: weak soil, medium soil, and hard soil.

Moreover, Shiraz has been placed in a high seismic risk region, and a base acceleration of 0.3 g is recommended for it in the Iranian code of practice for seismic-resistant design of buildings. The major faults in the Shiraz region and its vicinity are Sabzposhan, Kohenjan, Sarvestan, Karehbas, Bezain, Soltan, and Qalat [19]. These faults have caused the relocation of many water channels to balance the water supply possibilities. For example, the Kore Bes fault has caused the deviation of the Qara Aghaj river and the rise of the Mishvan and Shurab rivers [12].

The old fabric of Shiraz, located on a natural terrace, is characterised by medium soil whose deposits act as a filter when the bedrock earthquake acceleration is transmitted through them. As shown in Fig. 1, choosing a natural terrace in the middle of sedimentary soil for historical settlement sites in Shiraz illustrates human ingenuity in improving climatic resilience by reducing the risk of earthquakes and increasing the possibility of accessing water simultaneously.

3.2 Order of Air (Wind) and Its Effect on the Old Fabric of the City

Shiraz climate situations reveal a high alteration in temperature patterns between day and night, a high fluctuation of temperature during the day, a prevailing wind breeze with various speeds in different seasons from the western north to the eastern south, and continuous local winds from the sides of the mountain to the plain and vice versa. Nevertheless, the weather moderation of Shiraz is the key to the presence of gardens in the north and western north of Shiraz, among the Zagros mountains [14, 29]. On most days, the wind blows from north to south with low intensity, from west to east,

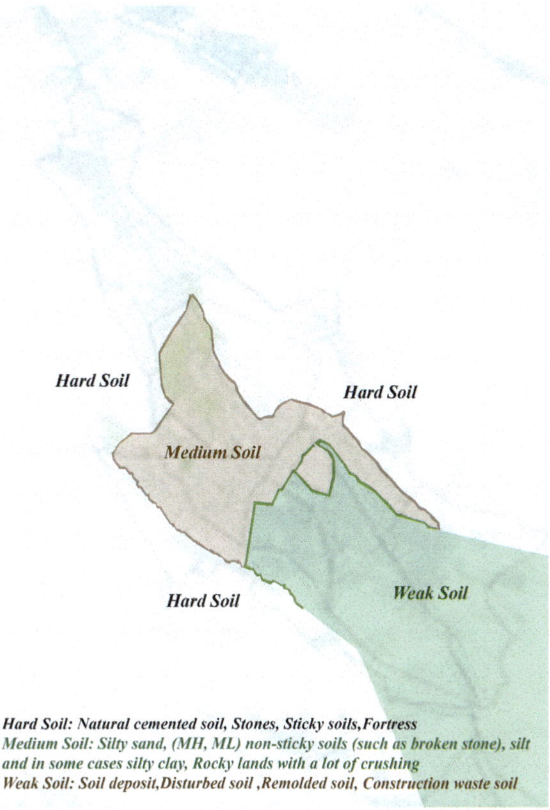

Fig.1 Shiraz types of soil. Redrawn by Ekhtiari adopted and adapted from Mosalanejad et al. [37]

or from west south to east north with higher intensity, and rarely from east north, east, and east south at low speed [14]. Bioclimatic indexes of Shiraz reveal a wide range of variety from warm to very cool conditions. during the year. According to Baker's method, Shiraz has mild alterations in human bioclimatic conditions due to cool, winter weather during the colder months of the year. Springtime brings with it a shift in the environment from one of relative stimulation and pressure to one of gentle and moderate. Summertime conditions are bearable, while springtime conditions are ideal for human comfort due to the gentle, moderate temperature [50].

Different environmental conditions that are tolerable help people perceive and experience a variety of spaces and environmental behaviors. Climatic strategies in selecting settlements in Shiraz contribute to the experience of a moderate climate, with mild to warm summers, cool winters, and mild springs and autumns.

3.3 Order of Light and Its Effect on the Old Fabric of the City

The historical city of Shiraz is located at 29.591768° north latitude and 52.583698° east longitude with a height of 1540 m above sea level. The average annual temperature is 17.3 °C, the maximum temperature rises to 43.2 °C in warm seasons, and the minimum temperature drops to 14 °C in cold seasons. As shown in Fig. 2, according to Sadeghi and Bahadori [49], the thermal surface simulation of Shiraz reveals three sections: high surface temperature (with red and yellow color), low surface temperature (with dark blue and blue color), and medium surface temperature (with light blue and green color) [49].

Residents' attitudes toward different kinds of places of sunshine in Shiraz have been illustrated in three sections: sunny places (Shiraz-Ishe, Abunasr hill, Bemo, including Sabzposhan, Chehlmaqam, and Babakohi mountains), shaded places (Nasar-Derak) and places with mild and pleasant sunshine (in Persian Khosh-Makān) as lively places [26, p. 19]. The historical fabric of Shiraz is located on a medium thermal surface and Khosh-makān.

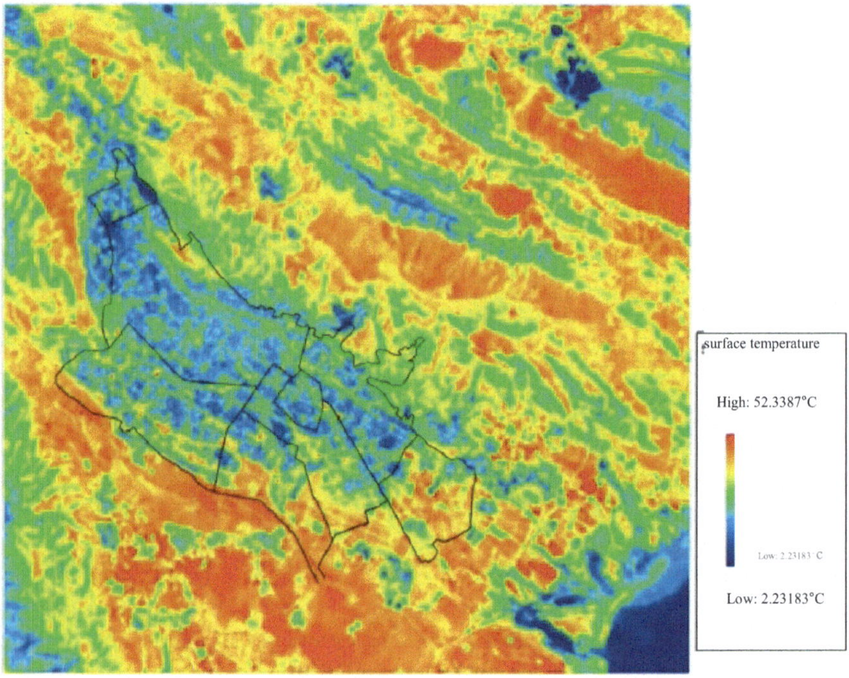

Fig. 2 Thermal surface simulation of Shiraz [49]

3.4 Order of Water and Its Effect on the Old Fabric of the City

Generally, four kinds of land can be identified based on the moisture in Shiraz plain (Fig. 3):

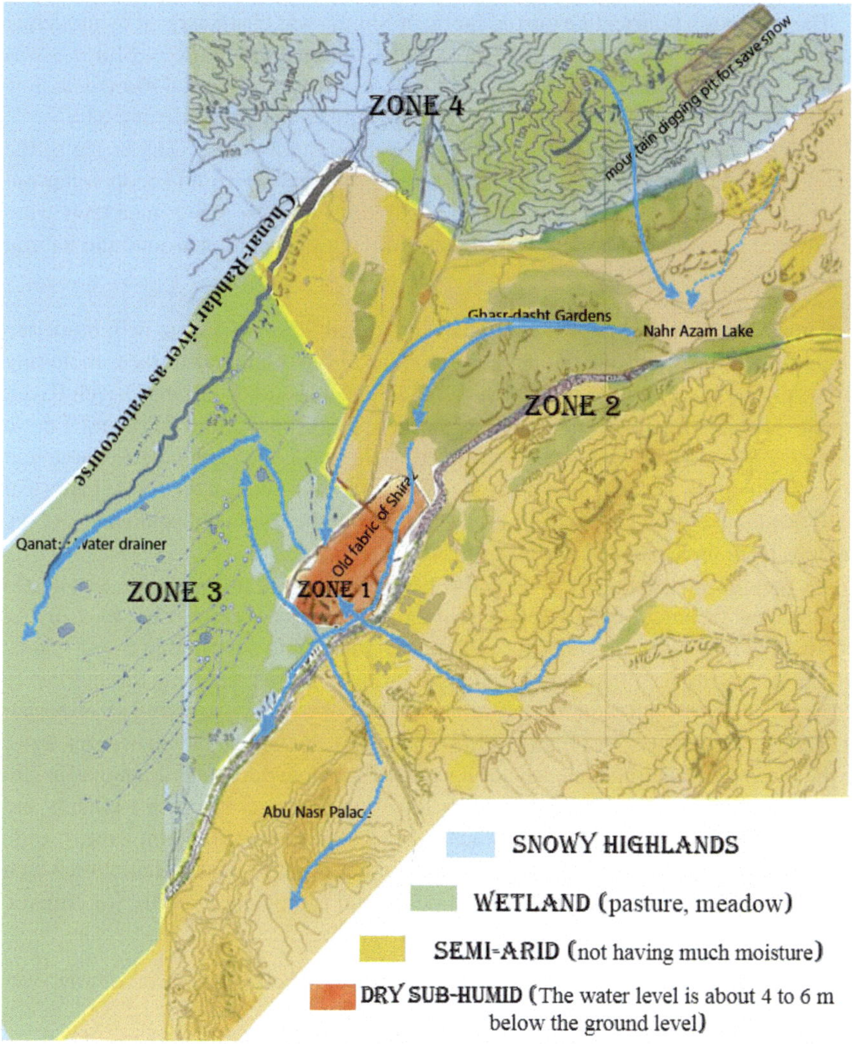

Fig. 3 Four kinds of land regarding moisture content and water circulation management in Shiraz plain Redrawn by Ekhtiari adopted and adapted from Javaheri, 2001 [22]

- Snowy heights and shaded places, Nasar (e.g. Derak, Qlat Mountain, Dehshaikh, and Red Mountain) used to serve as snow storage, which made ice available in the summer seasons and underground water.
- Arid lands or xeric shrublands were characterised by an inadequate supply of water, like Pahandej, Abu Naṣr palace (Qaṣr-i Abu Naṣr), and lands of the north and west-north of Shiraz.
- The dry-mesic prairie, like the old fabric of Shiraz, was characterised by moderate levels of moisture. The soil moisture of the site—the old fabric of Shiraz—was not too high to soak everything, but the level of water (about 4–6 m) was low enough to make access to water for drinking and courtyard gardening easy.
- In the mesic prairie-like meadow in the southeastern part of Shiraz plain, the underground water table used to rise so high that people could scoop water out of the well with a bowl in wet years. The lands of these green meadows were covered by water in the winter, and water-loving plants grew around the natural pond [26, pp. 22–23].

Choosing a good settlement site for the historical fabric of Shiraz with moderate levels of moisture indicates human ability. The site is a natural terrace, including proximity and accessibility to needed resources, such as the Khorram-Darreh River, which is the destination of many torrents originating from Tang-e Sa'di and Tang-e Allah o Akbar, which would virtually feed Maharloo Lake. The nearest mountain is a 15-min walk across Chahar Bagh Avenue, called Jahan-nama. The historical settlement site reaches the agricultural sectors from the south and southeast and the gardens from the north and northwest.

To provide this place for living, the people of Shiraz used a basic watershed management technique called water circulation. Therefore, their priority was to avoid irreversible loss of biodiversity through conservation and sustainable use of lands, which are essential to livelihood development. The Iranian ancients did not easily waste the water, which required laborious work to obtain. Constructing an aqueduct or drainage channels from a high ground table water, encouraged them to carry water to the lower surrounding area. Retaining water from aqueducts and drainages in situ contributes to a sustainable landscape by lowering the water levels in wetlands and redistributing it to arid areas, thereby achieving a more balanced soil moisture profile [27, p. 4]. The specific water management technique is called water circulation which ensures the equilibrium between human needs and ecological integrity for climatic resilience. Water circulation is fully illustrated with:

- Snow storage in northwest mountains such as Derak helped to preserve snow over the summer [26, p.116]
- The water from the snowmelt and fifteen aqueducts originating from the north gathered in Azam Creek in the northwest of Shiraz.
- The water of Azam Creek from a place called Maghsam Aala was divided into branches and flowed into streams
- One of these streams irrigated some mills in northwest gardens located in Ghasr-e-dasht and eventually reached the dry river; the other supplied the consuming

water of some neighborhoods in the Shiraz heritage fabric (i.e. Lab-e-Ab, Bala Qaft, Qavam, Shazde, and Darvaz-e Isfahan) [25, pp.187–189]

- Water cisterns, such as Vakil- addressed the need for rainwater harvesting to divert water runoff away from a residential area to prevent inundation [25, pp. 136–139]
- The source of Rokn Abad aqueduct was on the slopes of the north mountains and valleys of Shiraz, irrigating the west and northwest gardens and landscapes around upstream of the Khorram-Darreh river [25, pp. 187–189]
- The southeast area of Shiraz plain was drained with subsurface drains that induce excess water to flow through the soil to open wells and drains -Qanat- for irrigating the farmland in the south of the suburb [26, p. 22, 27, p.]. Furthermore, it supplied the rivers and streams of the south of Shiraz plain, such as the Chenar-Rahdar river basin [26, p. 22][1]
- Saadi aqueduct situated in the eastern region effectively irrigated the eastern section of the Shiraz plain and its adjacent downstream areas. This infrastructure extended its benefits to the typically arid southeastern region, subsequently channeling water to Maharloo Lake to regulate water levels according to varying land use requirements. This system played a crucial role in fostering a sustainable agricultural methodology.
- Some dams—such as the Qoran gate dam located in the southeast around the most vulnerable parts of the city is the self-closing flood barrier (SCFB) for flood protection and heavy rain prevention [15, pp. 136–139].

According to Fig. 3, garden zones in Shiraz were a part of dry lands or xeric shrublands characterised by an inadequate supply of water. However, through the use of watershed management and water circulation, the moisture from wet, upstream lands made these dry lands moist and arable. Consequently, gardens in Shiraz emerged as man-made rather than natural endowments reflecting harmonisation between climate and architecture in the Shiraz plain. Therefore, the relative humidity of the air around Shiraz was provided. As time elapsed, human life had a moderating effect of environmental stressors. As a result, the northwest part of Shiraz is called a garden as man-made green land. Shiraz was signified by the eco-friendly gardens and farms, as well as some protected areas of vegetation inside it, because of its traditional watershed management and improved soil conservation, which provide fertility and various types of plants to grow [26, p. 203]. Moreover, it revealed the special moderate weather and the traditional moderate culture. The careful and efficient watershed management involved supplying drinking water from aqueducts' springs, which were stored inside the reservoirs, directly from the streams or surface wells [25, pp. 136–139].

The fast growth of the city in recent years has destroyed most of the gardens and farms around the city [53]. This reveals that land use plays a significant role in controlling spatial and temporal variations of soil moisture by influencing infiltration rates, runoff, and evapotranspiration, which are important to crop growth and vegetation restoration in semiarid environments.

[1] Drainage is the most effective method in Shiraz plain to intercept the water and divert it away from the property and prevent groundwater rising to floor level (javaheri3, p4).

4 Second Level: Understanding the Climatic Spatial Configuration

Given Shiraz's high annual temperature fluctuation, the square (rectangle close to square ratio) plan with the least amount of exterior surface is the best structural design. The reason is that it loses the least amount of heat in the winter and gains the least amount of heat from the sun and surrounding environment in the summer. Furthermore, it is more suited to use a cube-shaped, condensed, and compressed fabric with more shared walls between buildings to avoid temperature loss within the structure [14, 29]. A microclimate that produces air purification and passive cooling is created by an introverted plan with a central courtyard as an open area that is simultaneously used for plants and water in this cube. This microclimate harmonises many elements that reflect the cooling cycle and lessens the impacts of direct sun radiation and air heat on residents' activities [14, 29, 36]. The average dimensions of the courtyards are overall regulated according to the latitude, which is narrow enough to maintain various kinds of shaded areas—such as light shade, partial shade, full shade, and deep shade—during the heat of the day but wide enough to receive solar radiation—direct, diffuse, and reflected radiation. Therefore, varying degrees of solar intensity, the sun-to-shade continuum, and the needs and tolerances of various human behaviors and habits towards lighting and thermal requirements have already been addressed in the biotic structures of historical buildings of Shiraz by seasonal spaces (Winter section, Summer section), Veranda (Iwan), basement, pool house (Hozkhane), planted courtyard with trees, underground spaces as an earth coupling technique, Hashti, or entrance space of houses, works as a thermal filter between hot outdoor air and the cool atmosphere of the courtyard, different kinds of windows, Sunbreaker, and Portico [38, 53]. Consequently, various ambient temperatures in the range of thermal comfort in the spatial configuration of Shiraz vernacular architecture gain environmental perception, which results in climatic residence.

A successful style in terms of reducing overall energy consumption in Shiraz climate is the transition zone, which is defined as semi-open or semi-closed spaces that are in-between closed and open spaces. These spaces, such as Veranda (Iwan), semi-open corridors in-between courtyards and rooms (such as Shahneshin), can control their surroundings' temperature by shaping different daily and monthly shadows [28]. Furthermore, this flexible spatial configuration provides natural ventilation for closed spaces and creates a wide variety of spatial configurations, such as multiple entrances for each space and selective connectability and disconnectability for residents to adapt to changes [13]. For further elaboration, studies conducted by Barzegar and Heidari [3] revealed the depth at the entrance of residential houses in the historical fabric of Shiraz, which has caused the formation of different monthly shadow heights. Consequently, it addresses the heating and cooling requirements. Therefore, the entrance space areas, which are much more connected to the exterior ambient environment, play a decisive role in the external thermal comfort of users. Moreover, these entrance zones reduce the thermal shock of changes in conditions, allowing time for adaptation [3].

The passive methods ensure the thermal comfort of the indoor environment by providing healthy and sufficient energy [40]. The buildings of the old fabric of Shiraz illustrate two kinds of air infiltration, consisting of the free circulation of fresh air on the roof level and limiting air infiltration and energy loss on the building level, which ranges from maximum urban porosity to the city's disaggregation in vertical organisation. The diversity based on geometrical design optimises natural ventilation in harmony with the natural air in Shiraz plain and alternate climatic comfort in the range of the comfort zone in variable quality, such as air of elegance, gusty winds, the breeze, and strong winds [10, 44]. The results of Zare Mohazabie and his associates indicated that most of the rooms of old residences such as Shahneshin (the main sitting chamber) are in the acceptable range of thermal comfort for eight months without mechanical ventilation [59]. Historical houses reveal diverse spatial configurations. However, some of the rooms act as independent respiratory organs and usually have no connection with other spaces. These kinds of rooms have only small windows opened to the protected central courtyard area, with less harsh conditions and a more favorable environment, or limited vents as wind catchers (mainly Khowrak, partially Badgir) on their roofs above living rooms to obtain fresh air [10]. Therefore, the buildings prove resilient to not spreading or dispersing viruses. Furthermore, larger windows, such as sash windows, lattice doors, and rose windows, play a significant role in the movement and rotation of the fluid at a significant volume in some other rooms, causing dynamic behavior and developing resilience.

Some other rooms are connected to each other with the possibility of transporting air from one room to another [10], which offers temperature variety and multi-directional daylighting. Different kinds of spatial configurations produce peaceful settlements that meet the residents' requirements and culture, which is rooted in their compatibility and adaptability to the environment. Generally, semi-open spaces—Takhtgah or corridors—separate the various kinds of rooms, which play an important role in room ventilation and indoor air quality because of their responsibility for providing healthy and purified air. Despite changing lifestyles for rapid evacuation and emergency response as well as the access of rescue teams, changing the placement of the historical fabric of Shiraz from the past (a residential area surrounded by gardens and agricultural lands changed to a central fabric located in the core of the city), destruction of heritage buildings, and changing the formation of some historical neighborhoods subsequently fabric, which caused not well-connected narrow and irregular street networks [54], studies reveal significant findings in terms of resilience to seismic risks, heat risks, and flood risks. Despite all these drawbacks, traditional neighborhoods have moderate levels of performance in terms of resilience to flooding because of their moderate levels of block and lot size and their comparatively high fractions of open and green spaces, which contribute to mitigating flood risks. However, traditional neighborhoods provided good access to the available green and open spaces because their compactness contributed to improved accessibility, and their narrow and irregular street networks were flanked by one- to two-story buildings used for temporary shelter and keeping the temperature low. Furthermore, nodes and links with high centrality values increase the accessibility of the organic network in traditional neighborhoods. These neighborhoods are accessible on foot, and their relatively high land-use mix can enable better access to facilities during the immediate aftermath of the earthquake. The well-preserved Sang Siyah, a traditional neighborhood, performs well in the face of some stressors, such as heat risk and flood risk, contributing to improving resilience because of its moderately sized open spaces that are well distributed across the neighborhoods (Table 1).

5 Third Level; Understanding the Climatic Components

Climatic components, such as the twisty and winding side alley and lanes, wind tower (Badgir), air-suction tower (Khowrak), openings, heavy material, ponds and trees, and pond house (Howz Khaneh), provide stack ventilation, radiative cooling, solar control, and evaporative cooling, which are named passive strategies for cooling. The twisty and winding side alleys and lanes in the organic architecture of the historical fabric of Shiraz behave as windbreaks, providing daily air circulation with a pleasant breeze. Wind towers (Badgirha) cause the penetration of cool summer breezes through cross-ventilation [51]. The air-suction tower (Khowrak: chimney-like tower) creates a vacuum suck at the top part by moving the heavy inside air to the base. Khowrak, a special ventilation system in Shiraz used for kitchens, leads to pressure drops and vacuums in them based on the wind current in a boundary layer near the vents. Furthermore, the wind tower (Badgir) and air-suction tower (Khowrak) polluted the outdoor air near the vents. Openings based on the spatial constituent elements such as location, color and material, size and shelter, and their orientation offer climate comfort for residents.

In addition, openings were the connection between outdoors and indoors, providing ample airflow and an outflow while simultaneously considering prevailing winds to be the most efficient and beneficial form of natural ventilation for indoor spaces. The number of windows that opened to the side corridors as an open-air corridor maximises favorable climate penetration into the indoor area. Another climatic strategy considered by the opening element is the usage of large openings with large windows for increasing temperature in the winter section and providing shading devices for openings such as shelters or louvers in the summer section. The colored-window panes of different sizes protect interior spaces from severe sun radiation and also prevent insects from disturbing residents [51].

Various apertures in Shiraz's historical architecture, such as sash windows (Orosi), lattice doors and windows, rose windows, and Panjdari, a five-window room can also affect the indoor condition of thermal comfort. Orosi, as a daylight-related component, provides three categories of different functions, namely daylight performance, thermal performance, and decorative role, which present a significant potential to address the requirements of visual comfort [24]. Orosi with different glass colors (colorful, red, blue, yellow, green, or no color) shows improvements in annual thermal comfort conditions and acceptable daylight performance [47].

The high fluctuation of weather during the day in Shiraz makes the use of heavy materials, such as brick, stone, and cement, with high heat capacity and thermal

Table 1 Climatic strategies for spatial configuration

Climatic strategies		Case studies or diagrams
1	Various state of connectivity (left) and various sky views (right) in the selected neighborhoods of old fabric of Shiraz et al. [54]	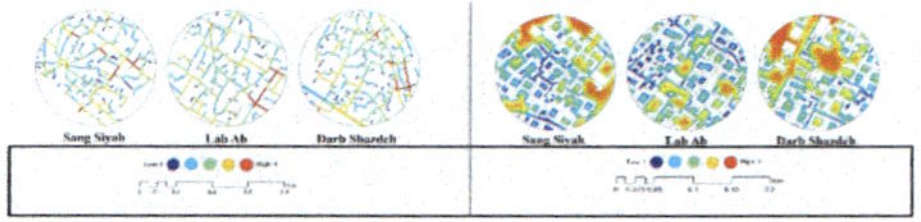
2	Various routes to a building	
3	Light diversity in the old lanes and alleys in historical fabric of Shiraz	Covered alley and shady alley, narrow alleys in old district of Shiraz, light and shadow simultaneously 

(continued)

Table 1 (continued)

Climatic strategies		Case studies or diagrams					
4	Entrances' depth and various shadow on outdoor thermal comfort						
5	Connectivity [39]	Rafael zadeh, Pahlavi	Sedaghati, Pahlavi	Zakeri, Pahlavi	Sheikh al-Islam, Qajar	Saadat, Qajar	Zinat al-Molk, Qajar
		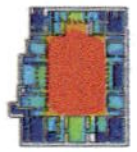	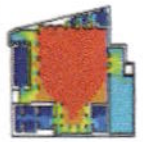	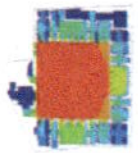		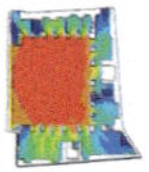	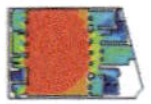
6	Different types of spaces	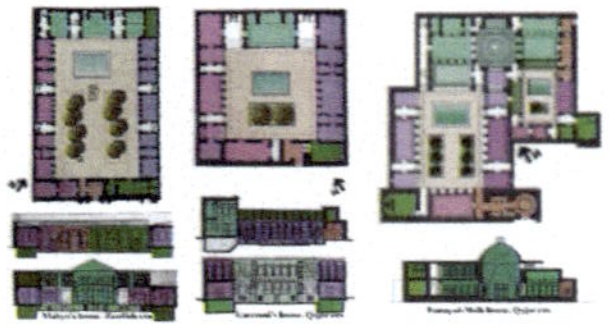					
7	Minimum exterior surface (close to square form),	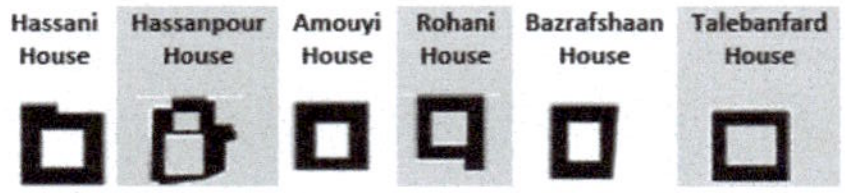					

Source The Author's own.

resistance, against the low thermal capacity of air in the courtyard. Therefore, the mass of the walls and floor of the courtyard serves as a reservoir of coolness by outgoing long-wave radiation at night and remaining cool by the following morning, consequently affecting the very low thermal capacity of the courtyard air that follows the surrounding surface [14, 29]. The microscopic and many pores of the mentioned materials, which are filled with air, change them to a material similar to a thermal insulator, besides their sun radiation absorption by their external surfaces [38].

Ponds, which both provide humidity and act as resources for low-water conditions, prevent severe heat during the day and severe cold at night. Therefore, plants and trees around the ponds can overcome unsuitable dry climate conditions and flourish [38]. Furthermore, dancing shadows of trees on ponds improve the quality of water.

The pond house in the traditional architecture of Shiraz uses the evaporative cooling of water by the pond and a fountain in the middle to cool the air of a space, with some platforms for sitting around it and frequently covered by high ceilings. Furthermore, by passing the air from the vents on the roof over the water fountain, cooled air is directed to other spaces around the pond house [38]. Using water and vegetation Water (Howz and fountains) was used under air vents combined with light catchers to humidify and dust the airflow that moves through [51] (Table 2).

6 Discussion and Implications

Based on the in-depth study, discussion, and implications have been generated in three areas: overall climatic integration of texture, climatic spatial configuration, and climatic components.

First level: overall climatic integration of texture: Notice to proper landscaping of the historical fabric of Shiraz demonstrates an integrated strategy to improve the microclimate regulation of green-blue-grey infrastructures for heat resilience, water resilience, and seismic resilience by historical settlement sites in Shiraz and green space management, which contributes to urban biodiversity, resources' provision, ecosystem services, and habitats for species, improving systemic connectivity at the urban level.

To sum up, the quality of designing the heritage urban fabric of Shiraz is based on the interaction between humans and nature, and it is not based on adverse effects on ecosystems, even though until now many modern architectures in the world provide human comfort at any cost, such as destroying their environment. Moreover, architecture and built environments are becoming part of the climate very quickly after its rise. The findings in this chapter prove that, in a heritage fabric, adapting to the wholeness of climate as a life system and life cycle is a common part of human culture. Iranian ancestors do deny bringing out destructive effects by damaging one part to improve the other part. Therefore, historical settlement sites in Shiraz reveal coexistence and harmony with nature to gather environmental and climatic quality to precisely manage human environmental comfort. Otherwise, urban life disturbs or

Table 2 Climatic strategies and components

Climatic components		Case studies or diagrams		
1	Entrance filter	Graph of entrance 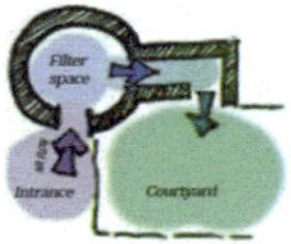	Ekhtiari et al. [11]  Khalilpasand house, Zandieh era  Zinat-al-molk, Forough-al-molk, Sharbatiha house, Gajar era  Jamshidi house, Pahlavi era	
2	Wind tower (Badgir), Air-suction Tower (Khowrak) [11]	 Khowrak, Nasr house, Gajar era	 Badgir, Jamshidi house, Pahlavi era	
3	Pond and trees	 Khalilpasand house, (Ekhtiari er.al. 2019)	Pond House (Howz Khaneh)	 Foruq-al-Molk,

(continued)

Table 2 (continued)

Climatic components		Case studies or diagrams
4	Various openings	 Foruq-al-Molk house  Naranjestan Qavam house  Zinat al-Molk house
	Ventilation and corridors	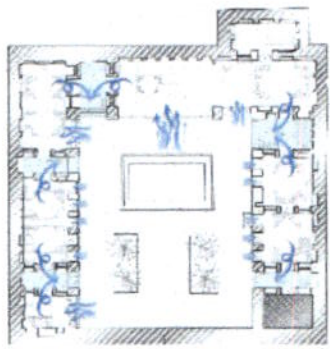 Khalilpasand house 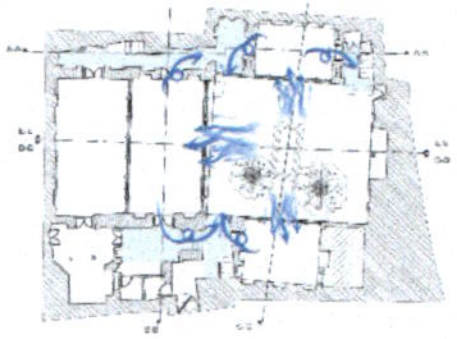 Jalalian house

Source Author.

disorders regional and global climates, as well as weather and geographic places. This is aggressively confronted with natural orders and makes it possible to damage itself and its wholeness system. Therefore, they have tried to make nature and built space as adaptable as they can preserve the climate and natural environment to achieve a desirable life. Thereafter, heritage architecture is the gathering, inspiration, and representation of the whole of climate. Thereby, we can perceive the climate by understanding and being in heritage architecture.

Hereby, inspiration and representation of the whole of the climates facilitate the unity between humans and existence. Moreover, in this case, the natural climate treats the man-made environment as a unified whole, referred to as the existential climate. Such a perspective allows for redundancy and buffer capacity to appear in both the natural climate and human culture through desirable architecture that symbolises lived spaces, such as the sustainable gardens of Shiraz. Furthermore, it illustrates the harmony between climate and architecture, which is central to the fame of Shiraz. That explains why Shiraz is distinguished by its eco-friendly gardens and farms, reflecting the region's unique moderate weather and traditional, balanced culture.

Second level: climatic spatial configurations: All of these arrangements improve climatic resilience in Shiraz. For instance, integrative factors provided by coordination of different spaces such as nodes or links with high centrality values, narrow and irregular street networks, an entrance transition zone, a courtyard, a transition zone as semi-open or semi-closed spaces (Iwan, corridors), and closed spaces (rooms). nested space. All transition zones improve transformative capacity. Moreover, nested space and a diverse range of views contribute to interconnectivity, and above all, multifunctionality enables houses to rearrange themselves according to changes without changing the main structure and form of the house.

The spatial configuration of Shiraz's historical fabric addresses some socio-cultural needs and desires. For example, urban fabric with alleys and lanes, which makes movement possible in any direction, allows the city not to have a non-defensible place. This metaphor describes intimate associations in society. Therefore, the social structure of Shiraz's heritage fabric is unified, and unlike many authoritarian cities, it is not divided into distinct centers of power and wealth.

Human needs and desires are the cultural characteristics that interact with climate. Thereby, these desires create urban architecture. Consequently, accurate interaction contributes to accurate architecture and vital spaces that form our existence. This is like our heritage urban architecture in Shiraz, which in turn contributes more to the system than the natural climate. Nowadays, buildings are designed to achieve or create a suitable atmosphere for only human comfort, but the historical fabric of Shiraz illustrates that designing a suitable atmosphere for humans and nature simultaneously is important for climatic resilience. For instance, coolers, on the one hand, help the interior space stay cool and, on the other hand, generate heat for the outside and environment. The example shows a self-destruction character that causes an increase in the environmental temperature. Furthermore, as mentioned before, self-regulation (self-organisation), adaptive capabilities (flexibility), and diverse categories have been revealed in the spatial configuration of Shiraz's historical fabric.

Third level: the climate components: The wind speed and temperature depend on each other for heat transfer by transaction. The continuous flow of wind inside roofed spaces can create a noticeable coolness on the shell of the buildings. Lack of airflow in the environment will increase the temperature and humidity by increasing the inside heat and humidity of the building in comparison with outdoor space, consequently creating a stuffy condition for the residents [1]. Therefore, the outdoor environment is affected by built-environment elements, too and the whole house behaves as a self-regulating thermal system [32]. These technologies utilise multifunctional resource quality to turn Iranian homes into paradise [48].

Consequently, the climatic component in Shiraz, from my point of view, provides more than passive strategies. They supply interactive strategies because of their interconnectivity with the natural environment and their effect on outdoor airflow or thermal comfort. They also prevent irretrievable damage to the environment by improving the microclimate. It is noted that creating heat balance between the human body and the surrounding environment is a significant comfort need offered by climatic components that supply interaction with the natural environment. Climate, similar to all complex entities, acts as a unit system and a field. It means every part affects another substance to change the whole [12]. Therefore, if climate as a natural environmental system tends to be durable and sustainable, it could concern every part and substance of the whole. The climatic component in Shiraz illustrated unity in diversity.

7 Conclusions

Recognition of architecture, climate, and culture as a whole, named existential climate, has taught us about having a desirable, sustainable, and durable life in urban texture. For instance, the natural environment needs to have some factors to make it possible for humans to live on the earth. Major parts of environmental factors and natural quantity, i.e., bisection winds swept through valleys and mountains, high-pressure air, faults, the flowing of fresh water through the ravine, etc., remain stable. These factors have been known as static environmental factors. The other parts of the environment are natural climate fluctuations, e.g., daily climate change, wind, and sunlight. This fluctuation extends beyond human comfort and tolerance and maintains vitality in life. There are dynamic environmental factors introduced in static and dynamic environmental factors that organise life systems [20]. Consequently, we need spaces in a static and dynamic manner simultaneously, to have sustainability and durability in life.

Nowadays, providing climate comfort at the cost of destroying the natural environment causes problems and makes sustainable architecture appear. As this architectural system is not compatible with the natural environment, the sustainable field of study appears in many sciences to solve environmental problems. Unfortunately, the construction method does not change. This field of study tries to help fix human architecture in nature, which has caused the natural environment to be destroyed in the past.

However, research [2, 8] proves that we need to gather the static and dynamic factors of nature and conserve them with designing urban architectures. This can be done by paying attention to natural and native features. Therefore, some original regulations in the architecture of the region must remain and we cannot transfer it or imitate it from one region to another region. In addition, we need to unify the orders of nature (water, soil, and air) with human needs and desires as a whole. This whole system contributes to sustainability and durability in life. When humans achieve harmony with the natural environment, a phenomenon greater than nature itself emerges, known as the "climate of existence". This is an illustration of a culture that gathers desirable orders for an accurate life. Thereafter, architecture shows not only human behavior and nature but also upgrades life knowledge and demonstrates wisdom by re-ordering our lives toward a more harmonious and Spirit-filled relationship with climate factors.

Furthermore, the implications of Iranian lessons for other regions in the age of global heating require digging into their underlying theories and practical technologies. The blind Imitation of Iranian technologies alone cannot address today's environmental problems. Knowing how to think and look at nature is also imperative.

References

1. Afzali K, Moarefzade A, Afzali K (2012) Particularity study of sustainable architecture in Iran traditional architecture. Int Sci J Arch Eng 1. http://architecture.scientific-journal.com
2. Ahmed CH, Rasul HQ (2023) Natural and architectural convergence: a model of nature-based strategies in the architectural design domain. Buildings 13:2015. https://doi.org/10.3390/buildings13082015
3. Barzegar Z, Heidari S (2017) A study of the functions of traditional house entrances' depth and shadow on outdoor thermal comfort in the historical context of Shiraz. J Arch Hot Dry Clim 5(5)
4. Berkes F, Folke C (eds) (1998) Linking social and ecological systems: management practices and social mechanisms for building resilience. Cambridge University Press, Cambridge, UK
5. Bosher L, Dainty A (2011) Disaster risk reduction and 'built-in' resilience: towards overarching principles for construction practice. Disasters 35(1):1–18. https://doi.org/10.1111/j.1467-7717.2010.01189.x Epub 2010 Aug 16. PMID: 20722695
6. Bourne L (1981) The geography of housing. Halsted Press, New York
7. Community & Regional Resilience Institute (2013) Definition of community resilience: an analysis. Meridian Institute
8. Couvelas A (2020) Bioclimatic building design theory and application. Procedia Manuf 44:326–333
9. Dashti S (2018) Designing a tourism complex in water stressed region with an approach to resilient design. Shiraz University

10. Ekhtiari et al (2022) An income on knowledge: Fars residential vernacular architecture, South of the Central Plateau of Iran (Fars). In: ICOMOS Iran scientific committee on vernacular architecture (CIAV-IRAN), NO 10 & 11/Summer & Autumn 1401/167 Pages
11. Ekhtiari M, Ashkani A, Rezayimanesh A (2019) Survey and analysis of historical building of Shiraz. Architecture center for education, culture and research, University of Qazvin
12. Ekhtiari M (2009) Quality of integral relationship between human needs and desire and architecture in all aspects. Dissertation, University of Tehran
13. Estaji H (2014) Flexible spatial configuration in traditional houses, the case of Sabzevar. Int J Contemp Arch 1(1):2014
14. Falamaki MM (2005) A review on urban repair "from Venicia to Shiraz." Faza Publication, Tehran
15. Farahani L, Lozanovska M (2015) The social life of historical neighborhoods: case study of a Middle Eastern city Shiraz. Int J Contemp Arch 39(3):176–187. https://doi.org/10.3846/20297955.2015.1088416
16. Field RD, Parrott L (2017) Multi-ecosystem services networks: a new perspective for assessing landscape connectivity and resilience. Ecol Complex 32:31–41. https://doi.org/10.1016/j.ecocom.2017.08.004
17. Folke C (2006) Resilience: the emergence of a perspective for social–ecological systems analyses. Glob Environ Chang 16(3):253–267. https://doi.org/10.1016/j.gloenvcha.2006.04.002
18. Ganor M, Ben-Lavy Y (2003) Community resilience: lessons derived from Gilo under fire. J Jew Communal Serv Winter/Spring:105–108
19. Ghodrati Amiri G, Razavian Amrei SA, Alaei H (2014) Uniform hazard spectra for the Iranian city of Shiraz. Asian J Civ Eng (BHRC) 15(6):817–829
20. Golami Y, Ekhtiari M (2015) Co-behaving of city and humans with nature in the way of human's sublimation and sustainability of city, case study Kashan. In: Paper presented at the 20th International congress in sustainability on architecture and urbanism, Masdar, Dubai
21. de Graaf-van Dinther R, Ovink H (2021) The five pillars of climate resilience. In: de Graaf-van Dinther R (ed) Climate resilient: urban areas governance, design and development in coastal delta cities. Rotterdam University of Applied Sciences, Rotterdam, The Netherlands, Palgrave MacMillan. https://doi.org/10.1007/978-3-030-57537-3
22. Holling CS (1973) Resilience and stability of ecological systems. Annual Rev Ecol Syst 4:1–23
23. Holling CS (2004) From complex regions to complex worlds. Ecol Soc 9:1–10. https://doi.org/10.5751/ES-00612-090111
24. Hosseini SN, Hosseini SM, HeiraniPour M (2020) The role of Orosi's Islamic geometric patterns in the building façade design for improving occupants' daylight performance. J Daylighting 7:201–221. https://doi.org/10.15627/jd.2020.18
25. Javaheri P, Javaheri M (1996) Challenge for water in ancient Fars, vol I. In: Iranian National committee on irrigation and drainage and water national museum of Iran
26. Javaheri P, Javaheri M (2001) Challenge for water in ancient Fars, vol II. In: Irainian National committee on irrigation and drainage and water national museum of Iran
27. Javaheri P, Javaheri M (2006) Challenge for water in ancient Fars, vol III. Fars regional water authority
28. Karim-zadeh J, Mahdi-nejad Darzi JA, Karimi B (2022) Climatic performance of traditional houses in the old texture of Shiraz using the thermal comfort approach case study: the Iwan (Veranda). J Iran Arch Stud 10(20):89–115. https://doi.org/10.22052/jias.2022.111875
29. Kasmaei M (2006) Climate and architecture. In: Ahmadinejad M (ed). Khak Publication, Isfahan
30. Kruk ME et al (2017) Building resilient health systems: a proposal for a resilience index. BMJ 357:j2323 https://doi.org/10.1136/bmj.j2323
31. Maghsoudi M (1998) Study of temperature changes in north coast of Persian Gulf. Roshd Amouzesh Geogr J 47:29–32. Retrieved 12 July 2015, fromhttp://www.noormags.ir
32. McCabe C, Roaf S (2012) The wind towers of Bastakiya: assessing the role of the towers in the whole house ventilation system using dynamic thermal modelling. Archit Sci Rev 56(2):183–194. https://doi.org/10.1080/00038628.2012.723398

33. Memarian G (1996) Introduction to Iranian residential architecture; introverted typology. Iran University of Science and Technology (IUST) Press, Tehran, Iran
34. Merriam-Webster (2015) Resilience. In: Merriam-Webster's collegiate dictionary. 11th ed., p 146, 165, 266
35. Mohammadi H, Ranjbar F, Moghabel M (2006) Assessment and comparison of climatic potentials for development the tourism activities in Chabahar and Shiraz. Tour Manag Stud 4 (11.12):41–56. 20.1001.1.23223294.1385.4.11.12.3..24
36. Moore F (1993) Environmental control systems: heating, cooling lighting. McGraw-Hill, University of Minnesota
37. Mosalanejad M (2021) Soil mechanics, classification types of soils in Shiraz city. Fars Construction Engineering Organization, Shiraz, Iran
38. Najafi N (2013) Iranian traditional architecture and energy saving (case study: Shiraz Ghajar houses). Int J Energy Environ 4(5):871–878
39. Nasiri Kianabad S (2019) Designing model of contemporary residential architecture in Shiraz based on space syntax analysis, case study: traditional architecture of Ghajar and Pahlavi houses. A master's thesis, Shiraz University (Under the direction of [Ekhtiari, Maryam])
40. Nematchoua MK, Tchinda R, Orosa, JA (2014) Thermal comfort and energy consumption in modern versus traditional buildings in Cameroon: a questionnaire-based statistical study. Appl Energy 114(C):687–699. https://doi.org/10.1016/j.apenergy.2013.10.036
41. Nikghadam N (2013) Patterns of semi-open spaces in vernacular houses of Dezful, Bushehr and Bandar-e-Lenge considering climate attributes [electronic version]. Honar-Ha-Ye-Ziba Memari-Va-Shahrsazi J 18(3):69–80
42. Norris FH, Stevens SP, Pfefferbaum B, Wyche KF, Pfefferbaum RL (2008) Community resilience as a metaphor, theory, set of capacities, and strategy for disaster readiness. Am J Community Psychol 141:127–150. https://doi.org/10.1007/s10464-007-9156-6
43. Nouhi K, Kamali G Mortazavi A (2008) Study and analyzing of heat tension effect on wheat in south province in Iran. Geogr Res J 88:74–51. Retrieved 12 July 2015, fromhttp://www.noormags.ir
44. Pesaran A, Kariminia S, Nazemi E, Toghyani S (2019) Evaluating the Existence of a design mechanism based on natural ventilation in the traditional architecture of cities in hot-arid climate without the use of windcatchers (case study: outer space of four-sided Qajar buildings in Shiraz). 7(9):101–119
45. Pirniakarim M (1999) Shiraz Gardens. Golestan-e-Honar 11:13–25
46. Pourjafar MR (2011) Growth of Bandar lenge and Kong. Tehran Times XVII(281):4
47. Raisi R, Vakilinezhad R,Khayam manesh M (2024) Daylight and thermal performance evaluation of Orosi; traditional colored window (case study: Kazeruni house in Shiraz). Int J Arch Eng Urban Plan 34(1). https://doi.org/10.22068/ijaup.839
48. Roaf S (2009) Resources quality cascades in traditional low energy technology: the Qanats and Badgirs of Yazd. In: 3rd International conference on smart and sustainable built environments (SASBE 2009), pp 15–19
49. Sadeghi AR, Bahadori Y (2021) Urban sustainability and climate issues: the effect of physical parameters of streetscape on the thermal comfort in Urban Public spaces; case study: Karimkhan-e-Zand Street, Shiraz, Iran. Sustainability 2021(13):10886 https://doi.org/10.3390/su131910886
50. Safaeipoor M, Shabankari M, Taghavi T (2013) The effective bioclimatic indices on evaluating human comfort (a case study: Shiraz city). Geogr Environ Plan J 50(2)
51. Saljoughinejad S, Rashidi Sharifabad S (2015) Classification of climatic strategies, used in Iranian vernacular residences based on spatial constituent elements. Build Environ 92 (2015)
52. Scotti Petrillo A, Prosperi DC (2011) Metaphors from the resilience literature: guidance for planners. In: Proceedings REAL CORP 2011 Tagungsband, 18–20 May 2011, Essen. http://www.corp.at ISBN: 978-3-9503110-0-6 (CD-ROM); ISBN: 978-3-9503110-1-3 (Print), Editors: Manfred SCHRENK, Vasily V. POPOVICH, Peter ZEILE

53. Shaeri J, Mahdavinejad M (2022) Prediction indoor thermal comfort in traditional houses of Shiraz with PMV/PPD model. Int J Ambient Energy 43(1):8316–8334. https://doi.org/10.1080/01430750.2022.2092774
54. Sharifi A, Roosta M, Javadpoor M (2021) Urban form resilience: a comparative analysis of traditional, semi-planned, and planned neighborhoods in Shiraz Iran. Urban Sci 5:18. https://doi.org/10.3390/urbansci5010018
55. Sharifi A, Yamagata Y (2016) Urban resilience assessment: multiple dimensions, criteria, and indicators. In: Yamagata Y, Maruyama H (eds) Urban resilience, advanced sciences and technologies for security applications. Springer International Publishing Switzerland. https://doi.org/10.1007/978-3-319-39812-9_13
56. Taghavi F, Mohamadi H (2007) Study of back period of extreme event to understanding environmental outcome [electronic version]. J Environ Stud 33(43):11–20
57. Tahbaz M et al (2008) Challenge of vernacular architecture and modern life style—case study in Iran. Paper presented at the 25th conference on passive and low energy architecture, Dublin
58. White I, O'Hare P (2014) From rhetoric to reality: which resilience, why resilience, and whose resilience in spatial planning? Eviron Plann C Gov Policy 32:934–950. https://doi.org/10.1068/c12117
59. Zare Mohazabie A, Shahcheraghi A, Shahin H (2022) Indoor environmental quality with an emphasis on thermal comfort in traditional houses, case studies: two Qajar houses in Shiraz. J Iran Arch Stud 5(9):85–100

Challenges of Climatic Modelling Persian Vernacular Architecture and Solutions to Overcoming Them

Rouhollah Rahimi

Abstract The rate, scale of the growth and variety of new data sources relating to the built environment are challenging. The modelling of vernacular buildings architecture with innovative programmes has the potential to provide more integrated design and analysis tools than were traditionally available. However, in an often traditional construction industry, the challenges is to be aware of what is on offer and identify which tools are practically best suited to a particular purpose. In this chapter, a range of functions of models used to analyse vernacular architecture are reviewed, prioritised, and categorised by use and usefulness. Optimisation tools for the reduction of energy consumption in buildings are widely used and here these tools, along with building energy simulation software are evaluated as potentially powerful tools to described and understand the climatic design characteristics of Iranian architecture. The limitations and capabilities of different optimisation algorithms within models in actually reducing the amount of energy consumption in the building are presented, with a view to understanding where the gaps in the use of optimisation and simulation algorithms exist, removing them or providing patches to those gaps to improve the accuracy of the models of the climatic functions of the often very complex performance of the traditional building in Iran. The study starts with a literature review of five widely used building energy optimisation and simulation programs. Their software criteria, applications, capabilities, and limitations were defined and compared. According to the opinions of optimisation experts in Iran, some challenges and potential solutions to overcome them were extracted by the Delphi method. The result of the research indicates that the lack of training and skill creation in the architectural community makes the use of these new tools very limited, so there is a need for optimisation experts as analysts. Therefore, it is necessary to take codified actions on the educational and professional levels to expand the knowledge of optimisation based on the needs of architects.

Keywords Climatic modelling · Persian vernacular architecture · Challenges · Solution · Iran

R. Rahimi (✉)
Department of Architecture, University of Mazandaran, Babolsar, Iran
e-mail: r.rahimi@umz.ac.ir

A. Cheshmehzangi and S. Roaf (eds.), *Persian Vernacular Architecture*, Urban Sustainability, https://doi.org/10.1007/978-981-96-1116-4_13

1 Introduction

Vernacular architecture refers to any type of architecture that belongs to a specific place. This includes architecture ranging from rural and urban buildings to large structures and monuments. According to the definition provided by Paul [14: 16], Vernacular architecture is :*architecture that grows from within communities and adapts itself over time to be compatible with social, climatic, and technological conditions, evolving in harmony with the values and ways of life of the cultures that produce it*". Vernacular architecture is architecture by and for the people, not imposed upon them [20]. Since the inception of architecture, Vernacular architecture has been striving to achieve harmony between nature and buildings based on local knowledge and experiences. Vernacular architecture utilises passive strategies dependent on the form and orientation of buildings, with a tendency to provide appropriate responses to climatic conditions, thereby ensuring human comfort [15].

Vernacular architecture in Iran possesses unique characteristics that not only consider environmental factors but also respond to the climatic needs of each region. The features of Vernacular housing can be observed in terms of physical, climatic, and cultural attributes in each area. Physical factors of indigenous architecture encompass all architectural elements influenced by the local climate. Indigenous housing, from a physical perspective, is influenced by the climate, and from a cultural perspective, it is influenced by the nationality, religion, and customs of its inhabitants. In this regard, understanding and examining the principles of indigenous architecture and its climate-appropriate features can contribute to improving the architecture of contemporary buildings [5].

The diverse and variable climatic and geographical conditions across this land have influenced architects to devise their best and most suitable adaptation methods to the climate and effectively utilise climatic conditions. These principles and patterns can now be integrated with new technologies in contemporary architectural structures, creating architecture that is both contemporary and sustainable while maintaining its unique identity. An appreciation of the valuable works of vernacular Iranian architecture across this vast land and in various scales demonstrates a comprehensive and purposeful approach to utilising climatic conditions in different structures. This approach, coupled with proper recognition and appropriate integration with contemporary architectural principles, can lead to the advancement of this architecture [9]. Among all environmental factors, controlling only physical environmental factors such as heat and light will lead to more desirable results in terms of thermal and visual parameters. In local and traditional buildings, the main function of the building envelope—while respecting physical environmental factors—is to provide thermal comfort by controlling the effects of climatic factors. Moreover, visual comfort is achieved by controlling natural light to understand the characteristics of traditional building in cases of heat, humidity, wind, and light, providing valuable experiences for current designs, which is the result of gradual evolution over a long period of time [16].

The main question and the final goal of the research is whether it is possible to define indicators for the application of modern modeling technologies in the direction of analysing Persian Vernacular Architecture while preserving the identity foundations of Iranian-Islamic architecture.

In the first stage, this chapter tries to create a theoretical framework and a method for conducting research. This is done by taking a general view and considering different ideas used in important projects of Vernacular architecture in Iran according to the main goals of this research. Secondly, in the following, the basic principles of native Iranian architecture in traditional Iranian buildings are analysed in order to achieve the main factors of desirability in these buildings and their application in contemporary architecture. In the next step, successful and famous examples of architecture built and implemented in different sites are selected, and considering these examples, the research hypothesis is qualitatively examined. A number of questionnaires have been prepared with questions from selected case samples and the main objectives of the research. Also, to check the research hypothesis, the main experts in the research field are interviewed. Based on the results, the main criteria—both qualitative and quantitative—will be evaluated and a model for the implementation of innovative modeling technologies in contemporary architecture will be presented.

In this way, by selecting case examples, doing the work Fieldwork, and surveying people using targeted questionnaires, indicators are obtained from the results of surveys qualitatively and quantitatively. An analysis and a model for the application of new technologies in Persian Vernacular Architecture are presented. Then, using a qualitative approach, it will be done with information modeling experts of Persian Vernacular Architecture in order to identify its challenges. The collected data are analysed with the theme analysis method and the most important challenges of using Persian Vernacular Architecture modeling and solutions to reduce or remove the identified challenges and obstacles are suggested.

According to the results of this research, the use of modeling technology considering the main characteristics of Iranian architecture is one of the necessities of contemporary architecture. Neither is meant to be considered separately. This research also suggests that in the modeling of Iranian buildings, the climatic and socio-cultural value of every part of Iran should be carefully considered. Considering all these factors, it is believed that a high degree of desirability can be achieved in today's architecture.

2 Background Knowledge

Numerous studies have been done on the issue of natural ventilation and its role in buildings. Alongside these efforts, Computational Fluid Dynamics (CFD) has become a very powerful and popular tool in building simulation and has been widely used to simulate indoor and outdoor airflow, heat transfer, containment transport [6] and in a wider sense, natural ventilation performance of buildings [3]. Van

Hooff and Blocken [18] categorise different methods to assess natural ventilation into three sections: full-scale measurements, reduced-scale measurements and CFD, and describe the advantages and disadvantages of each. As they have argued in their paper [18], one of these advantages is that CFD simulations easily allow parametric studies to evaluate alternative design configurations, especially when different configurations are all a priori embedded within the same computational domain and grid, which the most principal reason for using CFD simulations in this chapter.

Several studies have been done on Iran's vernacular architecture using CFD simulations on single buildings [2, 7]. For example, Montazeri and Azizian [10] have studied the performance of one-sided wind-catcher using CFD and Montazeri and Montazeri [11] have studied the performance of two-sided wind-catchers. At the international level, Mora-Pérez and Guillen-Guillamón [12] evaluated the façade opening distribution impact on natural ventilation performance in Valencian coastal region and Liu and Huang [8] tried to find out the natural ventilation values and limitations of "Yinzi" dwellings as well as designing an optimum ventilation strategy for such dwellings in their paper.

2.1 Computational Wind Engineering (CWE)

According to Blocken [1] "*in the last 50 years, Computational Wind Engineering (CWE) has undergone a successful transition from an emerging field into an increasingly established one in wind engineering research, practice and education*". As one of the many branches in this field, Computational Fluid Dynamics or CFD has become a very powerful and popular tool in building simulations and has been widely used to model indoor and outdoor airflow [6], especially in the last two decades, due to increased computing hardware capacity as an integral part of scientific research [4]. CFD method is applied in three main fronts [1]:

1. CFD simulation of pedestrian-level wind conditions around buildings;
2. CFD simulation of natural ventilation of buildings; and
3. CFD simulation of Wind-driven rain in building façade.

CFD is particularly suited for simulating natural ventilation conditions in indoor spaces [3];

A process that utilises natural outside air flow to both cool and ventilate a building passively, without the use of any electro-mechanical system [17]. CFD can be used to assess indoor natural ventilation by solving the interaction between the surrounding wind flow and the indoor airflow [18].

Modeling tools for Iranian vernacular architecture can be divided into two general categories: independent tools that only perform modeling operations and modeling tools that are built based on specific simulation software. The most widely used are Energy Plus, Spark, TRNSYS, MATLAB, GenOpt, ModeFrontier, BeOpt, Opt-E-Plus, Grasshopper plugin in Rhino and Dynamo plugin in Revit.

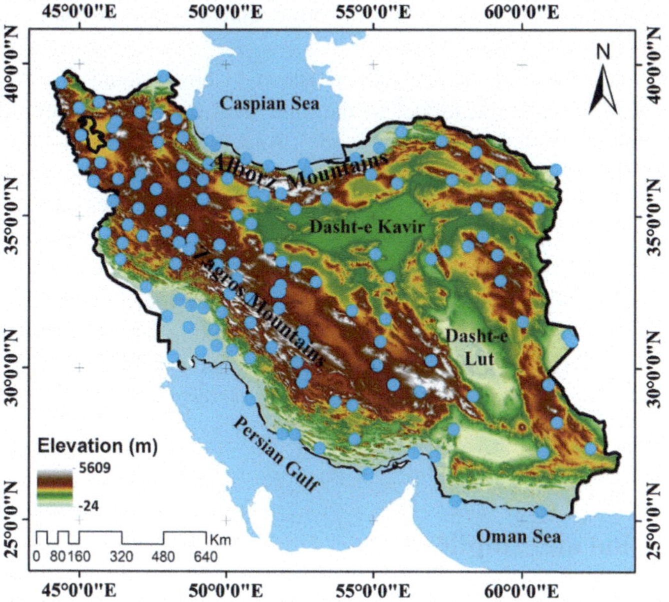

Fig. 1 Climatic zoning of Iran. *Source* Redrawn by the author and adapted from the original source available at Najafi and Alizadeh [13]

3 Methodology

This study relies on qualitative methodology consist of the investigation, library research and analysis of the outcomes. The method in this study consists of two sections based on descriptive methods and field studies along with a comparative approach. In the first section, library-based studies and documentation review are the main sources for expressing, analysing, describing, and interpreting the materials in this chapter. Also, the Delphi method was used in order to receive and analyse the opinion of experts in the field of Iranian native architecture. The second section contains a comparative analysis of the library- based findings in the first part with field studies and analyses the circumstances leading to the results in this study (Fig. 1).

4 Challenges and Solutions in the Use of Modeling in the Analysis of Persian Vernacular Architecture

In this research study, the main objective has been addressed by examining past studies and identifying the existing challenges in the use of modeling in the analysis of vernacular architecture. This primary goal is to determine the possibility of utilising modeling technologies to analyse Persian vernacular architecture while preserving its Iranian-Islamic identity foundations.

It can be mentioned obstacles such as speed of energy analysis; accuracy in transferring the architectural model to the analytical model; compatibility of the architectural model with the analytical model; Analytical model compatibility with simultaneous multi-branch calculations; Analytical model modeling speed; Coordination of inputs and outputs of the analytical model and also Weakness of suitable tools for modeling; limitations in the time and approach of the expert; and the need for complete knowledge of the problem such as constraints, objective function and detailed list of available solutions. In relation to the mentioned obstacles, solutions such as defining geometry in modeling software; Defining inputs of modeling software; generalisation of useful information in the decision-making stage; Evaluation and comparison of building energy performance; Interpretation of results; Smart repetition can be suggested (Table 1).

5 Discussion and Implications

Climatic modeling of Persian vernacular architecture presents several challenges due to the diverse climate conditions found throughout the region, as well as the intricate architectural features specific to Persian traditional buildings. Here are some challenges and potential solutions to overcome them:

Variability in Climate: The Persian region experiences a wide range of climatic conditions, from arid deserts to humid coastal areas. Modeling such diverse climates accurately can be challenging.

Solution: Use a combination of regional climate data, historical weather records, and advanced climatic modeling techniques to capture the variability in climate patterns accurately. High-resolution climate data from meteorological stations and satellite observations can also provide valuable inputs for the models.

Complex Architectural Features: Persian vernacular architecture is known for its intricate designs, including passive cooling systems, wind catchers (badgirs), and courtyard layouts, which may not be easily incorporated into traditional climatic models.

Solution: Collaborate with architects, historians, and local communities familiar with traditional building techniques to accurately capture the architectural features in the models. Incorporate detailed building material properties, orientation, and shading elements into the simulations to assess their impact on indoor comfort and energy performance.

Limited Data Availability: Data scarcity, especially regarding historical buildings and construction materials, can hinder the development of accurate climatic models.

Solution: Conduct field surveys and collect empirical data on traditional buildings, including building materials, construction techniques, and environmental conditions. Collaborate with local universities, research institutions, and heritage organisations to gather historical data and documentation on vernacular architecture.

Table 1 Challenges and solutions in the use of modeling in the analysis of Persian vernacular architecture

Variables	A physical criterion affecting climatic condition	Climatic conditions				Challenges	Solutions
		Mild and humid	Cold	Warm and dry	Warm and humid	Variability in Climate; Complex Architectural Features; Limited Data Availability; Scale and Resolution; Incorporating Traditional Building Techniques; Modeling Complex Geometries; Accounting for Microclimate Variations; Validating Design Performance; Accessibility and Usability; Cultural and Contextual Sensitivity	The scope of the software database and its ease of access; Ability to connect with other tools; User interface; Ability to graphically display and understand tool outputs (post-processing); Optimisation parameters; Ease of use for architects

Temperature	Orientation materials modeling the plan massing (form compression) height difference Porch Occupancy level Elongation The area of the central courtyard Opening area wall thickness roof form Canopy depth Kitchen place	East-west extension Use of materials with high heat capacity Roof = sloping and light roofs with thermal insulation Materials = (straw)	Use of materials with high heat capacity compact form Square plan flat roof Darker material	Use of materials with high heat capacity Dense and compact plan Low surface to volume ratio shading White material Kolak, Kharkhneh, Dorcheh	Use of materials with low heat capacity Transferring the air flow reduces the temperature and creates thermal comfort in the indoor environment = Kolak, Kharkhaneh, Dorcheh		

(continued)

Table 1 (continued)

Humidity		Ventilation East-west orientation Extensive and extroverted texture Not using the basement Terrace and porch	Below the passage Larger openings on the south area	Guiding the flow of wind and using clean energy in nature, in adjusting the temperature and bringing the temperature of the living space to a temperature that is comfortable for humans = wind girder, kolak, kharkhane, Dorcheh Central courtyard planting	Guiding the flow of wind and using clean energy in nature, in adjusting the temperature and bringing the temperature of the living space to a temperature that is comfortable for humans = wind girder, kolak, kharkhane, Dorcheh		

Scale and Resolution: Traditional climatic models may lack the spatial and temporal resolution required to capture microclimate variations and localised effects around buildings.

Solution: Utilise computational fluid dynamics (CFD) simulations and high-resolution modeling techniques to analyse microclimate variations around buildings. Incorporate site-specific topography, vegetation, and urban morphology into the models to account for localised climatic influences.

On the other hand, we question what are the challenges of climate design software for the design of vernacular architectural spaces? Climate design software for the design of vernacular architectural spaces faces several challenges due to the unique characteristics of traditional buildings and the complex interactions between architectural features and climatic conditions. Here are some of the challenges:

Incorporating Traditional Building Techniques: Vernacular architecture often employs passive design strategies and traditional building techniques that may not be adequately represented in conventional design software.

Solution: Develop specialised modules or plugins within existing design software that account for traditional building materials, construction methods, and passive cooling/heating strategies specific to vernacular architecture. This may involve integrating knowledge from architectural history and ethnographic studies into the software algorithms.

Modeling Complex Geometries: Traditional vernacular buildings often feature intricate geometries, such as courtyards, wind towers, and shaded arcades, which can be challenging to model accurately in conventional design software.

Solution: Enhance the geometric modeling capabilities of design software to accommodate complex architectural forms and irregular building layouts characteristic of vernacular architecture. This may involve incorporating parametric design tools or geometry optimisation algorithms tailored to vernacular building typologies.

Accounting for Microclimate Variations: Vernacular architecture responds to local climate conditions and microclimate variations, which may not be adequately captured by standard climate data sources used in design software.

Solution: Integrate microclimate modeling capabilities into design software to simulate localised weather patterns, solar radiation distribution, and airflow dynamics around vernacular buildings. This may require coupling design software with computational fluid dynamics (CFD) simulations or utilising high-resolution climate datasets for specific geographic regions.

Validating Design Performance: Evaluating the performance of vernacular architectural designs using climate design software requires validation against empirical data and real-world performance metrics.

Solution: Develop standardised validation protocols and benchmarking datasets for assessing the performance of climate design software in simulating vernacular building performance. This may involve conducting field measurements, post-occupancy evaluations, and performance monitoring of traditional buildings to validate software predictions.

Accessibility and Usability: Design software tailored to vernacular architecture should be accessible and user-friendly for architects, designers, and builders with varying levels of technical expertise.

Solution: Prioritise user interface design and usability testing to ensure that climate design software for vernacular architecture is intuitive and easy to use. Provide comprehensive documentation, tutorials, and training resources to support users in effectively applying the software to their design projects.

Cultural and Contextual Sensitivity: Vernacular architecture is deeply rooted in local cultural traditions, social practices, and environmental contexts, which may not be adequately addressed by generic design software solutions.

Solution: Incorporate cultural heritage considerations and contextual sensitivity into the development of climate design software for vernacular architecture. Engage with local communities, architects, and heritage conservation experts to ensure that the software respects and reflects the cultural identity and socio-environmental context of traditional building practices.

By addressing these challenges, climate design software can better support the design and evaluation of vernacular architectural spaces, enabling architects and designers to create sustainable, climate-responsive buildings that preserve cultural heritage and harmonise with local environments.

6 Concluding Remarks

Built environment's ventilation consumes a large portion of energy in today's world, which consequently leads to a great amount of pollution, especially in less-developed countries with less clean technologies. At the same time, vernacular architecture has come up with numerous effective techniques and elements to use natural wind induced ventilation to provide a comfortable setting for dwellers and reduce the energy consumption. Different types of climate divisions are prepared with different goals. Therefore, in any type of division according to certain components in line with the goal, some factors are not taken into consideration. Therefore, a division may be suitable for one purpose and not suitable for another purpose. Many climate divisions have been made by botanists and geographers based on elements such as vegetation, average monthly temperature and annual rainfall. Considering that in architectural decisions and designs, based on the type of climate, decisions are applied in the design of building components and mechanical systems, the existence of suitable climatic zoning and the correct identification of the climatic characteristics of the building location, from It is very important.

In this chapter, the types of climatic divisions of Iran are considered in relation to the building and architectural goals, and considering the shortcomings in these divisions, the need to create climatic zoning according to the appropriate thermal comfort range in the building. Benefits of natural ventilation have been proposed. Among the shortcomings of existing division systems in building design applications, we can point out the lack of attention to the difference in the range of thermal comfort and the amount of wind driving forces in different areas. Considering the importance of using natural ventilation as one of the most efficient methods of building cooling, this study proposes a new method to create a new climate zoning of Iran according to the range of adaptive thermal comfort and based on the potential of natural ventilation.

It should be noted that the potential of natural ventilation to create cooling depends not only on the climatic characteristics but also on the characteristics of the building. Due to the effect of the building characteristics on its thermal behavior, the potential of using ventilation in an office building is different compared to a residential building. In addition to the difference in the amount of natural ventilation potential according to the type of use, the optimal times for the use of ventilation are also different. In this way, in order to achieve more accurate results, it is possible to create climatic zoning according to different uses (residential, educational, administrative, etc.). Applying this method to Iran, by presenting a realistic picture of the existing potentials of natural ventilation, is a useful help to architects to design buildings benefiting from natural ventilation and will have the minimum need for mechanical cooling systems and reducing energy consumption.

Based on a study conducted by Welle et al. [19], the barriers to energy efficiency are as follows:

1. Long analytical model preparation times;
2. Inaccurate conversion of architectural model to analytical model;
3. Inconsistent conversion of architectural model to analytical model;
4. Missing or invalid data in the architectural model;
5. Inconsistent analytical models for multidisciplinary analysis;
6. Long analytical model simulation times;
7. Poor coordination of analytical model outputs/inputs;
8. Inconsistent performance metrics [19: 295].

According to the achievements of various researchers regarding the limitations of simulation tools and the needs of architects, these needs are divided into six categories:

1. **The scope of the software database and its ease of access**

One of the points that is used in the efficiency of modeling software is the comprehensiveness of their database. This database can play a decisive role in the proper definition of model variables and solve the user's needs in this regard in the software itself. The issue of database scope and ease of access is divided into the comprehensiveness of the database and the intelligent and automatic use of available information in various modeling and design solutions.

2. **Ability to connect with other tools**

The ability to receive and send information between the modeling tool and other software, including geometric modeling software and other simulation tools, is another criterion that has a significant impact on the efficiency of the modeling tool.

3. **User interface**

In order to use climate modeling tools as best as possible, especially for architects, it is necessary to have a user-friendly environment for building modeling and defining its parameters.

4. **Ability to graphically display and understand tool outputs (post-processing)**

Providing graphic outputs helps to understand and analyse the results of the modeling process, so having the ability to create different charts and display different charts has a noticeable effect on the software's efficiency.

5. **Optimisation parameters**

Regarding the modeling parameters, there should be the ability to change, the ability to calculate parallel, as well as the ability to optimise cooling, heating, lighting, natural ventilation and air conditioning systems, and the possibility of changing variables by the user.

6. **Ease of use for architects**

The meaning of simple use for architects is the lack of knowledge and skills beyond what is normally available to architects, including the ability to program computer and adjust and select modeling software for specific purposes.

In conclusion, increasing the capabilities of modeling software to help decision-making; creating awareness of the application and method of using modeling software; Understanding the difference between the climate modeling process and the architectural design process and how to simplify the model in the early stages of design and coordinate with the architect's needs at the beginning of the design process are the solutions to remove the obstacles in the climate modeling of Iranian vernacular architecture. Also, I suggest that due to the advancement of technology and the development of the application of optimisation software, it is necessary to use architects with optimisation expertise, who are proficient in optimisation software specific to historical buildings, to help the project design team as analysts; Because various variables such as the orientation and form of the building, the amount and intensity of sunlight and rainfall, and the materials used in the building are important in the optimisation analysis.

A key issue is that the very sophisticated methods for passive heating and cooling in traditional buildings might be incorporated into modern buildings if they are understood and can be modelled to work well in new buildings too. But this may require an architectural and engineers' team to create. It is a good use for the analysis to understand how traditional buildings work then architects can simply apply the principles of their working to new designs, not model them. Further work can be done to enhance the chapter aligned with the lessons learnt from Persian Vernacular Architecture due to the variety of climate in Iran. These lessons can help people adapt to our rapidly heating climate in the world while something that has been missing when we teach/educate/practice architectural history, vernacular architecture, sustainable buildings, climate-resilience, etc. Western models can be touch on the sophistication of systems like Wind towers, Qanats, Badgirs (Windcatchers) and etc. Also, it can be trying to understand how traditional buildings work using models developed for very different traditional buildings in very different climates and societies (privacy) to help us design modern buildings and use them in every city for a wide range of twenty-first century purposes.

References

1. Blocken B (2014) 50 years of computational wind engineering: past, present and future. J Wind Eng Ind Aerodyn 129:69–102
2. Calautit JK, Hughes BR (2014) Measurement and prediction of the indoor airflow in a room ventilated with a commercial wind tower. Energy Build. 84:367–377
3. Chen Q (2009) Ventilation performance prediction for buildings: a method overview and recent applications. Build Environ 44:848–858
4. D'Agostino D, Congedo PM (2014) CFD modeling and moisture dynamics implications of ventilation scenarios in historical buildings. Build Environ 79:181–193
5. Eslami Mahmoodabadi S, Mofidi Shemirani SM, Habib F (2021) Investigating the principles of indigenous housing architecture of hot and dry climate of Iran with the formal impact approach of the building in promoting thermal comfort of building users. Islam Art Stud 17(40):21–39
6. Hajdukiewicz M, Geron M, Keane MM (2013) Formal calibration methodology for CFD models of naturally ventilated indoor environments. Build Environ 59:290–302
7. Kobayashi T, Sandberg M, Kotani H, Claesson L (2010) Experimental investigation and CFD analysis of cross-ventilated flow through single room detached house model. Build Environ 45:2723–2734
8. Liu S, Huang C (2016) Optimization of natural ventilation of "Yinzi" dwellings in western Hunan based on orthogonal experiment and CFD. Buildings 6:25
9. Molanaei S, Soleimani S (2016) Insight into the valuable elements of Sistan local architecture in relation to climatic factors of sustainable architecture. Mon Sci J Bagh-e Nazar 13(41):57–66
10. Montazeri H, Azizian R (2008) Experimental study on natural ventilation performance of one-sided wind catcher. Build Environ 43:2193–2202
11. Montazeri H, Montazeri F, Azizian R, Mostafavi S (2010) Two-sided wind catcher performance evaluation using experimental, numerical and analytical modeling. Renew Energy 35:1424–1435
12. Mora-Pérez M, Guillen-Guillamón I, López-Patiño G, López-Jiménez P (2016) Natural ventilation building design approach in Mediterranean regions—a case study at the Valencian coastal regional scale (Spain). Sustainability 8:855
13. Najafi MS, Alizadeh O (2023) Climate zones in Iran. Meteorol Appl 30(5). https://doi.org/10.1002/met.2147
14. Oliver P (2006) Built to meet needs: cultural issues in vernacular architecture, 1st edn. Routledge. https://doi.org/10.4324/9780080476308
15. Silvayeh S, Asefi M (2019) Assessment of factors influencing form in vernacular architecture; in cold and mountainous climate. JRIA 7(3):17–35
16. Soltanzadeh H, Ghaseminia M (2022) Climatic building envelope employed in vernacular residential architecture in Golestan-Iran. Int J Arch Urban Dev 6(1):53–64
17. Tong Z, Chen Y, Malkawi A (2016) Defining the influence region in neighborhood-scale CFD simulations fornatural ventilation design. Appl Energy 182:625–633
18. Van Hooff T, Blocken B (2010) Coupled urban wind flow and indoor natural ventilation modelling on a high-resolution grid: a case study for the Amsterdam ArenA stadium. Environ Model Softw 25:51–65

19. Welle B, Haymaker J, Rogers Z (2011) ThermalOpt: a methodology for automated BIM-based multidisciplinary thermal simulation for use in optimization environments. Build Simul 4:293–313
20. Zarghami I, Sadat SA (2017) Indigenous technology and architecture. Shahid Rajaee Teacher Training University, Tehran, Iran

Analysis of Traditional Iranian Vernacular Architecture in Response to Sunlight: An Approach to Sustainable Development

Mahdiyeh Tabatabaei

Abstract The geographical uniqueness and extreme climatic conditions of Iran predispose the country to great potential for energy conservation through the exploitation of solar energy. This free, available, and renewable resource important for environmental preservation has found its utilisation from the past in traditional Iranian architecture to optimise energy use and provide thermal comfort. Important architectural elements, including central courtyards, proper building orientation relative to the sun, and window dimensions, among others, have been all taken into account for designing around the different climatic conditions of different regions. For instance, this research compared traditional buildings' solar strategies in two selected cities with two climatically different climates: hot-dry (Yazd) and cold-dry (Tabriz). It is shown how these strategies can effectively reduce energy consumption and improve residents' quality of life. The research is based on qualitative and analytic approaches to the examination of solar architectural strategies in traditional Iranian buildings. Data were collected from library sources using comparative methods. This involved the selection of samples of hot-dry and cold-dry climate traditional architecture, which were analysed for elements and techniques. The results indicate that coupling architectural design with the climatic parameters of any region would indeed save a significant amount of energy while providing a desirable level of living comfort for the residents. In the case of a hot-dry climate like Yazd's, thermal comfort has been ensured through wind catchers, materials with high thermal capacity, and compact buildings. In contrast, the compactness of design, use of high thermal mass materials, and capturing solar radiation in Tabriz's cold-dry climate during winter months are beneficial. Results from the current study emphasise the application of climate-responsive design in realising sustainable living environments. Modern architects and designers, by adapting some principles from traditional Iranian architecture, are able to create modern buildings that are energy-efficient and environmentally friendly. This traditional knowledge, when integrated with modern design practices,

M. Tabatabaei (✉)
Department of Architecture, Field of Architecture and Design Cultures, University of Bologna, Bologna, Italy
e-mail: mahdiyeh.tabatabaei2@unibo.it

A. Cheshmehzangi and S. Roaf (eds.), *Persian Vernacular Architecture*, Urban Sustainability, https://doi.org/10.1007/978-981-96-1116-4_14

can lead to the creation of buildings that are not just sustainable but also resilient in confronting the challenges posed by climate change.

Keywords Passive solar design · Traditional Iranian architecture · Sustainability · Energy efficiency · Climatic adaptability · Renewable energy

1 Introduction

Relying on their great civilization and extensive experience in living in regions with a variety of climates, Iranians have been pursuing the use of natural resources, including water, wind, earth, and fire, for making environmental comfort for centuries. In this way, sustainable architecture unique for its time has emerged. Moreover, Iranians have been pioneers in sustainable design thanks to the employment of various architectural styles and natural principles.

In the context of current global challenges related to climate change and the energy crisis, the relevance of traditional Iranian architecture has risen. Its compatibility with the environment and ability to save natural resources make it a valuable model for modern sustainable development in architecture. This chapter, therefore, looks into the design and architectural strategies in the traditional structures of the hot-dry and cold-dry regions of Iran. This study focuses on how passive elements of architecture, such as central courtyards, wind catchers, thick walls, and small windows, contribute to energy efficiency and optimisation. Moreover, strategies such as good building orientation in relation to the sun, choice of building materials suitable for the climate, and controlling indoor temperatures to minimise reliance on mechanical systems ensure that buildings naturally adapt to the prevailing climatic conditions.

The fact that many modern buildings are highly dependent on energy-intensive systems, leading to rising energy costs and environmental impacts, is the issue this study addresses. It is likely that such dependence could be reduced and better energy efficiency achieved through the study and implementation of traditional architectural practices.

The research investigates if regional variances in solar architectural strategies have different environmental and energy-efficient results. Is it possible to apply those ancient principles innovatively in modern buildings? These are the questions that we aim to answer in the following sections through detailed examinations of passive solar architectural elements and their adaptation to different climatic conditions. We will clarify the subject with detailed case studies, analyse data using SWOT tables, and assess the challenges of applying these principles to modern architecture. Finally, conclusions will be drawn, and a vision for sustainable modern architecture will be presented.

The structure of this study is comprised of a few central elements. We begin by analysing theories and concepts regarding solar architecture and sustainable design principles. Following that, the research will present its analysis of examples of traditional Iranian architecture in various regions. Collected data will be analysed through

SWOT analysis, and conclusions and recommendations for modern applications will follow.

Studying and learning from solar architecture in Iran is particularly significant at this time when the world is seeking sustainable solutions to address climate change and rising energy costs. This approach not only addresses widespread climate challenges but also proves imperative in promoting energy-efficient construction in modern communities.

2 The Role of Sustainable Development in Solar Architecture

2.1 Vernacular and Traditional Architecture

In the past, before architecture was shaped by specialised architects with technical knowledge, it was an integral part of daily life and human creativity. This type of architecture emerged without specific regulations. Over time, it evolved into a richer and more regulated form of architecture. Therefore, vernacular architecture represents a more advanced form of popular architecture, shaped by the climate, culture, and building materials of a specific region [29]. In Iranian vernacular architecture, buildings are in complete harmony with nature and the environment and are constructed based on national-local culture and identity. The form, materials, and structure of these buildings are designed to maximise the use of natural elements and renewable energy sources. The architectural traditions that have developed over the centuries ensure the sustainability of vernacular architecture.

Traditional Iranian architecture, with over 6000 years of history, spans a vast area from Syria to northern India and from the Caucasus to Zanzibar [24]. Many scholars of traditional Iranian architecture believe that the first classification of buildings was based on two general principles: introversion and extroversion. This classification was influenced by various factors such as culture, society, climate, and regional weather conditions. Given that Iranian culture and society lean more towards introversion, and a large part of Iran's territory lies in hot-dry or cold regions, most traditional buildings are designed based on the principle of introversion. Extroversion is more commonly seen in areas with high humidity [31].

As mentioned earlier, vernacular architecture, which is based on an experiential rather than a scientific approach, meets people's needs in relation to natural factors and their material and spiritual desires. It is born out of their participation and inspired by their daily lives, thus belonging to its time. Vernacular architecture is devoid of ostentation and external ornamentation, without a specific style, and retains the characteristic of improvisation [10]. Therefore, from its inception, vernacular architecture has had features that indicate a close relationship with the climate and nature.

2.2 The Principles of Sustainable Development in Architecture and Their Role in Iranian Vernacular Architecture

Sustainable development in architecture is the creation of buildings and living environments that people not only need but also emphasises the protection of the environment and the conservation of natural resources [32]. Sustainable design works on the efficiency of resources, reduces energy consumption, and maintains the durability of structures. Norman Foster, a reputable architect, associates' sustainability in architecture with the quality of thought and ideas related to the environment [21]. Some of the key principles in sustainable architecture are climate-responsive design, the use of renewable energy, and the enhancement of human comfort [1]. In order to categorise a building under the sustainable category, it has to be bound by principles, which include energy conservation, climate harmony, reduced use of new resources, meeting occupants' needs, site compatibility, and a holistic approach [9]. The application of these principles needs to be participants in a comprehensive process of designing the built environment while considering the respect towards the natural world.

Sustainability is a call for the protection of the natural environment as well as the balancing act of the ever-growing needs of humanity. In this respect, the basics of traditional architecture come from ancient methods and cultures that give respect to nature and the environment. Iranian vernacular architecture is very well justified by sustainable principles in design. This architecture, especially in its traditional forms, emphasises the possibility of maximising sunlight in wintertime and providing proper shade during the summertime. This enhances natural ventilation for better comfort of the inhabitants. In traditional Iranian architecture, energy conservation is stressed, as are non-fossil fuel uses and renewable energy sources. It harmonises with the local climate and uses regional materials that are renewable and recyclable. The designs and practices in traditional architecture not only preserve national identity but also address the daily and seasonal needs of the residents.

Furthermore, traditional Iranian architecture incorporates structural elements, which might be classified as arches, domes, and minarets, among others, which also do not only offer aesthetic value to the house but are some of the reasons the houses last and may be sustainable in nature. Such an architecture has been highly successful in the use of renewable sources of energy, such as wind and solar energies, through special strategies of design including, for instance, wind catchers and central courtyards that play a critical role in both energy consumption optimisation and the provision of thermal comfort. The same goes for traditional Iranian architecture, which enables adaptation to environmental and climatic conditions by providing solutions to creating a comfortable living environment while assuring an optimised use of energy. With intelligent spatial organisation and elements such as seasonal rooms, natural ventilation, and solar control, the traditional Iranian architectural model provides a modern architecture model whereby energy consumption is reduced and cultural identity is preserved [32].

In summary, the principles of sustainable architecture can deeply be traced to traditional Iranian architecture. It was the architecture that respected nature and benefited from natural resources, which led to the creation of habitable and sustainable environments. These principles will offer invaluable lessons for contemporary architects willing to design harmoniously with the environment.

3 Architectural Strategies in Dealing with the Sun: Adaptation in Hot-Dry and Cold-Dry Regions

Studying vernacular buildings in each climate reveals that all these structures are designed and constructed based on climatic principles to maximise the use of natural energies and combat extreme temperatures [29]. In Iran, buildings are designed according to different climates with the aim of heating in winter and cooling in summer. This type of architecture focuses on energy conservation and human comfort in the most efficient way possible. The goal of this type of architecture is to use passive solar systems, as was common in traditional Iranian architecture [17].

In this context, this section emphasises how solar energy is utilised in the two mentioned climates, which is employed in traditional Iranian architecture. These systems enable us to provide innovative and effective solutions for sustainable design and dealing with climatic changes by drawing inspiration from the past.

3.1 Analysis of Climates and Architectural Strategies of Vernacular Architecture in Iran

3.1.1 Diverse Climate Zones of Iran

There are several climate zones in Iran due to this extensive geographical diversity and varied topography, though several studies have, therefore, tried to classify the climate of Iran into varied zones. For instance, Najafi and Alizadeh [22] identify six main climatic regions in Iran, and these include temperate and humid, cool and semi-humid, cold and semi-arid, warm and semi-arid, cool and dry, and lastly hot and hyper-arid zones. Further, Khatibi and Saberi [15] used multivariate statistical analysis to divide Iran into 26 bioclimatic zones unique to the country. Another study is by Raziei [27] where the author makes use of the classification system known as Köppen-Geiger, which indicated ten distinct types of climates, which included, among others, the hot desert, cold desert, Mediterranean, warm semi-arid, and cold semi-arid, with an extent of reaching more than 98% of the country.

Climate change has also had far-reaching impacts on Iran and has contributed to the development of arid and hyper-arid zones, whose expansion is projected to increase in the future [26, 33]. According to the world's bioclimatic classification,

Iran lies within three macrobioclimates, ten bioclimates, as well as three bioclimatic variants, which are equivalent to the major phytogeographical regions of the country [5]. From the architectural point of view, the climate in Iran can be broadly characterised in four categories: temperate and humid, hot and humid, very cold with long winters, and extremely hot and dry summers. Moreover, findings from recent research also discovered that the high altitudes, which are colder in the regions of Iran, experience increased heat, while the level of precipitation has generally reduced in most parts of Iran.

The climate of Iran is typically divided into four major categories: hot-dry, cold-dry, hot-humid, and temperate-humid. Of those, the hot-dry, with the cold-dry climates, are the most important since for these regions, adaptation and management of solar radiation is a key factor.

3.1.2 Research Methodology

The workflow of the present study includes the above phases to systematically examine and analyse the design strategies applied in the selected climates. This research mainly uses descriptive and analytical methods with an applied goal, and its key aim is to identify historical patterns and successful strategies in Iranian vernacular architecture, ultimately providing logical reasoning for its revival in contemporary architecture. The design strategies of the buildings initially in cold-climate areas were analysed. For cold and dry, Tabriz was selected, while for the hot-dry climate, Yazd was selected. The texture and structure of these two cities were analysed in terms of their climate and adherence to any architectural principles that are most suitable in the case of solar radiation.

After a general comparison, the proposed strategies and principles are evaluated in a more refined manner with SWOT analysis to gather insight into the adaptation of buildings with the climates of regions and principles ready for practical solutions for modern projects. This comprehensive approach allows us to derive sustainable architectural solutions that are peculiar to the unique climatic conditions of any given region and which guarantee application in contemporary settings (Fig. 1).

3.2 Investigating Solar Utilisation Strategies in Cold-Dry Climate

3.2.1 Climatic Conditions of Tabriz

As analysed by Shaghaghi and Mofidi [29], in the region of Tabriz, winters are cold and long; during most months, the ground is usually covered with ice and snow. The scanty rainfall of this season withers summers because it comes in the form of snow, given that the western mountain ranges of Iran block any moist Mediterranean air.

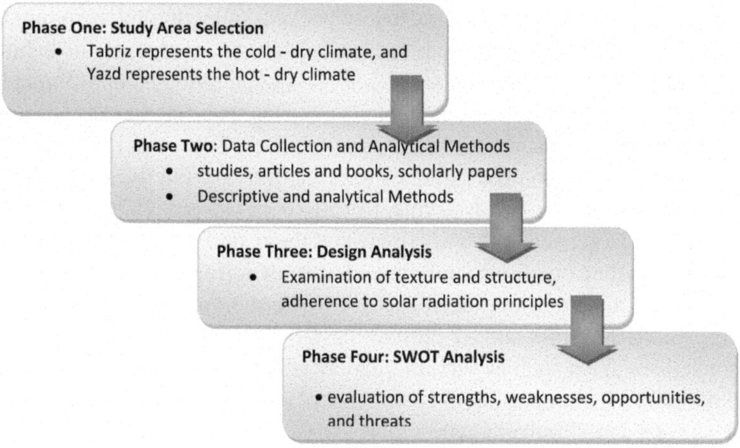

Fig. 1 Workflow

According to the findings of this research, the climatic characteristics of Tabriz are such that its winters are cold, dry, and lengthy with less fluctuation in the temperature between day and night for winter but more in summer because of strong solar rays. This results in a greater rise in temperature in the daytime during the summer season and with dry air, the temperature difference is quite pronounced between day and night. Given the significance of the solar radiation angle concerning the shadow formation, particularly in winter, it can be calculated that considering the geographical latitude of Tabriz ($\lambda = 38 - 05°$), the minimum angle of solar radiation is 27–28° in early January, while the maximum angle of solar radiation is 75° in early July [29].

3.2.2 Design Elements and Strategies for Cold-Dry Climate

The harsh climatic limitations have led to new design solutions. In cold climates, maximising solar radiation is essential due to the severe cold, and creating comfort conditions is important both inside and around buildings.

Buildings in cold climates should be spaced apart to avoid casting shadows on each other and to create medium-width sidewalks and streets. Streets should be oriented to benefit from sunlight for most of the day and avoid being in the shade. Streets are categorised into two types:

- The first type includes those that are primarily north-south and east-west, dividing the city into a grid pattern.
- The second type follows the land's topography, climatic conditions, and geology, potentially deviating from the north-south axis and running diagonally from northeast to southwest or northwest to southeast [20].

Urban centers in cold regions are densely packed, reflecting the response to the cold periods in these areas and providing better physical conditions for habitation. These strategies include reducing the surface-to-volume ratio, using grouped buildings to minimise heat loss, reducing the city's exposure to winter winds, minimising shading, and creating warm indoor and outdoor spaces [29].

Traditional buildings in cold climates have central courtyards, with other parts arranged around these courtyards. Typically, the rooms located on the north side of the courtyard are larger than other sections, and the main living hall is also situated on this side to benefit from direct sunlight and warmth during the cold winter months. The south side of the building is used less due to the short and mild summer season. Therefore, southern rooms and, if present, eastern and western rooms are used as storage or service areas such as servant quarters or bathrooms [32].

In cold regions, it is important to ensure that plantings do not create shade during the cold seasons and simultaneously block undesirable winds. Deciduous trees are suitable for providing summer shelter against sunlight and enhancing sunlight absorption in winter. Intermittent planting is preferable to straight planting as it helps to break the wind. A belt of evergreen trees as a windbreak from northwest to southeast, and planting deciduous trees on the east, west, and south sides of the building to benefit from sunlight, is recommended [14].

A south-to-southeast orientation is suitable for habitation as it creates spaces that attract sunlight during the early morning hours. This orientation is also very suitable for the main facade and entrance of the building. Factors that favor this orientation include:

- Utilisation of sunlight
- Increased heat absorption
- Protection against prevailing northern winds.

Avoid placing spaces in the eastern and northern directions of the building to minimise the use of shaded and cold areas. Western spaces are used for very cold periods or regions, allowing maximum ventilation in the afternoon while avoiding direct sunlight. Most habitats in cold-dry regions are selected in these directions, and experiments show that these orientations are highly effective.

Buildings should be compact and individual, positioned next to each other to avoid layouts that create wind tunnels and frost. Buildings should have an east-west orientation, and the distance between buildings should be such that maximum sunlight absorption is achieved with minimal combined shadow. The size of open spaces (courtyards) should be adjusted based on the need for sunlight absorption and wind protection, and they should be primarily located on the south to east sides of volumes to maximise sunlight exposure and avoid creating central courtyards that cause shading.

The predominant annual wind direction is east to west and northeast to southwest. Therefore, minimal openings should be placed on the eastern and northeastern sides of buildings to prevent cold wind penetration. Additionally, walls and windows should be double-glazed, and using windbreaks such as coniferous trees in this direction is recommended.

To optimise the use of solar radiation, buildings should have an elongated shape with a larger southern facade (with a deviation of 12.5° towards southeast to 12.5° towards southwest) and a smaller east-west facade. In examining the lateral surface area and plan density of buildings, it is found that all ancient residential buildings have compact plans with reduced lateral surface areas. Elongating the building in the northeast to southwest direction is recommended for cities like Tabriz [13]. Roofs, due to their orientation and relatively large surface area, are important sources of heat gain in buildings. Intense solar radiation generates a lot of heat during the day, increasing the temperature under the roof surface. Experiments show that even if the surrounding air temperature is only 32°, the conductor under the roof spaces can raise the temperature above 65°. Using dark colors and matte surfaces on building exteriors and urban arteries is recommended for better sunlight absorption [29].

The overall strategy for energy conservation in cold regions is driven by the unique climatic conditions, characterised by cold and long winters and dry summers. These conditions necessitate designs that can maximise solar gain in winter and mitigate the negative impacts of cold winds (Table 1).

These strategies not only contribute to reducing dependence on fossil fuels and energy consumption but also align with sustainable development goals and environmental conservation. Ultimately, for urban design in Tabriz, principles such as dense layout, planting coniferous trees, and using dark colors and matte surfaces should be employed to achieve the best results.

3.3 Investigating Solar Utilisation Strategies in Hot-Dry Climate

3.3.1 Climatic Conditions of Yazd

The general climatic characteristics of the plateau plains are as follows: hot and dry summers, cold and dry winters, very low rainfall, very low air humidity, minimal vegetation, significant temperature differences between day and night, and dust-laden winds in desert and semi-desert areas. Due to low humidity and distance from the sea, the temperature difference during the day is significant [13].

Yazd Province, located in the central plateau of Iran, largely falls within the hot and dry climate zone. The region's remoteness from extensive water bodies, low rainfall, and high evaporation rates are key factors contributing to its dryness. Yazd experiences significant temperature variations during the day and night due to its low humidity and distance from the sea. Additionally, severe desert winds carrying sand and dust create a harsh environment for human habitation [28]. The climate in Yazd, which is in the central plains of Iran, is hot and dry, with big changes in temperature between day and night, little rain, low humidity, and frequent desert winds that bring dust and sand. These weather conditions make it very hard to live in Yazd. As a result, Yazd's weather conditions make it challenging to plan sustainable buildings.

Table 1 Design strategies for optimal solar energy utilisation and building protection in cold-dry climates

Strategy	Description
Optimal building orientation	Buildings should be oriented towards the south to southeast to exploit sunlight and absorb more heat, maximising solar gain during the cold season
Spacing between buildings	Proper spacing should be maintained to prevent shadowing between buildings, ensuring that each building receives adequate sunlight throughout the day
Use of central courtyards	Central courtyards should be designed to harness direct sunlight during winter, with surrounding buildings arranged to maximise warmth and minimise shading
Planting deciduous trees	Deciduous trees provide summer shade and enhance sunlight absorption in winter. They should be planted on the east, west, and south sides to optimise solar gain
Street and sidewalk design	Streets and sidewalks should be oriented to receive maximum sunlight throughout the day, avoiding shading from adjacent buildings or structures
Planting evergreen trees	Evergreen trees should be used as windbreaks from northwest to southeast to reduce the impact of cold winds on buildings and outdoor spaces
Use of dark colors and matte surfaces	Dark colors and matte surfaces are recommended for building exteriors and urban spaces to improve sunlight absorption and heat retention, particularly on roofs
Reducing surface-to-volume ratio	Grouped buildings with reduced surface-to-volume ratios help in minimising heat loss and increasing internal warmth, essential in cold-dry climates
Compact building design	Buildings should be compact and closely spaced to avoid creating wind tunnels and frost pockets, enhancing thermal comfort
Adjusting open space sizes	The size of open spaces (e.g., courtyards) should be optimised for maximum sunlight absorption and protection from cold winds, primarily on the south to east sides
Minimising openings in windward directions	Openings on the eastern and northeastern sides should be minimised to prevent cold wind penetration, enhancing insulation

(continued)

Table 1 (continued)

Strategy	Description
Double-glazed windows and walls	Use double-glazed windows and walls to reduce thermal exchange and protect against heat loss, improving energy efficiency in cold climates
Elongated building shape	Buildings should have an elongated shape with a larger southern facade and a smaller east–west facade to benefit from solar radiation while minimising wind exposure

3.3.2 Design Elements and Strategies for Hot-Dry Climate

The hot and dry climate significantly influences architecture, impacting all buildings, interior architectural elements, urban fabric, and the lifestyle of people. Consequently, the spatial and physical configurations, considering wind direction and solar radiation, play a crucial role in enhancing spatial quality and providing environmental comfort [30]. In light of these conditions, urban and rural textures in these areas are highly dense, and urban spaces are entirely enclosed. Alleys are narrow, irregular, and sometimes covered with vaults. Buildings are interconnected, and the layout of living complexes is based on the direction of the sun and wind, helping urban spaces better adapt to adverse climatic conditions [13].

Buildings are designed to be entirely inward-looking and enclosed, featuring central courtyards, basements, porches, and windcatchers. Room heights are relatively high, and the arches are often semicircular or domed. Furthermore, thick and massive walls control heat exchange throughout the day and night [13]. Central courtyards and shaded areas are key elements in traditional architecture, particularly in hot and dry regions, where they improve thermal performance and create favorable microclimatic conditions. The design of central courtyards, considering factors such as proportion, orientation, geometry, and the features of openings, can reduce energy consumption and enhance both indoor and outdoor temperatures. Shading elements like galleries can also improve the thermal performance of buildings by 30–60%, facilitating natural ventilation and reducing humidity. This design approach is crucial as the central courtyard, with its vegetation and water pools, plays a vital role in air conditioning and thermal comfort for the inhabitants. Warm air in the central courtyard rises at night, replaced by cooler air. In the morning, the courtyard air slowly warms up and retains its coolness until the sun's rays directly hit the courtyard [6]. Traditional water cisterns are exemplary of the ingenuity of historical design, demonstrating a sustainable approach to managing scarce water resources and ensuring thermal comfort in arid and semi-arid regions. These structures play a crucial role in storing and maintaining cool water through advanced architectural and engineering techniques. By preserving stable thermal layering and utilising natural ventilation, these cisterns provide cool and high-quality water during the hot seasons. The underground design of the cisterns allows for the cooler water to remain in the lower layers, even in high ambient temperatures, with water temperatures in these lower layers remaining below 10 °C during the summer. This ability to maintain cool water underscores the effectiveness and relevance of traditional cisterns in contemporary environmental design, particularly in regions with harsh climatic conditions.

Moreover, the shape and orientation of buildings are critical in reducing solar radiation inside the building. Tall and narrow buildings with a longer length in the east-west direction can reduce the surface area exposed to direct sunlight, thereby reducing heat ingress. Thus, the interior spatial arrangement of the building can improve thermal conditions. In hot and dry areas, main rooms should be oriented to minimise direct solar exposure, while less important spaces can be positioned in warmer directions to act as thermal insulators for the main parts of the building.

Additionally, dome and arched roofs are chosen because the angle of solar radiation hitting different parts of these roofs varies. Part of their surface remains shaded in the morning and evening. The curved form of the roofs is suitable for the dispersion and exit of solar radiation waves at night, aiding in nighttime cooling. This design choice is essential as roofs have the largest surface area exposed to sunlight and release the most energy [11]. These roofs have a small parapet wall to provide privacy and security, which also offers thermal comfort. Experiments have shown that in buildings with domed roofs, indoor air temperature remains in the comfort zone, but in buildings with prefabricated concrete roofs, the indoor air temperature exceeds the outdoor temperature [4].

Furthermore, massive walls, about one meter thick, are essential elements of vernacular houses in hot and dry climates. Due to their heat storage properties, these walls slowly transfer heat to the interior. A delay factor in these materials causes heat to be absorbed a few hours after solar exposure and then released slowly. Therefore, this feature reduces temperature fluctuations inside the houses. Native materials such as mud, adobe, brick, and stone have high thermal resistance, large heat capacity, and the ability to absorb sunlight through exterior surfaces. Additionally, low vapor pressure in this climate allows temperatures to rise to 27–28 °C with minimal ventilation. This phenomenon is observed in the basements of houses in Yazd, where residents rest in the afternoons [11].

Small windows are usually placed at the top of walls and just below the roof. As a result, in hot and dry climates, large windows are less common unless shaded or oriented appropriately to prevent sunlight penetration into the interior. Exterior walls have fewer windows, with most opening to the interior courtyard [4]. In hot and dry climates, construction materials with specific thermal properties play a crucial role in enhancing thermal comfort and reducing the need for artificial cooling systems. Clay walls, commonly made from mud bricks, are widely regarded as ideal building materials for such environments due to their unique thermal qualities, including low thermal conductivity and high thermal capacity. These walls absorb heat during the day and gradually release it at night, helping to maintain a more consistent indoor temperature and reducing temperature fluctuations [12].

One key advantage of clay walls is their low thermal conductivity, which minimises heat transfer from the outside to the inside of the building. This characteristic is particularly beneficial in hot and dry regions where outdoor temperatures can be extremely high [2]. Furthermore, the natural materials used in these walls, such as straw and compacted earth, improve thermal performance, reducing the need for energy-intensive cooling systems [3]. Traditional Iranian systems, like wind catchers (badgirs), are among the most effective examples of passive cooling methods in Iranian architecture. Wind catchers use natural airflows to cool buildings in hot and dry regions without requiring external energy sources. These systems reduce indoor temperatures and increase thermal comfort by harnessing wind and water evaporation (Table 2).

In summary, the optimal design strategies for hot-dry climates, as outlined in this table, emphasise the importance of orientation, material selection, and traditional architectural elements to enhance thermal comfort and energy efficiency. By leveraging features such as thick walls, small windows, dome or arched roofs, and central courtyards with shading elements, these strategies reduce heat ingress and facilitate natural ventilation. The integration of high thermal capacity materials and passive cooling systems like wind catchers further minimises the reliance on artificial cooling, making these design principles essential for sustainable architecture in hot and dry regions.

4 Use of Traditional Elements and Techniques in Enhancing Energy Efficiency and Sustainability in Modern Architecture

SWOT Analysis is a simple strategic management tool that identifies and assesses the Strengths, Weaknesses, Opportunities, and Threats associated with any organisation, project, or even a geographic area. Just considering this point and doing a SWOT analysis on the cold-dry and hot-dry climate of Iran allows one to indicate the factors that greatly affect both sustainability and energy efficiency in architecture.

Strengths

The use of traditional Iranian materials, such as the earth, adobe, and pozzolanic substances, including Sarooj, is very important in improving the efficiency of energy use, as well as sustainability, in buildings due to their good thermal and mechanical properties. The inclusion of these types of materials in modern designs can be very effective in the effective control of indoor temperatures and saving a great deal of energy used in heating or cooling. For instance, the addition of pozzolanic material to mortars and plasters leads to the improvement of thermal properties while at the same time contributing positively to the mechanical strength of buildings. The progressive

Table 2 Optimal design strategies for solar energy utilisation and building protection in hot-dry climates

Strategy	Description
Optimal building orientation	Buildings should be oriented to maximise natural shade and minimise direct solar exposure. A north-south orientation with a slight east or west deviation is suitable. Additionally, buildings in hot and dry areas should be oriented east-west to minimise sun exposure on the main facades
Use of small windows	Small windows positioned high on walls and near the ceiling to minimise heat ingress
Thick walls	Thick walls, about one meter in thickness, for heat storage and temperature fluctuation reduction. They slowly transfer heat to the interior, reducing indoor temperature fluctuations
Dome or arched roofs	Roofs with small parapet walls for privacy and thermal comfort. Dome or arched roofs are particularly effective in shading parts of the roof during morning and evening, aiding in nighttime cooling
Central courtyards with shading elements	Central courtyards as the main space for ventilation and temperature control, often featuring shading elements like galleries and overhangs to reduce solar heat gain by 30–60%, facilitating natural ventilation, trapping cool air, and improving microclimatic conditions
Clay walls color	Natural brown or gray, minimise sunlight reflection by absorbing heat during the day and releasing it slowly at night, reducing temperature fluctuations and balancing indoor comfort
High thermal capacity materials	Use of mud, adobe, brick, and stone with high thermal resistance and capacity to absorb heat during the day and release it slowly at night
Interior space arrangement	Main rooms should be oriented to minimise direct solar exposure, while less important spaces should act as thermal buffers
Dense urban and rural layout	Dense and enclosed urban and rural layouts with narrow, irregular alleys, sometimes covered with vaults, help in reducing heat exchange and improving thermal comfort
High ceilings	Relatively high ceilings for better ventilation and reduced heat in the interior spaces, allowing hot air to rise and maintain cooler temperatures below
Water cisterns/ qanats	Maintain a stable thermal layering, allowing cooler water to remain in lower layers, even in high ambient temperatures. These structures also support natural ventilation and sustainable water management
Wind catchers	As passive cooling systems in the hot and dry regions of Iran, they play an effective role in reducing energy consumption and providing natural ventilation, significantly improving thermal comfort

developments also make it possible to encapsulate these materials with state-of-the-art technologies such as Phase Change Materials so as to further minimise heat transfers and enhance energy efficiency.

Windcatchers are counted as one of the most remarkable elements in traditional architecture in hot and dry regions of Iran. They have historically functioned as passive cooling and ventilation systems. Studies have found that these systems reduce energy consumption and improve thermal comfort, thus making them effective models for sustainable design in modern architecture. New windcatcher designs, including those with rotating and adjustable tops that may automatically respond to wind velocities, have been found to reduce temperature dramatically and raise thermal comfort. They additionally enhance air change and therefore the efficiency of ventilation, thereby indirectly decreasing the demand for electrical cooling.

Another essential component of traditional Iranian design are central courtyards, which are crucial for controlling the microclimate within structures. These courtyards aid in lowering interior temperatures in the summer and raising them in the winter by offering shade and natural ventilation. These courtyards may be used in contemporary design to create green, open areas within buildings, improving the quality of life and lowering energy use.

The use of indigenous techniques and climate models developed from traditional Iranian architecture may provide significant improvements in thermal comfort and energy economy. These models may be used in contemporary designs to minimise energy consumption and create more sustainable surroundings by making the most use of natural light, shading, natural ventilation, and wall and roof insulation [16].

Another important aspect of traditional architecture that supports indoor temperature control, better than the flat or sloping-roofed buildings, is the dome-shaped buildings. They build more warmth in cold seasons and reduce heat in warm seasons. This could be well exploited in modern architecture as one of the ways to reduce the use of mechanical heating and cooling systems.

In Iran's hot and arid areas, the restoration of conventional water management methods such qanats and cisterns can assist lower reliance on outside water supplies and strengthen community drought resilience. Approved as one of the most efficient and sustainable approaches of water management, these systems may be used with new technology in modern architecture to guarantee sustainable water supply, maintain cultural legacy, and enhance quality of life in arid and semi-arid areas. Rainwater collecting systems have been demonstrated, for example, to provide over 75% of the water required for non-potable purposes like toilet flushing in highly populated places like Spain. Apart from supporting sustainable water supply, rebuilding and incorporating these systems into urban infrastructure improves environmental aesthetics and protects cultural legacy.

Weaknesses

Strong winds that could compromise thermal comfort and building performance can affect both cold and hot-dry regions. Buildings in cold areas are often subjected to strong winter winds, so strong windbreaks and careful construction orientation are rather important to reduce heat loss. In hot-dry areas, desert winds bringing dust and sand may enter structures, so specialist design elements like thin, insulated walls and few openings on windward sides—which might not be possible in all current designs—become essential.

Urban planning is complicated by the necessity for compact, inward-facing designs in hot-dry regions to control sun gain and ventilation as well as by thick urban layouts in cold climates to lower heat loss. These design criteria may restrict the adaptability and freedom of metropolitan areas, therefore making it difficult to combine the demands for contemporary urban growth, comfort, and energy economy.

Although conventional materials with great thermal qualities—such as adobe, mud bricks, and thick walls—may not necessarily be useful or readily accessible in contemporary building The use of these materials also demands certain construction skills and knowledge, which may be absent in modern building procedures, therefore limiting its applicability on a broad basis.

While conventional cooling solutions such as windcatchers and the use of large walls are efficient, combining them with current technology might prove difficult. Adapting windcatchers to fit contemporary HVAC systems or making sure historic materials satisfy current building rules and standards, for instance, calls for careful design and creativity that could raise complexity and expenses.

Threats

Rising frequency of natural catastrophes such floods and earthquakes, as well as climate change, may severely strain old buildings and call for fresh approaches for their preservation and management. Furthermore, the predisposition toward modern lives and modern wants might contradict with conventional ways, therefore promoting unsustainable behaviors and maybe compromising cultural values.

Opportunities

In arid and semi-arid areas, the use of conventional systems such as water reservoirs and qanats offers a sustainable answer for water and energy management. Low-energy and ecologically friendly design is in more demand as environmental problems and the depletion of energy supplies become more well known. This scenario presents chances to incorporate historic tools and materials into contemporary projects in order of importance. Furthermore, the integration of cutting-edge technologies such Phase Change Materials (PCM) and Computational Fluid Dynamics (CFD) simulations helps to improve the efficiency of conventional systems like windcatchers, thereby opening fresh directions for current architectural development.

5 Concluding Insights

This chapter has discussed the important role of Iranian vernacular architecture in addressing current sustainability challenges. The study has examined the design strategies used in buildings located in cold-dry and hot-dry climates, specifically in Tabriz and Yazd cities. It has emphasised the successful incorporation of traditional elements such as windcatchers, dome-shaped roofs, thick walls, and central courtyards into modern architectural practices. The SWOT analysis we conducted has given us a better understanding of how these traditional techniques can help improve energy efficiency and promote environmental sustainability in modern architectural designs. The strengths that have been identified include the thermal and mechanical properties of traditional materials such as adobe, earth, and pozzolanic substances.

Additionally, traditional Iranian architecture also benefits from passive cooling and ventilation systems.

However, the study also recognises the difficulties and dangers presented by current climate change, the growing number of natural disasters, and the transition to modern lifestyles. These factors necessitate the adaptation and innovation of traditional methods to meet current building standards while preserving cultural heritage. However, with the increasing global focus on protecting the environment and using energy efficiently, there are great opportunities to bring back and incorporate these traditional strategies into modern architecture. Using advanced technologies like Computational Fluid Dynamics (CFD) simulations and Phase Change Materials (PCM) can improve the performance of traditional systems such as windcatchers and water reservoirs.

The main point of this chapter is to emphasise the importance of using historical architectural knowledge to build sustainable, resilient, and culturally enriching environments in the midst of contemporary difficulties. The findings from this study provide a basis for future research and development in sustainable architecture. It highlights the significance of combining traditional and innovative approaches to tackle the urgent environmental issues we face today.

References

1. Asadi A (2007) Sustainable architecture and primary bio-systems. Iran J Arch Urban Plan 12
2. Bahobail M (2014) The mud additives and their effect on thermal conductivity of adobe bricks. JES J. https://doi.org/10.21608/jesaun.2012.112711
3. Binici H, Aksogan O, Bodur M, Akça E, Kapur S (2007) Thermal isolation and mechanical properties of fibre reinforced mud bricks as wall materials. Constr Build Mater 21:901–906. https://doi.org/10.1016/J.CONBUILDMAT.2005.11.004
4. Bonine ME (1980) Aridity and structure, desert housing. In: Golavv G (ed), New York, no 82
5. Djamali M, Akhani H, Khoshravesh R, Andrieu-Ponel V, Ponel P, Brewer S (2011) Application of the global bioclimatic classification to Iran: implications for understanding the modern vegetation and biogeography. Ecol MediterrEa 37:91–114. https://doi.org/10.3406/ECMED.2011.1350
6. Dunham D (1961) The courtyard house as a temperature regulator. Ekistics 11(64):181–186. http://www.jstor.org/stable/43613511
7. Ebrahimi M, Kalantari K (2003) Examining the balance between environmental necessities and development needs in sustainable development. J Environ Sci Technol 5(2)
8. Falamaki M (1975) Popular and native architecture
9. Ghiasvand J (2006) The interaction of architecture and new (sustainable) energies. Rah Sakhteman J 9
10. Gioni L (1977) Institutional dimension of urban sustainability in southern Italy. In: 30th Congress of the European regional science association, Zurich, vol 122, no 1
11. Haji Qasemi K (2004) Ganjnameh encyclopedia of Islamic architectural works of Iran, vol 14th Houses of Yazd. Rozaneh Publications, Tehran
12. Hanna C, Aly A, Shebl S, Abdallah A (2022) An empirical study on the thermal behavior of rice husk in eco-friendly brick for external walls of buildings. JES J Eng Sci. https://doi.org/10.21608/jesaun.2022.141836.1144
13. Kasmaei M (1989) Climatic zoning of Iran—educational buildings. Organization for development, renovation, and equipping schools of Iran, Tehran

14. Kasmaei M (1995) Climate and architecture. Cactus Publishing, Tehran
15. Khatibi R, Saberi M (2020) Bio-climatic classification of Iran by multivariate statistical methods. SN Appl Sci 2. https://doi.org/10.1007/s42452-020-03500-9
16. Lotfabadi P, Hançer P (2019) A comparative study of traditional and contemporary building envelope construction techniques in terms of thermal comfort and energy efficiency in hot and humid climates. Sustainability 11(13):3582. https://doi.org/10.3390/su11133582
17. Mahmoudi MM, Niavi S (2011) The trend of technological development in climate with a sustainable development approach. Naqsh-e-Jahan 1:36–51
18. Memarian G (2005) The impact of religious culture on the formation of houses in the neighborhoods of Muslims, Zoroastrians, and Jews in Kerman. Cult Stud 3(2):23–43
19. Mofidi Shemirani SM (2008) Stability and dynamics of passive systems. Andisheh Nameh 1, collection of interdisciplinary urban and architectural articles, Center for Urban and Architectural Studies and Research, Ministry of Housing and Urban Development
20. Mofidi M (1998) Climatic urban design, Ph.D. thesis, University of Sheffield, United Kingdom, p 7
21. Mozayeni N (1997) Sustainability in architecture from the perspective of Norman Foster. Fine Arts J 28
22. Najafi M, Alizadeh O (2023) Climate zones in Iran. Meteorol Appl 30. https://doi.org/10.1002/met.2147
23. Parhizgar G (2003) Culture and architecture. Hadith-E-Zendegi J 11
24. Pope A (2010) Architecture of Iran (Translated by Gholamhossein Sadri Afshar). Akhtaran Publishing, Tehran
25. Pourvahidi P, Ozdeniz M (2013) Bioclimatic analysis of Iranian climate for energy conservation in architecture. Sci Res Essays 8:6–16. https://doi.org/10.5897/SRE11.518
26. Rahimi J, Ebrahimpour M, Khalili A (2013) Spatial changes of extended De Martonne climatic zones affected by climate change in Iran. Theoret Appl Climatol 112:409–418. https://doi.org/10.1007/s00704-012-0741-8
27. Raziei T (2021) Comparing the Köppen-Geiger, Feddema, and UNEP climate classifications: an application to Iran. Res Square. https://doi.org/10.21203/RS.3.RS-508056/V1
28. Seflaei A (2004) Study of Climatic Conditions of Yazd. Iran Archit J
29. Shaghaghi S, Mofidi Shemirani SM (2008) The relationship between sustainable development and climatic design of buildings in cold and dry regions (case study: Tabriz). J Environ Sci Technol 10(3 (38)):105–120. https://sid.ir/paper/458612/fa
30. Tavasoli M (2002) Urban and architectural design in Iran's hot and dry climate. Center for Urban Studies and Architecture Research Publishing, Tehran
31. Torabi F (2018) Examining the climatic elements of Iranian architecture based on static solar architecture elements for reducing energy consumption in residential houses. Arch Knowl 1(6). https://sid.ir/paper/520905/fa
32. Zandieh M, Parvardi Nejad F (2010) Sustainable development and its concepts in residential architecture in Iran. J Hous Rural Environ 130
33. Zareiee A (2014) Evaluation of changes in different climates of Iran, using De Martonne index and Mann-Kendall trend test. Nat Hazards Earth Syst Sci Discuss 2:2245–2261. https://doi.org/10.5194/NHESSD-2-2245-2014

Preserving the Past, Designing the Future: Unveiling Climate-Resilient Design in Persian Vernacular Architecture for Modern Sustainability

Elham K. Hassani

Abstract Contemporary Iran preserves a rich tapestry of Persian vernacular architecture. Climate change makes this long tradition pertinent as a source of climate-resilient design, but in fact, its relevance was always in view because its origins are pragmatic and its principles a direct response to the actual conditions people experienced within their everyday. Persian vernacular architecture's historic context and cultural underpinnings include the strategies used by traditional master builders to adapt buildings to the region's varied climates. Their holistic approach combined resilience with cultural identity to set neighborhood patterns, building layout and orientation, and the use of locally sourced materials. As an illustration of their relevance, consider that the rotary wind catchers used in the Bluewater Mall in Kent, UK, trace back, despite an overlay of new technologies, to the Iranian *badgir* found in Kashan's traditional *Agha Bozorgh school and mosque* type. The underlying principle is the same in both. The chapter reviews the climate-resilient design principles found in Persian vernacular architecture and city-making, and their lessons for cities with similar conditions as they try to reduce the energy used in construction, improve overall building performance, and deliver settings at higher densities that meet current needs and foster a sense of belonging and place. It draws on the author's extensive research on Kashan, a traditional city located in central Iran, with an emphasis on the inherent resilience and potential for adaptive reuse of its historic buildings and fabric. Engaged in architectural and planning practice in Iran, including the redevelopment of Tehran and other cities, the author has actively considered how these traditions can be applied sustainably to meet current needs.

Keywords Persian vernacular architecture · Climate-resilient design · Sustainable architectural practices · Energy consumption reduction · Master builders of Iran

---

E. K. Hassani (✉)
Department of Architecture and Design, Faculty of Architecture, University of Sapienza, Rome, Italy
e-mail: Elham.k.hassani@gmail.com

A. Cheshmehzangi and S. Roaf (eds.), *Persian Vernacular Architecture*, Urban Sustainability, https://doi.org/10.1007/978-981-96-1116-4_15

1 Introduction

Given the increasing severity of climate change, it is essential to include sustainability in contemporary architectural methods. Persian vernacular architecture serves as an excellent example of climate-resilient design concepts throughout history. Through the analysis of these traditional approaches, modern architects can gain vital knowledge to guide current sustainable construction strategies. Persian vernacular architecture showcases its ingenuity through the use of indigenous materials, mindful spatial arrangement, and adaptable strategies to combat the challenging climatic conditions of the region. This architectural style provides valuable insights for creating environmentally sustainable and climate-resilient structures. Traditional Persian architecture demonstrates a harmonious relationship between human settlement and the surrounding environment [18]. Contemporary architects can respond inventively to current environmental challenges by examining and reevaluating historical design concepts. This approach promotes energy efficiency and helps to decrease carbon footprints [23]. Emphasising the significance of drawing lessons from history to develop a sustainable future, it demonstrates how contemporary advances in constructing resilient, sustainable, and culturally vibrant living environments can benefit from a knowledge of tradition.

1.1 Purpose and Scope of the Chapter

This chapter considers Persian vernacular architecture from the perspective of climate-resilient design in order to discover important ideas and examples that can be applied to contemporary urban settings facing comparable climate-related challenges. The chapter connects historical knowledge with current sustainability concerns by analysing traditional Persian architectural principles. It provides practical ideas for improving energy efficiency, reducing carbon footprints, and promoting culturally dynamic urban environments. These principles reflect centuries of work by Iran's traditional master builders as they addressed the varied climate conditions found in the region. Their long evolution not only guaranteed the long-term sustainability and comfort of the constructed cityscape and buildings, but also preserved a strong cultural heritage. As the chapter demonstrates, the principles they put in place are pertinent to contemporary architects and urban planners faced with the same climate conditions, whether their work has a traditional component or not. This is why you find historic and contemporary examples of their application to cities and buildings in Europe.

Focusing on the traditional central Iranian city of Kashan, the chapter discusses such sustainable aspects of vernacular Persian architecture as the orientation of buildings, their use of materials, the layout of their Indoor and outdoor spaces, and their use of climate-responsive elements like wind catchers (Badgir) and subterranean courtyards (Godalbaghcheh). The chapter also discusses the broader urban morphology and building typology of Kashan to show how they reinforce each household's desire to live comfortably and sustainably with the climate.

This chapter includes a number of multifaceted contributions:

1. Contextual Analysis: This involves a thorough investigation of the environmental, cultural, and socio-economic factors that influenced the development of Persian vernacular architecture. Comprehending this is essential for identifying how these elements might be incorporated into contemporary design to improve sustainability and resilience.
2. Design Principles: The chapter highlights essential design principles from Persian traditional architecture that are relevant to current environmental challenges. The ideas include passive cooling methods, optimised use of local materials, and the creation of microclimates to enhance thermal comfort and minimise the consumption of energy.
3. Case Study Insights: By conducting a thorough examination of Kashan, this chapter demonstrates the practical implementation of these traditional design principles. This involves a comprehensive examination of how particular architectural elements and urban layouts have been historically used to create places that are resilient to climatic challenges.
4. Contemporary Applications: The chapter suggests methods for adjusting and integrating these traditional ideas into contemporary architectural and urban design. This includes the use of modern technology and materials that may replicate the advantages of traditional methods while adhering to current building standards and aesthetic preferences.
5. Sustainable Urban Development: The chapter provides concepts for sustainable urban development drawn from Persian traditional architecture and city-making. It suggests ways to incorporate these concepts in contemporary projects to create resilient urban environments.

The chapter bridges between the extensive architectural legacy of Persia and the pressing needs of Iran today as an arid and semi-arid region. It shows comprehensively how this legacy can be used to improve and enrich contemporary Iranian (and other) cities by designing urban spaces there that are both resilient and culturally significant.

1.2 Methodology and Approach

This chapter examines Persian vernacular architecture from the standpoint of resilience to climate change and then considers its implications for contemporary sustainable buildings. Drawing on architectural history, empirical research, and practical applications, it connects traditional knowledge with contemporary innovation. It draws on the author's extensive experience with urban regeneration in Iranian and Western European contexts, and her Ph.D. dissertation research on Kashan, a traditional central-Iranian city. Her research combined historical analysis, morphological and typological studies, and review of actual conditions, applying an extensive knowledge of the theoretical foundations and practical implementations of Persian vernacular architecture to contemporary urban conditions. Kashan was chosen as a prominent exemplar because of its well-preserved architectural legacy and its pertinence to the study's emphasis on arid and hot climates. Her fieldwork in Kashan involved the use of direct observation, photographic recording, and other qualitative approaches. It provided valuable insights into the architectural characteristics and urban arrangements that are legacies of Persia's vernacular traditions, with particular emphasis on climate-resilient concepts that are relevant to today's cities and buildings, whether their contexts are visibly historic, as with Rome, or reflect the underlying conditions imposed by climate change and other environmental factors. To ensure objectivity and reliability, her research drew on multiple sources of data. Her observations were cross-verified by comparing them with historical documents and by consulting with contemporary experts. In addition, her results were peer-reviewed by relevant researchers and practitioners in Iran, Italy, and Sweden (Fig. 1).

1.2.1 Research Design

The research design incorporates a multi-method approach that encompasses:

1. Literature Review: A comprehensive examination of the current body of literature on Persian vernacular architecture, climate-resilient design, and sustainable architectural practices was undertaken. The sources include scholarly articles, historical documents, and contemporary studies.
2. Case Study Analysis: The city of Kashan was chosen as the main case study because of its well-preserved architectural heritage and its significance in desert and hot climates. A thorough analysis was made of particular structures and urban layouts to demonstrate the efficacy of design strategies applicable to climate-resilient design in current contexts.
3. Fieldwork: The process of directly observing and recording architectural elements in Kashan included photographing buildings, sketching layouts, and documenting the specific materials and methods used in the construction process. The purpose of the fieldwork was to document accurately the details of traditional planning and design methods, patterns, and elements to determine how effectively they perform in the specific local environment.
4. Morphological and Typological Studies: The urban morphology and building typology of Kashan were analysed utilising cadastral and topographic maps, aerial photographs, and AutoCAD LISP software. This facilitated the tracking of the development of urban forms and their adaptations to different climates.

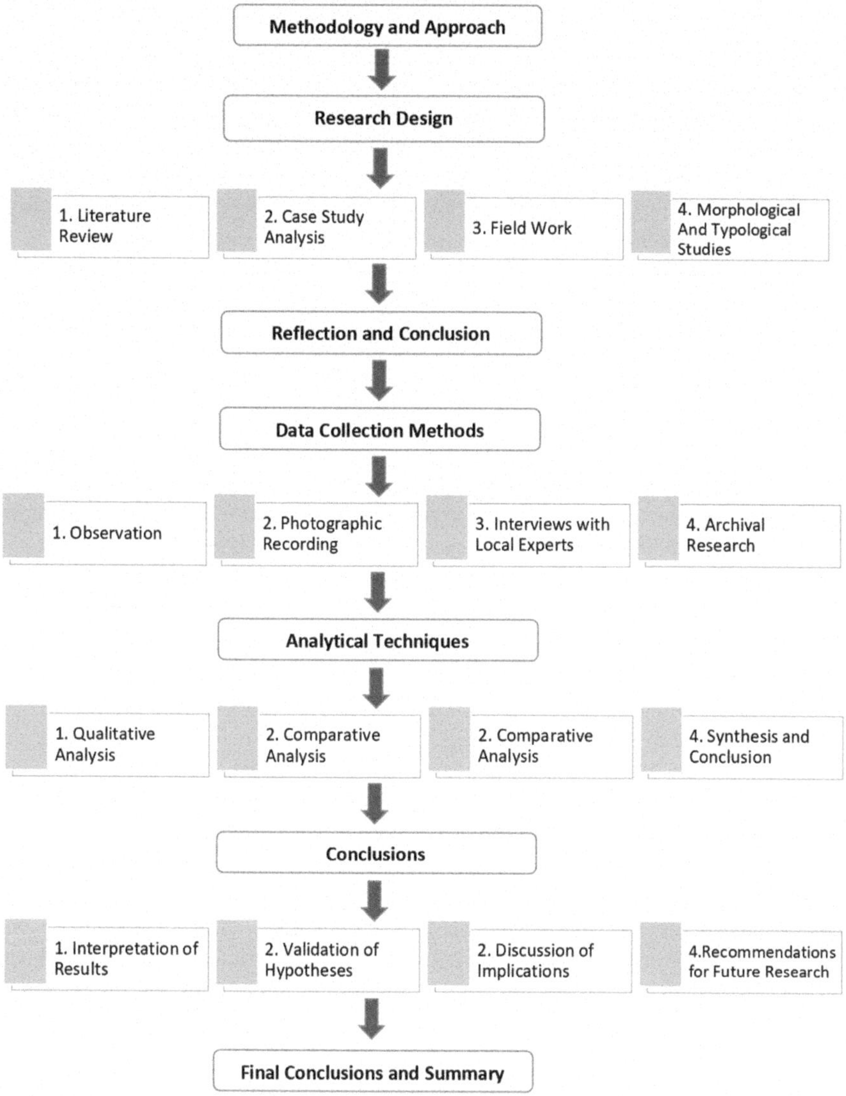

Fig. 1 This diagram and accompanying explanation provide a clear and structured overview of the methodology used in the study: methodology and approach, data collection methods, analytical techniques, and conclusions. *Source* Drawn by the author

1.2.2 Data Collection Methods

1. Observation: A systematic observation of architectural characteristics, such as courtyards, wind catchers (Badgir), sunken courtyards (Godalbaghcheh), and

other elements, was made, focused on their contributions to thermal comfort and energy efficiency.

2. Photographic Recording: Detailed photographic documentation of buildings and urban spaces provided the visual data necessary to understand traditional architectural characteristics and their role in regulating climate.
3. Interviews with Local Experts: Conducted to gain insights into the practical and cultural aspects of Persian vernacular architecture. The experts consulted included architects, historians, and the inhabitants of specific buildings.
4. Archival Research: Historical documents, maps, and records from local archives were examined to gain insight into the historical background and evolution of architectural practices in Kashan and other traditional Iranian cities.

1.2.3 Analytical Techniques

1. Qualitative Analysis: Thematic analysis was used to identify fundamental design principles and methods in Persian vernacular architecture that enhance climate resilience. The themes included passive cooling methods, the use of local resources, spatial organisation, and multiple additional methods.
2. Comparative Analysis: The study compared traditional design principles with contemporary sustainable architectural techniques to identify similarities and possible opportunities for integration. This included comparing the results of case studies with contemporary examples of climate-resilient design.
3. Geospatial Analysis: Geospatial tools and software, such as LISP, were used to map and analyse the geographical distribution, orientation, and dimensions of buildings. This facilitated understanding broader urban patterns and their corresponding climate adaptations.
4. Synthesis and Conclusion: The multiple analyses discussed in this chapter show conclusively Persian vernacular architecture's ability to mitigate Iran's arid climate, particularly the summer heat. The underlying principles of Persian vernacular architecture are relevant to sustainable design today, and a synthesis of traditional architectural knowledge with current technologies has the potential to generate more resilient cities in other hot arid climate zones.

Elements such as wind catchers can also be found in European vernacular architecture, albeit in a different form, which speaks to the principle that underlies both. These and other elements take new forms in current architecture and urbanism that reflect the higher performance possible with new materials and technologies, reducing the need for thick walls, for example, and automating the control of air flow. Yet these new forms can be traced back to the insights that led Persia's traditional master builders to develop passive solutions to the same climate.

1.2.4 Conclusions

1. Interpretation of Results: The research shows that Persian vernacular architecture is crucial to urban climate resilience. A careful study of Kashan shows the flexibility and sustainability of ancient building traditions. These techniques give insights into managing hot and dry conditions and new solutions for contemporary sustainable building design. The findings show a substantial association between ancient approaches and climate resilience concepts, showing that these historical practices are still relevant to modern environmental issues.
2. Validation of Hypotheses: The study confirmed that traditional Persian vernacular architecture includes climate resilience concepts relevant to current sustainable construction methods. The research used historical analysis, empirical data, and expert consultation to show that Kashan and similar examples of traditional and vernacular architecture were successful in their original contexts in enabling people to settle comfortably in climate-challenged locations. The consistency of the results across sources and peer evaluations supports this idea, showing that traditional and vernacular knowledge can inform current architectural practice.
3. Discussion of Implications: This finding has major implications for sustainable architecture. The research implies that combining traditional architectural expertise with current design approaches could make cities more robust, especially in harsh climatic conditions. It also allows architects and urban planners to combine historical insights with technical advances to construct ecologically friendly and culturally appropriate structures. The results also underscore the need to maintain and research traditional architecture's suitability for current needs.
4. Recommendations for Future Research: To understand Persian vernacular architectural ideas' adaptability and effectiveness, future studies should apply them to regions with similar climates and different cultures. Comparative studies of vernacular architecture and its reflection or lack of reflection in contemporary architecture in a range of "desert urban" contexts would also improve the knowledge base. Interdisciplinary study, involving environmental science, urban planning, and cultural studies, can clarify how to integrate traditional and current approaches to sustainable architecture. Further empirical research will be needed to confirm the suitability and effectiveness of these ideas to the goal of developing more resilient towns and cities.

2 Background Knowledge

2.1 Significance of Persian Vernacular Architecture

Traditional builders in Iran employed a range of methods to provide thermal comfort in different cities, especially in hot and arid locations, in order to adapt to particular climatic conditions. An in-depth examination of sustainable design and building practices in Iran reveals how traditional housing effectively addresses the constraints imposed by local culture and environment. These approaches may be incorporated into modern design frameworks [33]. Persian vernacular architecture has a wealth of historical and cultural importance. It is not just concerned with architectural forms but also demonstrates a profound comprehension of the environmental context, the needs of the community, and the importance of cultural identity. This section explores this historical context, emphasising the inherent connection between architecture and its environment throughout Persian history. It highlights the significance of preserving and acquiring knowledge from traditional architectural methods, particularly in the face of present-day environmental issues.

2.2 The Role of Traditional Master Builders in Shaping Climate-Resilient Design

Traditionally in Iran, a "builder" (Memar) was taught in a master-and-apprentice relationship in which elders passed along strategies and techniques, gained from centuries of experience, to adapt buildings to a variety of climatic conditions while maintaining sustainability and comfort.

2.3 Iran's Environmental Contexts

Iran is located in western Asia and is bordered by the Caspian Sea, the Persian Gulf, and the Gulf of Oman. The mountain ranges have long had a significant impact on its economic and socio-political history. Their expansive valleys define their agricultural and urban regions [26]. Iran is geographically characterised by four distinct climatic zones: 1. The Caspian Sea coastline has moderate and wet climates. 2. The coastline of the Persian Gulf and Oman Sea has hot and wet climates. 3. Regions characterised by mountains and cold weather 4. Arid and hot climates are found in desert regions [31]. Given that Iran has much lower average rainfall compared to other regions of the world, the weather in most of its regions is characterised by hot and arid conditions [20].

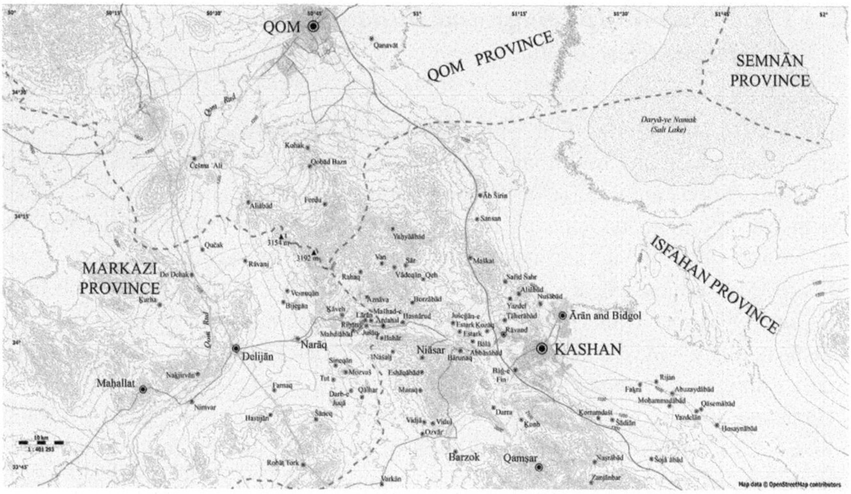

Fig. 2 The Kashan area. Map data © OpenStreetMap contributors; relief contour based on open-access information from NASA Earth Science Data (EOSDIS) *Source* Amanat [3]

3 Examination of Kashan as a Traditional Iranian City

3.1 Geographical Location and Climatic Characteristics

Kashan is a traditional Iranian city subject to Iran's predominant climate. It exemplifies Persian vernacular architecture's climate-resilient design approaches as exhibited by its architectural typologies, sustainable construction techniques, and growth patterns. From Tehran, the south/southwestern route passes through a region bordered by the western boundary of the central Iranian desert strip and the northwestern-to-southeastern Zagros Mountains. These mountains exceed 4000 m above sea level. Kashan lies between their alluvial fans and the Dasht-i Kavir[1] [26]. Kashan's settlement reflects its fertile oasis location, watered by several qanats and springs, with mountainous, mountain foothill, and plain geomorphologies (see Fig. 2) [36]. The geographical location and climatic conditions of Kashan have had a significant impact on its urban growth and architectural practices. The city's advantageous geographical position, situated between mountainous terrain and vast desert landscapes, together with its proximity to crucial water resources, has facilitated the development of sustainable and resilient construction methods. These elements show the inherent connection between Kashan's built environment and its larger setting and climate, providing a case study in the resourceful and adaptive nature of Persian vernacular architecture.

[1] KASHAN dates back to prehistoric antiquity. Its position on a fertile plain watered by springs, the most notable of which is the spring of Fin, resulted in the establishment of a city at Tepe Sialk, on the contemporary city's southwestern suburbs [19].

3.2 The Urban Morphology and Building Typology of Kashan City

Gaube's 2018 study found that while Kashan added postwar suburbs, its historic district maintains many traditional patterns and elements. Kashan's older residential neighborhoods are a maze of twisting pathways, arches, and dead ends. Urban growth was spontaneous, causing this inconsistency. Urban pathways are narrow [26]. Houses are compact and have merged walls [4]. This reduces air contact on wall surfaces, reducing heat exchange between buildings. Narrow allies with high walls limit wind speed and create shade pathways. High walls protect buildings from harsh heat, dusty winds, and cold winds in winter. Their façades have few openings, usually a single main entrance [21].

Cadastral and topographic maps were studied and aerial photographs from the archives were used to assess the evolution of plots and blocks, routes, and other aspects that diachronically and synchronously shaped the city. The historical urban fabric of Kashan was built of basic building units with courtyard layouts that were transformed using the same type and modified to meet the evolving requirements of the inhabitants.

A LISP application in AutoCAD was used to analyse plot and block sizes along nearby routes, as well as plot regularity and irregularity (trapezoidal forms, e.g.). In the most recent phase of redevelopment, Kashan imposed hypergrids and hyperblocks to assist mobility in the historic district. Analysing its overlapping historical layers, routes, bands of pertinence, and plots and blocks documented how its geometric patterns changed over time (Fig. 3 illustrates this evolution). These patterns reveal that, despite informal growth and city impositions, the climate characteristics have long been the main factor shaping Kashan's urban fabric [26].

In summary, Kashan's urban morphology and building typologies demonstrate an intricate relationship between ancient traditions and contemporary interventions. The organic development of its historic districts, distinguished by compact houses with shared walls, small passageways, and high protective walls, was primarily a response to climatic factors. These features collectively contribute to a built environment that efficiently reduces the impact of heat, wind, and cold. Recent redevelopment has sought to enhance movement within the historic area. Analysing these changes over time, it is evident that Kashan's urban structure is a dynamic combination of traditional and contemporary practices, adapting older patterns to meet current requirements. This ongoing interaction between historical precedents and contemporary adaptation highlights the resilience and creativity inherent in Kashan's urban setting. It echoes the long evolution of "tradition" by master builders, working empirically with climate and culture as unfolding contexts, and learning from the experiences of their peers and predecessors.

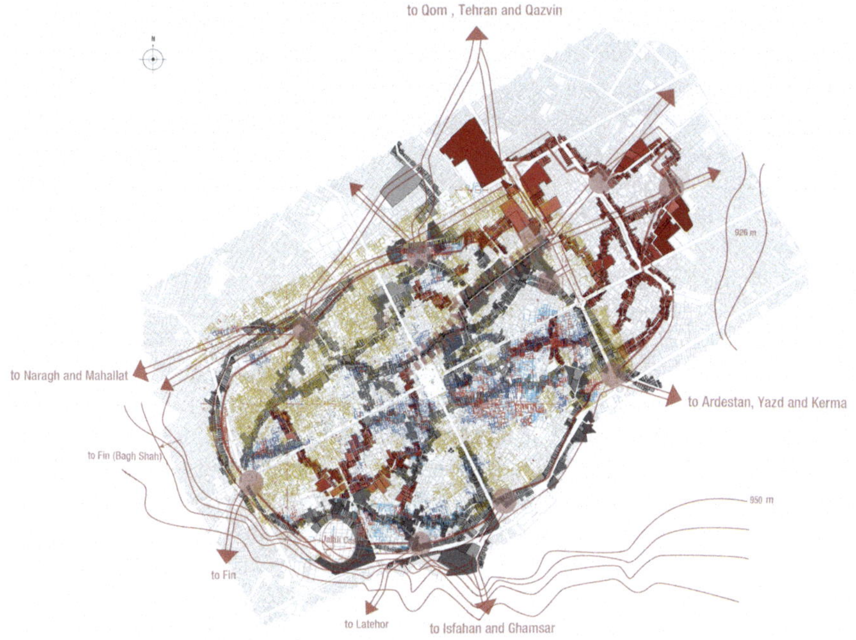

Fig. 3 Kashan city development from eighteenth to twentieth century A.D.: Afsharid, Zand, and Qajar. Different colored rectangles depict the buildings' alignments. The latest overlaid city levels are red, whereas the earlier development periods are grey. Plots along matrix and restructure routes are regular and irregular. The author accelerated plot and block regularity and alignment analysis using LISP. This investigation was done on an AutoCAD map from Kashan's municipality with hand drawings from the city's Organization of Cultural Heritage, Handicrafts, and Tourism [26]

4 Strategies for Climate-Resilient Design

Climate-resilient design is an integrative approach that aims to create buildings capable of sustaining and adjusting to the environmental effects of climate change. In Iran and other regions contending with arid and semi-arid conditions, traditional strategies for addressing them can be combined with the latest materials and technologies to produce a resilient urbanism that is culturally resonant. This section discusses the fundamental elements of climate-resilient design present in Iran's long vernacular traditions of building and city-making, identifying the principles that can be applied to the current urban contexts of Iran and other regions with similar climates.

An in-depth understanding of local climatic patterns is the foundation of design that can withstand the impacts of climate change. This entails closely examining past weather data, forecasting future climate scenarios, and identifying the potential for environmental events such as excessive temperatures, unpredictable precipitation, and natural disasters. These insights are essential for informing design choices that improve building durability and comfort. Persian vernacular architecture has historically exhibited a profound relationship with the surrounding environment, resulting in more resilient buildings and settlements [39].

Incorporating locally sourced materials is a crucial element of designing structures that can withstand climatic challenges. The materials used, such as adobe, clay, and locally gathered wood, are well-suited to local conditions and permit passive temperature control and structural stability. The exceptional thermal insulation properties of the thick earthen walls are evident in traditional Persian buildings, which are able to mitigate the region's extreme temperatures. The natural qualities of these materials argue for their continued use [17].

Passive design concepts greatly contribute to improving the resilience of structures. These concepts optimise the use of natural light, ventilation, and thermal comfort while minimising energy usage. Many passive design techniques can be found in Persian architecture, including wind catchers (badgirs), courtyards, and domes. These architectural features effectively control internal temperatures and airflow, providing a basis for designing new buildings that are energy-efficient and highly responsive to changing environmental conditions [38].

Employing adaptive reuse and retrofitting techniques for existing buildings plays an essential role in the development of climate-resilient design. Enhancing and altering historical buildings with contemporary technology and materials can greatly enhance their durability and effectiveness by strengthening structural components to resist earthquakes, for example, or using contemporary insulation methods to improve thermal efficiency. Persian architecture preserved cultural continuity while responding to environmental conditions. Maintaining this balance is part of resilience, particularly if embedded carbon is counted as "net zero." [45]

Effective water management is essential for achieving resilience, especially in dry environments. Traditional Persian architecture often incorporated intricate water management systems, such as qanats (underground aqueducts) and ab anbars (cisterns), to catch, store, and distribute water. Modern adaptations of these systems can conserve and manage scarce water resources equally effectively, while adding grey water and other recycling techniques [7].

Another essential element of climate-resilient architecture is to design of buildings that can be easily renovated or enlarged. Structures with flexible and modular designs can adjust to changing needs and conditions, extending their useful lives. Modular courtyard buildings in Persian vernacular architecture exemplify the capacity to adapt spaces to accommodate evolving household requirements and environmental conditions. The capacity to adapt is crucial to construct durable and resilient structures, and one reason for the persistence of tradition in many Iranian cities [24].

Incorporating the knowledge and traditions of local people is essential for successful climate-resilient design. Community involvement assures that designs are both culturally appropriate and functionally efficient. Throughout its history, Persian architecture has been guided by community knowledge, which ensures that buildings are appropriately designed to fit local environmental and social conditions. This collaborative approach fosters a feeling of shared responsibility in the community, making its built settings part of its social capital [42].

Climate-resilient design incorporates these essential elements to address contemporary environmental problems and promote long-term sustainability and cultural preservation. The following sub-sections discuss specific examples of how these strategies were incorporated into Persian vernacular architecture, using Kashan as its main example, and how they can be applied in new forms to contemporary sustainable buildings and cities.

4.1 Central Courtyard

In hot, arid areas, traditional settlements responded directly to their harsh external environment. A building's structure protects the micro-climate of its courtyard(s) from the severe macro-climate conditions outside its walls [22]. In Kashan, the majority of buildings have a semi-introverted design approach. Typically, rooms are organised around the courtyard, with a semi-open space like the Iwan playing a significant role in connecting enclosed areas. The presence of a large central water pool and the inclusion of resilient plants contribute to maintaining a moist environment, enhancing the beauty of the setting. In Iranian culture, courtyards serve as the primary centers of social interactions [26].

The central courtyard in the evolution of urban planning and architecture serves as a demonstration of humanity's ingenuity in responding to environmental challenges. Considering the challenges posed by climate change, cities are striving to develop urban designs that are resilient. In this pursuit, the courtyard architecture of Kashan provides valuable insights and serves as an enduring model for constructing sustainable and living places that foster a connection between people and the environment, as well as among individuals.

4.2 Sunken Courtyard (Godalbaghcheh)

A sunken courtyard is a central courtyard situated lower than ground level. Occasionally, it was integrated into the central courtyard to give it two levels [43]. The reasons for using Godalbaghcheh were to provide more shade and help maintain a cool atmosphere during the day. It also provided convenient access to the underground canals used for irrigating the courtyard's plants and gardens, also a factor in providing shade [37]. Another advantage of using two-level courtyards was to divide the home into two distinct sections, a summer area and a winter area [11]

Incorporating underground courtyards showcases a sophisticated approach to residential design that effectively addresses both climatic challenges and functional needs. The innovative characteristic not only improved the comfort and sustainability of traditional residences but also demonstrated a profound comprehension of environmental adaptability in architecture.

4.3 Badgir (Wind Catcher)

Wind catchers are structures used in hot climates to facilitate ventilation in buildings. Their openings are positioned in the direction of the cooler prevailing breezes. The multiple "faces" of the Badgir feed individual, chimney-like air shafts that connect directly to the lowest floor of the home. The Badgir's openings efficiently transmit even the smallest air movement through ducts located beneath the walls. Typically, a water pond below the ducts humidifies and cools the air coming from the Badgir. Its design varies, but the square form is the most typical. The Badgir is separated by two partitions positioned diagonally along the length of its shaft. It can also be used in pairs or multiples on the rooftop of buildings. Aside from its function as a ventilation system, it is also used as an ornamental feature in certain structures [35].

The Badgir is a remarkable architectural development that successfully combines functionality with visual attractiveness. Through the use of natural winds and the incorporation of water features, it efficiently provides cooling and ventilation to traditional buildings in Iran and other countries. The diversity of its form and function demonstrates the creativity of traditional architecture in creating sustainable and pleasant living spaces.

4.4 Iwan (Verandas), Hashti and Corridors

Iwans are important settings in hot, dry environments. Semi-open iwans offer shaded and comfortable living spaces throughout the day. Usually facing south and east, accessed by three enclosed passageways, they are pleasant and shaded on hot summer afternoons. As mentioned, a single entrance door connects houses in this climate to adjoining public spaces [8]. Visitors enter the Hashti, an octagonal area that acts as a thermal screen against severe weather [41]. The Hashti links to the Central courtyard by a corridor. Together, they act as a thermal barrier that keeps the courtyard's cool air from dissipating [32].

In brief, the design components of iwans, Hashti, and corridors together provide a successful and visually appealing solution to the environmental issues posed by hot and dry areas. These architectural elements not only improve the comfort and livability of traditional dwellings but also showcase a profound comprehension of environmental adaptability and passive cooling strategies.

4.5 Vaults, Domes, and Air Vents

The harsh climate and scarcity of precipitation in hot and arid cities necessitate the construction of roofs in arched vaults and domes, using adobe and mud. Vents expose these volumes to the airflow generated by the wind, lowering their temperature. Vaults and domes shade the roof, reducing the amount of sunlight absorbed across its total coverage [1].

The use of vaults, domes, and air vents reflects a sophisticated approach to addressing environmental issues. These elements not only provide efficient cooling and ventilation but also improve the stability and aesthetic attractiveness of buildings. This combination of form and function speaks to the creativity and adaptability inherent in traditional Persian architecture.

4.6 Cellar (Sardab) and Basement

A subterranean cellar with a ceiling situated several meters below grade is called a Sardab. Its floor, walls, and roof are much cooler than above-grade structures, so thermal comfort is passively achieved. Some historic buildings have a subterranean Qanat waterway underneath them. In this case, the Sardab either accesses it directly, or the Qanat water flows into a tiny pond located in a covered area, the Hozekhane. It forms a central pool, daylit from above, that raises the water level and is lit through the roof. The pond and Badgir vents in the Hozekhane make it a cool oasis in contrast to the dry midday heat outside [35].

The Sardab and basement designs exemplify the practicality of traditional architecture in constructing comfortable living areas under severe climatic conditions. Below-grade construction using water elements provides a pleasant retreat from the harsh conditions above ground.

4.7 Wall Thickness, Height, Material, and Color

Using insulating materials in the walls and roofs slows down the transfer of heat, reducing its impact on the internal environment. The heat flow within the wall is directed towards the inside surface, helping to keep it cool during the day. At night, the procedure is reversed [12]. The use of internal corridors and taller walls around the courtyard on the first level also contributes to a pleasant microclimate [2]. The mud, clay, and brick used in these structures have a significant climatic benefit due to their high thermal capacity, which effectively reduces temperature fluctuations between day and night [22]. The bright color of exterior surfaces reduces the solar radiation impact on buildings and internal temperature, even without the use of air conditioning systems [30].

These elements work together to provide stable and pleasant interior spaces, revealing an advanced understanding of thermal dynamics and energy efficiency. Pragmatically using readily available, recyclable or reusable building materials anticipates what we think of now as sustainable construction methods, in alignment with the natural environment.

Fig. 4 Three different construction orientations (Rons) in Iran. *Source* Adapted from Mohammad Karim Pirnia's book and redrawn by the author (2024)

4.8 Orientation of the Construction

Building alignment to the geographical direction is an important urban morphological factor. In traditional Iranian architecture, "Ron" indicates the building's geographic orientation. Isfahani, Qiblah, and Kermani Rons, each fitting within a hexagonal area, have been identified in Iran. These orientations can prevent energy losses. Kashan buildings are built northwest to southeast, following the Isfahani Ron (see Fig. 4) [41]. The meticulous consideration of building orientation in traditional design demonstrates an in-depth knowledge of environmental and energy efficiency. Traditional architects improved the sustainability and comfort of their buildings by aligning structures based on certain geographical orientations, which optimised natural light and thermal conditions. Integrating building design with the natural environment is an enduring strategy, as these historic buildings demonstrate.

5 Agha Bozorgh School and Mosque

Built in 1851, the Agha Bozorgh School and Mosque stands along a semi-circular path lined with several distinctive structures. Within the urban structure of Kashan, a mosque often functions as a nodal point for routes. This mosque is aligned in a counter-clockwise orientation from the north, in keeping with the surrounding urban layout. The mosque incorporates a subterranean school and has an axial layout, with a courtyard and two prayer halls. The hall at the entry is given more importance owing to its structural characteristics and the presence of a dome[2] aligned with the central axis. Due to the absence of an Iwan, the cross-axis is almost imperceptible.

[2] In traditional Iranian architecture, the domed hall is called Gonbad Khana.

The central dome[3] space complex is configured in several directions around a central point. The courtyard is marked by its triple-height form,[4] accentuated by a sunken floor. The two wind towers surrounding the entryway serve as counterparts to the two minarets on top of the Pishtaq.[5] They exemplify the classic Qajar design, which uses a precisely symmetrical entrance-Mihrab axis (see Fig. 5). The subterranean school is comprised of a Shabistan,[6] which is an underground space, and a courtyard with rooms located at opposite corners [19].

Adaptive reuse is seen today as an important strategy in resilient urban regeneration, given the embedded carbon content of existing structures. Claims of "net-zero" often focus only on the running costs of a new, energy-efficient building without considering that the cycle of tear-down and replacement is fundamentally unsustainable from a carbon standpoint.

This mosque demonstrates how adaptive reuse supports resilience, not only preserving the historical structure but also revitalising it to meet modern requirements. The mosque's spacious courtyard, originally a peaceful area for reflection and worship, has been redesigned to accommodate a diverse range of cultural events, exhibitions, and social gatherings while preserving its sacred atmosphere. The traditional classrooms have been sensitively updated to facilitate current education. The revitalisation of the underground areas of the mosque is also noteworthy. The subterranean chambers, first intended as protected areas from the hot desert climate, have been transformed into exhibition areas and small lecture halls. The exterior of the mosque has also undergone adaptive reuse. The wind towers, historically used for natural ventilation, have been restored for this purpose. The mosque's library now has a digital archive aimed at preserving historical documents. By respecting the historical value of the site and adapting it to meet modern requirements, the mosque continues to be a vibrant and pertinent component of the community [26].

The Agha Bozorgh School and Mosque is an excellent example of how historical preservation and contemporary functionality can be successfully combined. This architectural masterpiece's adaptive reuse preserves its historical and cultural importance while meeting the requirements of the community in contemporary times. Given the cultural importance of historical buildings, their sensitive and sustainable redevelopment sets an important precedent for other cities.

---

[3] The central dome is a motif of Iranian architecture, having a double roof. In this example, it is composed of two shells separated by a great distance, generating a large gap between them filled with brick brackets called Khashkhashi or Parra (Parra meaning spoke) that emphasize their radial arrangement. The dome is supported by a tall windowed drum and a transitional element with the typical Persian ribbed pendentive popular in the Qajar era in the central part of Kashan's domed structures.

[4] In traditional Iranian architecture, it is called Sahn.

[5] In Islamic architecture, a Pishtaq is a rectangular frame that surrounds an arched opening and is often associated with an Iwan.

[6] The Shabistan, or nave, is a section of a mosque that is covered by a ceiling. They are confined spaces that are connected to the mosque's courtyard by a passageway and contain single and parallel columns on one side.

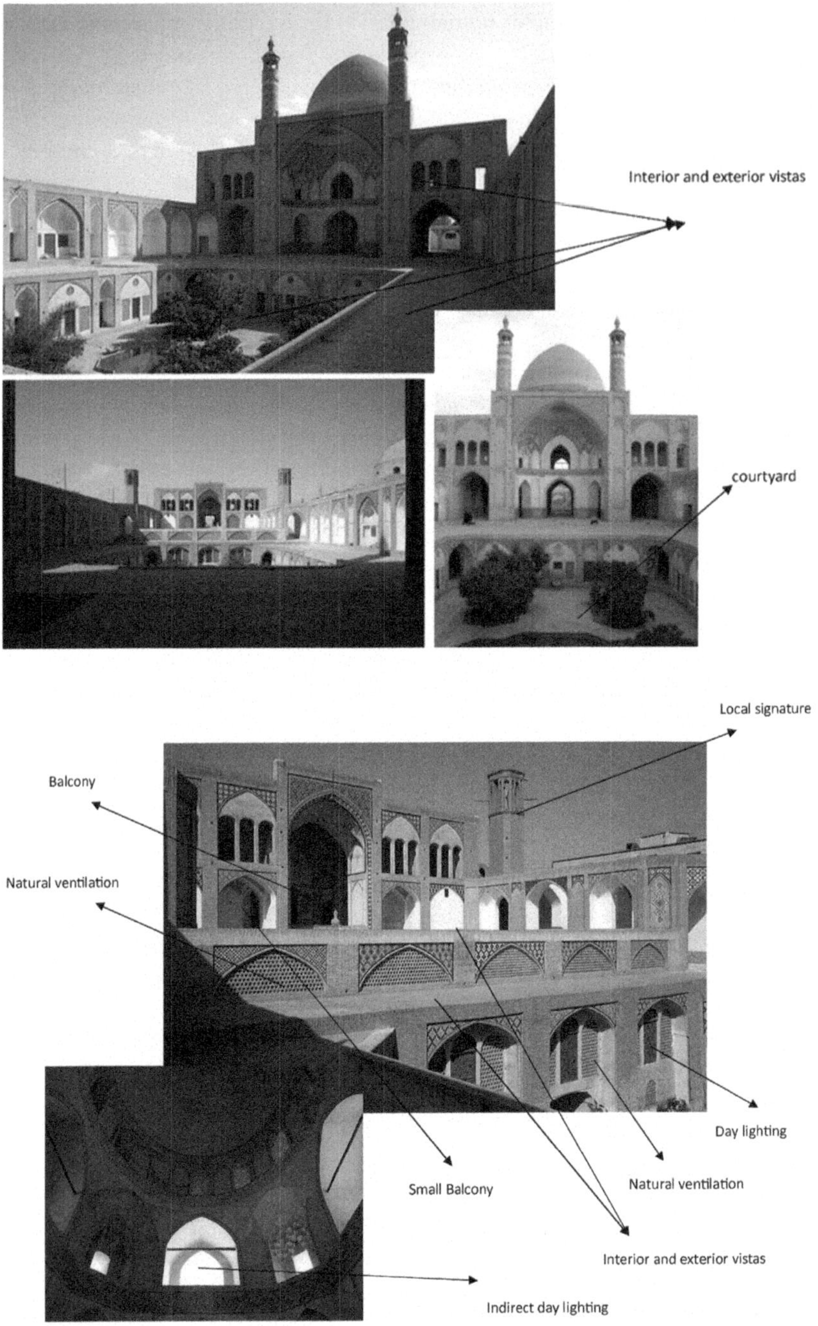

Fig. 5 Traditional Iranian elements were utilised while designing the Agha Bozorg School Mosque. *Source* Hassani [26]

6 Discussion and Implications: In Search of Cultural Fit and Optimal Performance

The architectural techniques used in Persian vernacular architecture provide valuable insights for constructing climate-resilient buildings that blend seamlessly with the surrounding environment and cultural legacy. The urban structure of Kashan, distinguished by dense building layouts and interconnecting walls, demonstrates how traditional designs efficiently reduce heat transfer and optimise climatic comfort. Where current options build on the same underlying principles, we can emulate tradition without replicating it. Green roofs, innovative water recycling systems, solar panels, materials with a high thermal mass, and insulation like aerogel can achieve the same thermal advantages as traditional buildings while meeting current tastes and regulations.

Computer-aided energy modeling and fluid dynamics simulations optimise building performance by combining new materials and systems creatively with traditional plans, forms, and elements. Dynamic facades, automated shading devices, and smart ventilation systems can improve the passive cooling and heating strategies we find in Persian vernacular architecture. This section considers the synthesis of traditional and modern architectural approaches, emphasising the application of learning from Persian vernacular architecture to contemporary practices.

6.1 *Bridging Tradition and Innovation: Modern Applications of Traditional Principles*

The principles that guided Iran's tradition of city-making and building form apply to sustainable architecture and urban development elsewhere. Cities have in common what nature imposes on them as they seek to reconcile socio-economic imperatives with environmental conditions, often now with the added stresses of temperature rise and shortages of potable water, for example and others.

This section provides examples of how to marry traditional approaches to architecture and city-making to current approaches that make use of the same underlying principles. The examples reference the elements of Iran's tradition discussed in the previous sections, to demonstrate their relevance to modern cities' need for ecologically responsible, culturally rich, and socially useful solutions that are tailored to each city's specific situation. Iran's traditions differ from those of other regions located in hot and arid climate zones, but all of them reflect human efforts to find comfort and shelter in the face of broadly comparable conditions. History suggests that we learned from each other, that tradition is a lineage that draws widely and evolves constantly. In this spirit, what follows are field notes that describe a possible bridge and then give examples.

6.2 Central Courtyard

1. Implement green roofs and vertical gardens in central courtyards to improve insulation and promote biodiversity: Green roofs can mitigate the formation of heat islands in metropolitan areas and enhance air quality. The California Academy of Sciences in San Francisco has a large green roof that imitates the natural environment and helps control the building's temperature, while also providing a home for local animals [10]. 2. Integrate advanced water recycling systems with central pools, using solar-powered pumps and filtration systems, to preserve water quality and minimise water consumption: Greywater recycling can be used to irrigate plants. The Bullitt Center in Seattle employs a greywater recycling system to repurpose water for irrigation and toilet flushing substantially decreasing its water use. 3. Integrate automatic shading systems that can adapt to the time of day and season to optimise comfort and energy efficiency: The Al Bahar Towers in Abu Dhabi have a responsive façade equipped with automatic shading devices that adjust their position to block the sun's rays, thus minimising heat absorption and glare [12].

Integrating sustainable design features like green roofs, innovative water recycling systems, and adaptive shading mechanisms in central courtyards improves individual buildings' environmental performance and that of their neighbourhoods. By adopting these cutting-edge technologies, we are able to evolve the central courtyard concept as a dynamic, environmentally sustainable refuge from the discomforts of climate change in urban contexts.

6.3 Sunken Courtyard (Godalbaghcheh)

1. Implement geothermal heating and cooling systems under sunken courtyards to utilise the stable temperatures of the earth for effective climate control: One Angel Square in Manchester's geothermal heating and cooling systems use the earth's consistent temperature to minimise energy use. 2. Integrate rainwater harvesting systems, directing rainfall to subterranean storage tanks for subsequent use in irrigation and greywater systems: The Sidwell Friends School in Washington D.C. employs rainwater harvesting, which involves the collection of water in subterranean tanks to reuse it in irrigation and toilet flushing. 3. Employ light wells and reflecting surfaces to bring additional natural light to the sunken courtyards, diminishing the need for artificial lighting: Therme Vals in Switzerland uses strategically positioned light wells to naturally illuminate its subterranean spa areas [13].

These features not only improve the efficiency of energy and water consumption, but also enhance the visual and functional character of these areas. They show that sustainability can be achieved without sacrificing architectural quality. On the contrary, the future of urbanity rests on well-considered models for regeneration that combine a strong sense of place with resiliency.

6.4 Iwan (Verandas), Hashti, and Corridors

1. Install photovoltaic panels on the roofs of Iwans and verandas to provide sustainable energy while also offering shade: The Bahrain World Trade Center incorporates wind turbines into its architectural design, effectively harnessing renewable energy while also providing shade. 2. Utilise computational fluid dynamics (CFD) models to maximise airflow and eliminate the need for mechanical cooling by designing corridors and Hashti with improved cross-ventilation: The Eastgate Centre in Harare, Zimbabwe, employs natural ventilation techniques based on termite mounds, resulting in a substantial decrease in reliance on air conditioning [5].

By integrating solar panels, using computational fluid dynamics for natural ventilation, and reimagining conventional architectural features like Iwans, Hashti, and corridors, one can greatly improve sustainability and comfort in contemporary structures. These solutions not only decrease reliance on non-renewable energy sources, but also display a harmonious relationship between modern technology and traditional design principles. By balancing innovation and heritage, they will help ensure that urban growth is efficient, resilient, and visually appealing.

6.5 Vaults, Domes, and Air Vents

1. Use high-performance, low-emissivity glazing in dome structures to enable the passage of light while reducing the amount of heat gain: The Reichstag Dome in Berlin utilises high-performance glass to optimise the entry of natural light and minimise energy use. 2. Incorporate smart ventilation systems equipped with sensors and automated controls to regulate airflow according to temperature and humidity levels, replicating the natural cooling effect offered by conventional air vents: The New York Times Headquarters has an intelligent ventilation system that adapts according to current environmental circumstances in order to enhance interior air quality and optimise energy consumption 3. Implement reflecting coverings and cool roof technologies on domes to minimise heat absorption and enhance energy efficiency. California State University, Northridge, uses cool roof coverings to deflect sunlight and decrease the amount of energy needed for cooling its campus buildings [34, 44].

Integrating solar panels, using computational fluid dynamics for natural ventilation, and rethinking traditional architectural features like Iwans, Hashti, and corridors can significantly improve sustainability and comfort in contemporary structures. They exemplify the importance of balancing tradition and innovation in order to achieve better performance while maintaining the integrity of urban areas incorporating traditional structure and growth patterns.

6.6 Cellar (Sardab) and Basement

1. Integrate phase-change materials (PCMs) in cellars and basements to store and release thermal energy, thus improving passive cooling: The Federal Center South Building 1202 in Seattle employs PCMs to control temperature and minimise the need for heating and cooling 2. Employ energy-efficient LED lighting equipped with occupancy sensors to tune artificial lighting to actual demand: The Edge in Amsterdam employs energy-efficient lighting systems equipped with intelligent sensors to decrease energy use. 3. Improve the insulation in cellar and basement walls and floors to provide consistent interior temperatures while minimising energy use: The Bullitt Center in Seattle utilises high-performance insulation in its basement to provide stable temperatures and minimise energy use [13].

Integrating phase-change materials, energy-efficient lighting with occupancy sensors, and high-performance insulation in cellars and basements greatly improves building energy efficiency and thermal comfort. These novel solutions not only decrease the need for active heating and cooling systems but also encourage sustainable construction practices. These spaces are often ignored or underutilised, but tradition shows their value to contemporary sustainable building design.

6.7 Wall Thickness, Height, Material, and Color

1. Utilise contemporary, eco-friendly materials like rammed earth, hempcrete, and recycled materials that provide comparable thermal characteristics: The Hakka Tulou structures in China, built with rammed earth, exemplify the long-lasting nature and thermal advantages of using earth elements in architectural projects [13]. 2. Incorporate an adaptive insulation technology like aerogel that provides exceptional heat resistance while requiring less material thickness than traditional alternatives: The Energy House at the University of Salford employs aerogel insulation to achieve excellent thermal efficiency while maintaining a thin profile. 3. Implement innovative paint technology on the outside walls to effectively deflect solar radiation and minimise heat absorption: The West Palm Beach City Hall in Florida employs a cool paint layer on its exterior surface to effectively deflect sunlight and minimise the need for cooling, thereby reducing energy consumption [44].

Architects can enhance the thermal performance of buildings and have a positive environmental impact by adopting such eco-friendly materials as rammed earth and hempcrete. Today, especially for Western-style housing types found in Iran after 1950, incorporating advanced insulation technologies like aerogel, and utilising reflective paint solutions to mitigate heat gain are alternatives to the thicker walls rammed earth and hempcrete necessitate.

6.8 *Orientation of the Construction*

1. Perform comprehensive solar path analyses to optimise the orientation of the building for optimal solar gain during winter and shade during summer: The Masdar Institute in Abu Dhabi is specifically oriented to optimise solar efficiency and decrease energy consumption. 2. Employ dynamic facades with changeable shading mechanisms that adjust to annual variations in solar angles: The Al Bahar Towers in Abu Dhabi include a responsive exterior that adapts to maximise shade and capture solar energy [13]. 3. Employ energy modeling software to simulate and enhance the energy efficiency of the structure by considering its orientation: The NREL Research Support Facility in Golden, Colorado utilised energy modeling to enhance its design for optimal solar efficiency and energy performance [44].

The alignment of a structure is of utmost importance to its energy efficiency and total ecological footprint. Architects can design buildings to maximise solar energy and minimise heat gain during peak summer months by making comprehensive sun path assessments and applying advanced energy modeling tools. Integrating dynamic facades that respond automatically to changing sun angles in different seasons can further optimise building energy efficiency and performance.

6.9 *City-Making Using Traditional Patterns*

1. Look for opportunities to extend traditional patterns and concepts to current, multi-family contexts. Iran Lab's[7] Fatemeh Behfar and Maryam Merajoddin's "Cultural Roof" proposal in Bushehr, Iran, draws from the city's courtyards and rooftops. Bushehr is hot and humid, with tall buildings and narrow streets for shade. Rooftops in Bushehr have long been popular for hot summer nights and community gatherings. This concept intends to update this heritage by turning rooftops into cultural and musical spaces. The project aims to make Bushehr a more dynamic location to live and visit by turning rooftops into social and performance spaces. Bushehr's climate and environment shaped its architecture, and the project revived its use of rooftops as social gathering elements [9].

Applying traditional patterns of city-making to contemporary urban design presents a significant opportunity to enrich community life and increase resilience. Initiatives like the "Cultural Roof" project in Bushehr demonstrate how traditional practices can be creatively adapted to promote social engagement and cultural vitality

[7] This project was among so many submitted proposals to participate in the call for vision 23 at Iran Lab. The author has been collaborating as a postdoc researcher with this research lab since 2023. Iran Lab, in keeping with the PDTA Department's cultural project at Sapienza University of Rome, is part of the strand of cultural exchanges between scholars, researchers, PhD students and students in the fields of design, technology and urban regeneration, with the aim of coordinating, directing and enhancing research and studies on Iranian architecture, experimenting, where possible, with forms of contamination between the two cultures. For more information, please refer to the website: https://www.iranlab.it/

in contemporary contexts. This strategy not only recognises and values cultural identity, but also fosters a feeling of community and inclusion, enhancing the livability and attractiveness of cities for both inhabitants and visitors.

7 Case Study: Rotary Wind Catchers in the Bluewater Shopping Centre

In contemporary wind catcher design, the two ventilation principles of the wind tower and the passive stack are integrated into a single design. The stack is divided into two halves or four quadrants, with the division extending the entire length of the stack [25]. The roles of each half or quadrant in the wind catcher vary as the wind direction changes. This ensures that the wind catcher remains functional regardless of the wind direction. Since wind catchers do not have any moving parts, their maintenance requirements are minimal. One advantage of the wind catcher is that it takes in fresh air from the roof, where it is usually cleaner than if taken in at ground level. This is especially important in urban locations where the building is near a road [14, 15]. The Bluewater Shopping Centre in Kent is a recently built commercial complex located outside urban areas. The structures use wind catchers to ventilate their interiors. The design of the wind catcher's cowl is derived from the traditional structure of a Kent oats house. The houses were prevalent in the region and eventually became a distinctive architectural element in contemporary structures across Kent (Fig. 6) [6, 40].

The Blue Water Center's wind catchers are a pioneering example of integrating a traditional concept with modern technology. They show that wind catchers are applicable to southeast England's temperate climate. A total of 39 rotating cowl wind catchers are placed over the three malls. The wind catchers are components of a centralised air conditioning system that is automatically regulated by air handling units. Building Energy Management Systems (BEMS) provide a comprehensive

Fig. 6 Traditional rotating wind catcher, Oast houses and rotating wind catcher at Bluewater Shopping Centre, both in Kent, UK. *Source* Elmualim and Awbi [16]

control system that detects equipment malfunctions and sends alerts, while gathering data on exterior wind conditions, temperature, humidity, and inside environment parameters [16].

The Bluewater Shopping Centre's wind catchers derive from a different tradition than the Persian Badgir, but both were developed empirically on the basis of wind conditions and climate. Thus, the underlying principle is the same, although their form and appearance differ. The use of wind catchers in the Bluewater Shopping Centre is an example of the successful integration of traditional architectural concepts with contemporary technologies. By modifying a historical idea to comply with current environmental requirements, the wind catchers not only improve the quality and comfort of interior air, but also decrease energy consumption linked to mechanical ventilation systems. This case study shows how both traditional and innovative new solutions can emerge from the same climatic and environmental circumstances, based ultimately on the same underlying principles. Hybrid approaches to designing resilient and ecologically responsible urban development show particular promise in resolving apparent dilemmas such as the need to balance indoor air quality and comfort with sustainable building performance.

8 Concluding Remarks

A. **Key Findings and Contributions**

This review of Persian vernacular architecture focused on its climate-adaptive techniques, shows how traditional master builders integrated resilience with cultural identity using neighborhood patterns, building layouts, orientation, locally obtained materials, and other strategies. The Kashan mosque and school case study illustrate how traditional techniques can influence contemporary architecture, notably in adaptive reuse and sustainable design. It suggests that the strategies found in Iran's building and city-making traditions, contending with an arid climate, can be combined creatively with modern technologies and materials. Each traditional concept is in effect an invitation to current practitioners to synthesise and innovate. As the Bluewater case shows, this is equally true for other traditions and climates. Any number can play.

B. **Future Directions and Further Study Areas**

This chapter points to the importance of considering traditional architecture and city-making on a comparative basis, in order to catalogue shared and divergent principles and practices that are applicable to our environmental challenges. Specifically, we should explore how tradition can inform today's buildings' long-term efficiency and adaptability. We should also consider how those who plan, design, approve, and implement the built environment's transformation interact with each other and with the communities impacted by their activities. If the effects of climate change live up to science's baleful predictions, making common cause to contend with them will be crucial. Tradition reminds us that we have been here before, that each generation stands on the shoulders of earlier giants, as Isaac Newton put it, resolving the challenges life hands us.

Acknowledgements John J. Parman, a Berkeley-based writer-researcher, kindly provided editorial advice to the author as she developed this chapter.

References

1. Akbari H, Konopacki S, Pomerantz M (2012) Design fundamentals in the hot and humid climate of Iran: the case of Khoramshahr. Canadian Center of Science and Education 4:65–72
2. Alkhalidi A (2013) Sustainable application of interior spaces in traditional houses of the United Arab Emirates. Procedia—Social Behav Sci 288–299
3. Amanat M (2018) KASHAN iii. History to the Pahlavi Period, Encyclopædia Iranica. Online edition. Available at http://www.iranicaonline.org/articles/temp-kashan-history. Accessed on 23 Jan 2018
4. A'zami A, Yasrebi SH, Salehipoor A (2005) Climatic responsive architecture in hot and dry regions of Iran
5. Bahadori MN (1978) Passive cooling systems in Iranian architecture. Sci Am 238(2):144–154
6. Barnwell PS, Adams AT (1994) The house within: interpreting medieval houses of Kent. Royal Commission on the Historical Monuments of England. HMSO, London
7. Beckers B, Berhanu D, Brook GA (2013) Traditional water harvesting in semi-arid northwestern Ethiopia. GeoJournal 78(5):845–858
8. Bonine ME (1980) Aridity and structure: adaptations of indigenous housing in Central Iran. Desert housing: balancing experience and technology for dwelling in hot arid zones. University of Arizona, Office of Arid Lands Studies, Tucson, AZ, USA, pp 193–219
9. De Cesaris A (2024) The results of call for visions 23: new public spaces for contemporary Iranian cities. Aracne Publisher, Rome, Italy
10. Chen A (2004) The New York Times building: designing for energy efficiency through daylighting research. Science|Beat, Lawrence Berkeley National Laboratory
11. Cho S, Mohammadzadeh N (2013) Thermal comfort analysis of a traditional Iranian courtyard for the design of sustainable residential building. In: 13th Conference of international building performance simulation association, Chambéry, France, pp 2326–2333
12. Coles A, Jackson A (1975) Wind tower house in Dubai. AARP, London
13. Edwards B (2010) Rough guide to sustainability: a design primer. RIBA Publishing
14. Elmualim AA, Awbi HB (2002a) Wind tunnel and CFD investigation of the performance of wind catcher ventilation systems. Int J Vent 1(1):53–64
15. Elmualim AA, Awbi HB (2002b) Full scale model wind catcher performance evaluation using a wind tunnel. In: Sayigh AAM (ed) World renewable energy congress VII (WREC 2002), 29 June–5 July, Cologne, Germany. Elsevier Science Ltd.
16. Elmualim AA, Awbi HB (2004) Evaluating a control strategy for a hybrid air-conditioning and windcatchers ventilation system. In: CIB 2004 World building congress, Toronto, Canada
17. Fathy H (1986) Natural energy and vernacular architecture: principles and examples with reference to hot arid climates. University of Chicago Press
18. Foruzanmehr A, Nicol F (2008) Towards new approaches for integrating vernacular passive-cooling systems into modern buildings in warm-dry climates of Iran. Build Environ 43(4):631–646
19. Gaube H, Neglia GA, Petruccioli A, Rafipoor F (2018) Kashan, an Iranian city in change. EB-Verlag Dr. Brandt Publication, Berlin
20. Ghobadian V (1996) Analyzing Iran's traditional structures from a climatic perspective. Tehran University Press, Tehran

21. Ghobadian V (2003) Barresiye Eghlimiye Abniye Sonati Iran (Climatic analysis of the traditional Iranian buildings), 2nd edn. Tehran University Press, Tehran (In Persian)
22. Ghobadian V (2009) Sustainable traditional buildings of Iran, a climatic analysis, 1st edn. Islamic Azad University, Dubai campus, Dubai
23. Ghobadian V (2014) Climate and architecture. Tehran University Press, Tehran
24. Graham S, Thrift N (2007) Out of order: understanding repair and maintenance. Theory Cult Soc 24(3):1–25
25. Harris DJ, Webb RS (1996) Wind Towers—old technology to solve a new problem. In: The 17th AIVC conference. Optimum ventilation and air flow control in buildings, 17–20 Sept, Gothenburg, Sweden, pp 613–621
26. Hassani EK (2022a) Schools as community centers: transformation of schools as elements to regenerate the neighborhood in the historical urban fabric, case study: Kashan, Iran. Doctoral dissertation, Sapienza University of Rome, Italy
27. Hassani EK (2022b) "The effects of super grids and superblocks on the transformation of the historic urban fabric of Kashan city in Iran," ISUF 2022, Urban Redevelopment and Revitalization, Łódź–Kraków, Poland
28. Hassani EK (2021) "Investigating the role of urban morphology in resilient cities," ISUF 2021, Urban form for sustainable and prosperous cities, University of Strathclyde, Glasgow, UK
29. Hassani EK (2020) "Schools as elements to regenerate the communities of contemporary cities," ISUF 2020, Urban substrata and city regeneration, ISUF Italy, Rome
30. Karam M, Al-Obaidin M, Abdul Malek A (2014) Passive cooling techniques through reflective and radiative roofs in tropical houses in Southeast Asia: a literature review. Front Arch Res 283–289
31. Kasmai M (1993) Climate zoning in Iran, housing and residential environments. Building and Housing Research Centre, Tehran
32. Khalili M, Amindeldar S (2014) Traditional solutions in low energy buildings of the hot-arid region of Iran. Sustain Cities Soc 13:171–181
33. Leylian MR, Amirkhani A, Bemanian MR, Abedi M (2010) Design principles in the hot and arid climate of Iran: the case of Kashan. Int J Acad Res 2(5):278–283
34. MASS Design Group (2015) Purpose built case study: the California academy of sciences. Candid|Issue Lab
35. Maeiyat M, Movahed N, Maeiyat F (2009) Sustainable housing in semi-rural settlements. In: 45th ISOCARP congress. Tehran University Publications
36. Mirmiran H (2011b) Development and construction comprehensive plan of Kashan city, vol 2. Naghsh Jahan Pars Consulting Engineering
37. Moradi S (2005) Environmental control system, pp 128–134. Tehran: 34–41, 55–57
38. Olgyay V (2015) Design with climate: bioclimatic approach to architectural regionalism. Princeton University Press
39. Parry ML, Canziani OF, Palutikof JP, van der Linden PJ, Hanson CE (eds) (2007) Climate change 2007: impacts, adaptation and vulnerability. In: Contribution of working group II to the fourth assessment report of the intergovernmental panel on climate change. Cambridge University Press
40. Pearson S (1994) The medieval houses of Kent: an historical analysis. Royal Commission on the Historical Monuments of England, HMSO, London
41. Pirnia M-K (2010) Islamic Architecture of Iran. Soroush Danesh, Tehran
42. Pretty J (2003) Social capital and the collective management of resources. Science 302(5652):1912–1914
43. Soltan Zadeh H (2011) The role of geography on the formation of courtyards in traditional houses in Iran. Hum Geogr Res 43(74):69–86
44. U.S. Department of Energy (2021) Energy efficiency & renewable energy. Retrieved from [https://www.energy.gov/eere/office-energy-efficiency-renewable-energy]
45. Yung EHK, Chan EHW (2012) Implementation challenges to the adaptive reuse of heritage buildings: towards the goals of sustainable, low carbon cities. Habitat Int 36(3):352–361

Iranian Vernacular Architecture: An Alter-Na[rra]tive

Ali Rad Yousefnia

Abstract Iranian [Persian] vernacular architecture is often associated with warm temperatures and dry conditions in the literature. Historic design solutions and centuries-long technological advancements to tackle and solve extreme heat conditions with novel 'smart' and 'sustainable' cooling ideas are the mainstream narratives in the existing literature. While the major focus of existing literature is the hot, arid climates, this chapter addresses two emerging gaps in the scholarship. Firstly, little account is taken of the historical accounts of the very diverse geographical landscapes and climatic conditions in Iran. To address this gap, narratives, and references are drawn from historical texts, such as descriptions of the country taken from the Balkhi School of Geographers that date back to the tenth century and combined to create a continuous picture of the climate and historic geographies of Iran. As detailed historical discourses on the regional climates, divisions and terrains of Iran are discussed for its first gap, the chapter focuses on Azerbaijan as one of the regions that has been described by the scholars of the Balkhi school. Three distinct examples of vernacular architecture and related exemplars are discussed, emphasising the role of the regional meso-climatic influences on these settlements. Some historical information about each village is accompanied by a contemporary photographic record and site observations. These narratives aim to provide a broad-based understanding of diversity in the village typologies in the selected case studies. While 'vernacular' buildings are usually portrayed and perceived as fairly homogenous in the Iranian context, this chapter offers an alternative narrative based on emerging literature on Persian vernacular architecture described here to expand the general understanding of the heterogeneity of its architecture.

Keywords Geography · Balkhi school · Village settlements · Azerbaijan · Iran

A. Rad Yousefnia (✉)
UQ College (UQC), The University of Queensland, Brisbane, Australia
e-mail: a.radyousefnia@uq.edu.au

A. Cheshmehzangi and S. Roaf (eds.), *Persian Vernacular Architecture*, Urban Sustainability, https://doi.org/10.1007/978-981-96-1116-4_16

1 Introduction

Throughout history, climatic conditions compelled appropriate responses in the shaping of the built environment not only to protect occupants from the extremes of climate outdoors but also to maximise the comfort of people indoors. Along with the available local materials, perhaps climate was one of the significant factors in creating different types of traditional, ecologically responsive solutions [1, 2]. These solutions were developed and evolved in harmony with the knowledge and skills of local craftspeople, usually known as *Ostads*.

In the first lines of his book, *Architecture without Architects*, Rudofsky points out: "Vernacular architecture does not go through fashion cycles. It is nearly immutable, indeed, unimprovable, since it serves its purpose to perfection" [3]. In addition, Rudofsky also highlighted the continuous biased focus of Architectural History and related scholarship on the Western World, criticising the fact that there is only a *total* picture of the rest. Following that, he calls the rest 'vernacular, anonymous, spontaneous, Indigenous, rural' since there was no defined categorisation for non-Western architecture at that time. Within these frames, Western scholars writing on the late nineteenth-century rural, non-monumental buildings were often antiquarians and architects who focused on movements such as the Arts and Crafts styles, largely ignoring the non-Western architecture [4–6]. After sixty years of Rudofsky's important and provocative publication, it seems timely to question his assumptions [7]. Did the literature of different regions describe diverse building typologies and geographies? Using the under-represented primary sources from the tenth century and literature vis-a-vis examples from different villages in the Azerbaijan region, some conclusions to this question are drawn and discussed.

2 Definitions and the Origins of *Klima, Iqlim* and Climate

The fascination of humans with the earth–*geo*, and related writings and descriptions–*graphos*, formed the field of geography. As the earliest of these descriptions we know of, Herodotus (484–425 BCE) explained the geography, social structure, and history of the Persian Empire in the *Histories* [8]. In doing so, Herodotus referred to multiple geographical components of the Persian Empire, such as its immense size, diverse landscapes, floating rivers, climate and resources. Persians have always been considered the '*Other*' by Herodotus [9, 10]. This ideological 'Other' was the largest empire known at that time, which touched and controlled the landscapes of Central Asia, the Indus Valley, Anatolia and the Nile basin [11]. Since then, Persia evolved through the arrival and emergence of new religions and ideologies, thus heavily influenced by the invasions and wars that changed its face continuously.

In her seminal book *Frontier Fictions*, Kashani-Sabet once again reminded us about the importance of geography in understanding the land and the categorisation of the world in the scholarship that emerged during the tenth century throughout the Islamic Civilisation by Persian and Arab geographers. According to Kashani-Sabet:

> Long before the French Revolution, an anonymous geographer of the tenth century had already imagined the "boundaries of the world" [*hudud al-alam*] [12] as compromising distinct regions [*nahiyat*]. According to this unknown figure, each territory varied from another "First, by the difference of water, air, soil and temperature [*garma-va-sarma*]. Secondly, by the different of religion, law [*shariat*], and beliefs [*kish*]. Thirdly, by the difference of words [*lughat*] and languages. Fourthly, by the difference of kingdoms [*padshai-ha*] [13].

This anonymous geographer's attention to differences in water, air, soil and temperature—dividing it into warm [*garma*] and cold [*sarma*] climates gives us further insights into the geographical knowledge almost a thousand years ago on seasons and temperature differences. The era in which this anonymous geographer lived coincided with the golden age of science and art in the Islamic world, particularly in geography. Some of the most well-known geographers resided in the Balkhi School of Geography, often known as the 'classical' era of geography in the Islamic world. Four of the most prominent scholars from the Balkhi schools are Abu Zayd Ahmad al-Balkhi (? d. 322/934), Ibrahim B. Muhammad al-Istakhri (d. c. 350/961), Abu Dl-Qasim b. Hawqal (d. c. 380/990) and Muhammad b. Ahmad al-Muqaddasi (d. after 378/988) [14]. Among them, al-Istakhri played a significant role in depicting a holistic understanding and interpreting the known world until that time.

By relying on the collaborative works of the Balkhi school, Ibn Hawqal and al-Istakhri exchanged ideas and created some of the sophisticated and thought-provoking maps of the world and regions known to them. In *Masaalik va Mamaalek* [Pathways and Kingdoms], which dates back to 957/8, Ibn Hawqal mentioned: "Our plan is to describe and delineate on maps the various seas or oceans which surround the land, the inhabited and the desert islands, and every climate and region of the earth; affixing the name of each, so that it may be known in the maps; and confining ourselves to those countries which are the seat of *Islam*, and the residence of true believers" [15]. Al-Istakhri's division of province(s) originated in the idea of different climates, or in other words, *iqlim*. Depending on their size and differences, al-Istakhri composed maps as single or multiple regions (Fig. 1). A complete list of al-Istakhri's maps that could also be found in his manuscripts are: (1) World map; (2) Arabia; (3) Indian Ocean; (4) Al-Maghrib [North Africa]; (5) Egypt; (6) Syria; (7) the Mediterranean Sea; (8) al-Jazirah [Upper Mesopotamia]; (9) Iraq [Lower Mesopotamia]; (10) Khuzistan; (11) Fars; (12) Kerman; (13) Sind; (14) Armenia, Arran (Alvan), and Azerbaijan; (15) Jibal (central Persian mountains); (16) Daylam and its neighbours (Rayy, Tabaristan); (17) Caspian Sea; (18) Persian Desert; (19) Sijistan; (20) Khurasan; (21) Transoxiana [16].

One of the outcomes of the collaborative work and a major region discussed in the literature by Ibn Hawqal and al-Istakhri is Azerbaijan (Fig. 2). Tibbetts highlighted these exchanges and referred to Ibn Hawqal's direct lines as:

> al-Istakhri had drawn a map of Sind, but he had made some mistakes, and he had also drawn Fars, which he had done extremely well. For my part, I had drawn the map of Azerbaijan

Fig. 1 Map of the world, by al-Istakhri. Courtesy of Leiden University Library. Public Domain

> (Figs. 2 and 3) which occurs on the following page and of which he approved, as well as that of al-Jazirah, which he considered excellent. My map of Egypt, however, he condemned as wholly bad and that of al-Maghrib as, for the most part, inaccurate [17].

As a result of historical evolution, the definition of *Klima* originated in Ancient Greece, meaning inclination and slope, adapted by Balkhi geographers. Persian and Arab geographers evolved these ideas through the application of mathematics, adding geographical descriptions, and eventually establishing climate zones. Definitions of these kinds became the predecessors of climate classifications, such as Köppen in the late nineteenth century and later evolved to Köppen-Geiger in the 1960s.

Almost the same categorisation of regions could be seen in the works of well-known Russian scholars, such as Vasily Barthold (1869–1930). In his book *An Historical Geography of Iran* (translated and published in English in 1984), Barthold divided Iran into several regions [18], as shown in the map in Fig. 4. It is possible to relate how he was influenced by medieval scholars, such as the ones from the Balkhi school.

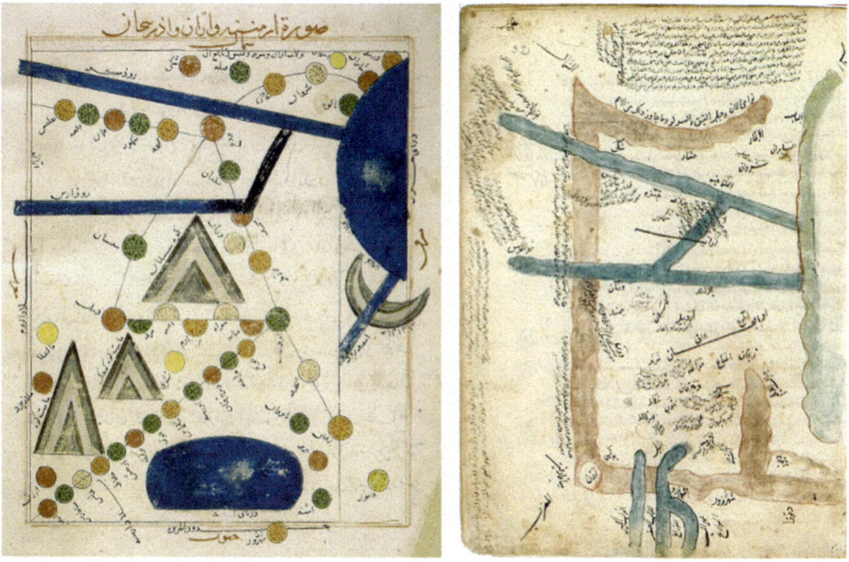

Fig. 2 Map of Armenia, Arran and Azerbaijan by al-Istakhri (left) and Ibn Hawqal (right). *Image Sources* Courtesy of Leiden University Library. Public Domain (left). Ibn Hawqal's Surath al-Ardh, BNF Paris, MS Arabe 2214. Public Domain (right)

3 Iran and the Climatic Sub-divisions

Modern-day Iran, as defined by its political borders, is between latitudes 24 and 40° N. The area is known as mid-latitudes, which creates a sweet spot for four seasons. While the southern parts of the country experience hot, dry summers with mild winters, the northern parts—bordering the Caspian Sea and the mountainous regions of Alborz and Azerbaijan are associated with different conditions. The seasonal changes of these areas are usually related to pleasant and mild summers but severe cold and snowy winters. Thus, the notion of four seasons also changes from one region to another.

Iran also exhibits a diverse range in terms of altitude. This diversity originates in its low-level seashores around the Caspian Sea and rapid increase in altitude, reaching two major mountain ranges. Some parts of the Caspian shores have been recorded around 30 m lower than the sea levels, and the peaks of the mountain ranges reach above 5500 m high. As its highest peaks, the order is Mount Damavand—5609 m, Alam—4850 m, and Sabalan—4811 m, respectively, which shape the three highest peaks in the Iranian Plateau. The juxtaposition of these different ranges in latitude and altitude creates various meso-climatic conditions over Iran's vast geography.

In terms of using modern-day classifications, Raziei employed and adopted the three most common climate classification methods in Iran: Köppen-Geiger, Feddema, and UNPEP climate classifications. The research found that:

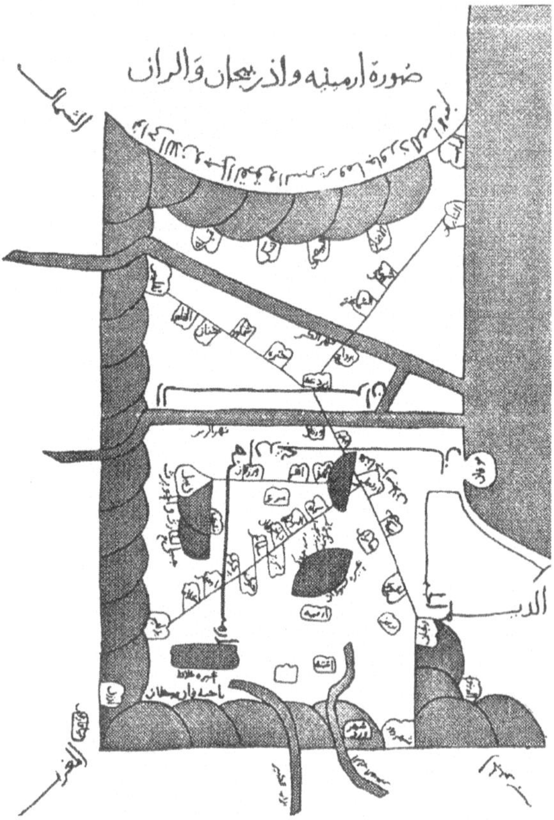

Fig. 3 Simplified version of Ibn Hawqal Azerbaijan map, showing all the cities and different regions from Dailam on the Eastern side to the Western mountains in the region. *Image Source* Wikimedia Commons. Public domain

> Iran encompasses ten out of the 31 possible climate types, from which "Bwh," "Bsk," "Csa," "Bsh," and "Bwk" are the five major climate types of Iran that cumulatively account for more than 98% of the territory. Likewise, from the 36 possible climate types of the simpler Feddema's classification, fourteen climate types were found in Iran, from which the climate types "dry cool," "semiarid torrid," "semiarid hot," "semiarid warm," "dry warm," "semiarid cool," and "moist cool" are the seven main climate types of the country that collectively occupied approximately 93% of the country [19].

The findings of this research attest to the diversity of the climatic conditions in Iran despite the difference in the land size compared to the Tenth century. In all models, the classified types are almost 50% or more of the overall typologies across all three methods. However, the diversity of climate is usually reflected in the existing literature, and there is an overwhelming attention to BWh (dry arid climate) and BWk (cold desert climate) regions, which are around 42% of the land mass of Modern-Day Iran (Fig. 5). However, further research and modelling suggest changes in the future, with the top findings being the decrease in the annual rate of precipitation, *vis a vis* the changes in its patterns, and the increase in the mean annual temperatures [20].

Following BWh and BWk climate zones, the second-largest climate classification within the borders of Modern-day Iran is BS, which includes dry semi-arid lands

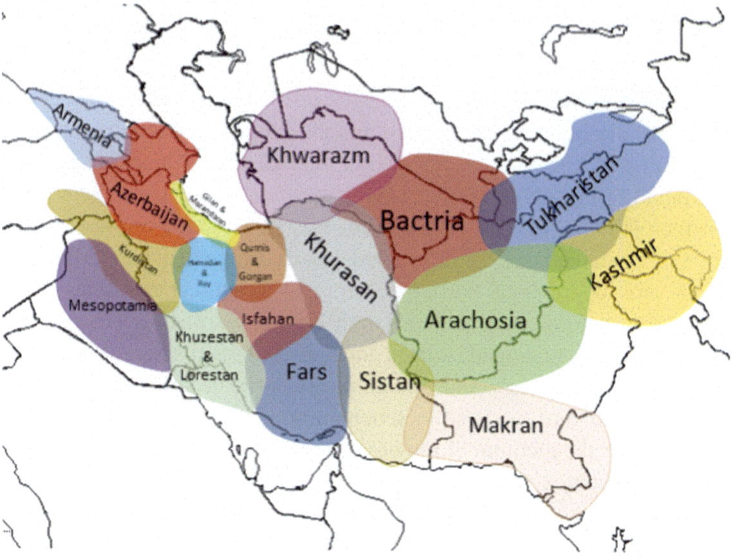

Fig. 4 The regions of Persianate Iran depicted by Barthold in an historical geography of Iran and their overlay on the current political borders. Illustrated by the author

known as steppes (BSk–23.69%, and BSh–15.70% account for 39.39% of the land mass). This is also the most relevant climate classification for the northwestern parts of Iran [21]. Within this classification, the following section highlights and explains four different typologies and how some of them have been influenced by geographical characteristics of their built environment vis a vis different altitudes, which created meso-climates within the region.

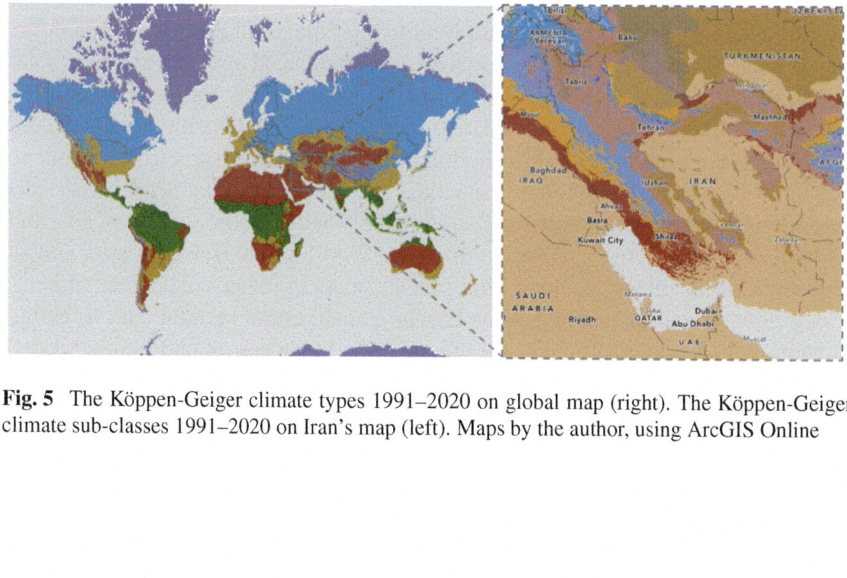

Fig. 5 The Köppen-Geiger climate types 1991–2020 on global map (right). The Köppen-Geiger climate sub-classes 1991–2020 on Iran's map (left). Maps by the author, using ArcGIS Online

4 Case Studies: Three Village Typologies, One Region

Iran's population used to be diverse in terms of their settlement types and mobilities. However, the numbers show a significant change in recent years since the urban population reached a 75% rate in Iran in 2022 [22]. Abrahamian divided the Iranian population into three segments as: tribesmen [*iliyati*], villagers [*dehqani*], and townsmen [*shahri*] [23]. As the past records show, many of these populations used to be self-contained, self-sufficient and usually isolated, due to the lack of navigable rivers and geographical barriers. During twentieth-century Iran, cities were intended to be transformed by Authoritarian high modernism approaches [24]. In contrast, most villages remained isolated and self-contained, operating much as they had in the nineteenth century and earlier, particularly in terms of their economic activities [25]. As one of the best documentation of village settlements, they have been recorded over the medium of movies in Iranian cinema.

In the Western context, the village is associated with the city in the same frame, and as AlSayyad says: "Today's cinematic city often reflects the nostalgic desires of postmodernity that translate in acts of replication and self-representation-as attempts to go back to the lost village, the family, and the community" [26]. However, in Iranian neo-realism cinema, the village is the self and starts to show itself as an independent entity without the need to be justified by relying on the meaning and existence of the city.

These isolated environments became the focus of avant-garde and pioneer scholars and intellectuals in cinema, such as Dariush Mehrjui. Mehrjui's fame came to recognition after one of his earliest movies was titled Cow [Gav] [27]. He depicted, documented and visualised the everyday lives of the villagers in an actual village settlement, giving rise to the genre later known as neorealism in Iranian cinema. What is important and in the interest of our discussion is an anecdote by Ali Nassirian, one of the leading actors in the movie. In an interview with Deutsche Welle (DW) World Service, Ali Nassirian explained the site selection of the movie. He explained that for the site selection, the production team went to the villages of Azerbaijan and documented them for the site selection. Visual materials produced from these villages were called 'shocking' [tekaan dahande] in terms of people's lives by Nassirian. According to him, the Ministry of Culture and Art strongly opposed making movies that would showcase these settlements [28]. The final selected site was somewhere closer to Qazvin, with minor alterations of a village named Buinak. Perhaps Nassirian's descriptions of the villages of Azerbaijan clearly indicate the quality of life and the built environment of the villages during the 1960s in Iran. However, after almost fifty years, these villages came in front of the camera in movies of another contemporary director and intellectual, Jafar Panahi. Panahi chose his film sites to be the authentic villages in Azerbaijan. By reflecting on his ancestral and familial roots, Panahi picked two almost untouched villages in the current-day Azerbaijan Province in Iran and used the existing houses and site for his movie to be shot there [29, 30].

Considering the climate, building materials, rich history and many other layers that compose these villages, the author visited three distinctive villages in the province

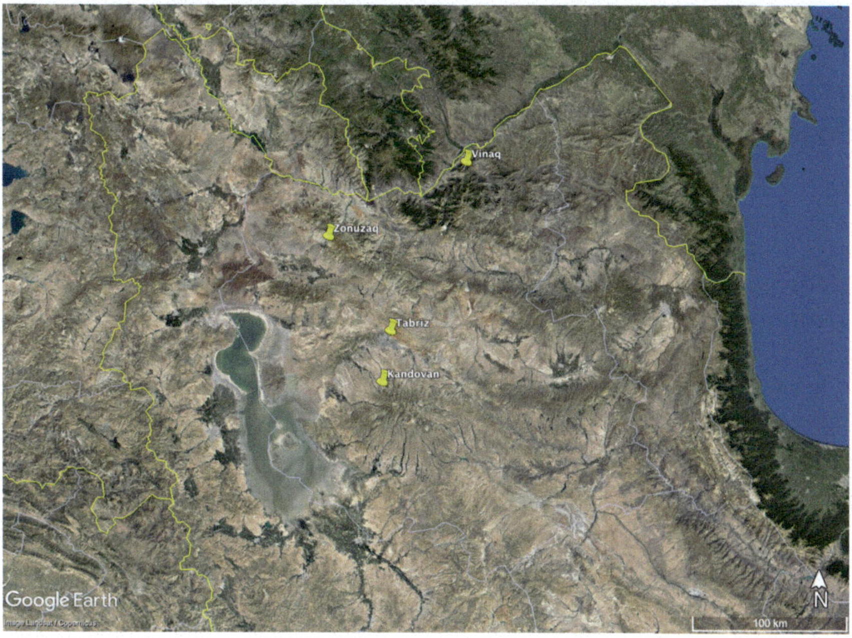

Fig. 6 Key map indicating the location of each village and their distance from the Centre of the Province. It also highlights the topography and altitude changes in the province. Map created by the author using Google Earth Pro

of East Azerbaijan in Iran, and each of them represents certain characteristics and qualities for the building techniques, materials and associated climates (Fig. 6). This section particularly focuses on these case studies by relying on the photography documentation of these villages between 2010 and 2015, and aerial views extracted from the Google Earth Pro.

4.1 Kandovan

Kandovan is situated in the foothills of Mount Sahand (3707 m) at an altitude of 2400 m. The most distinctive characteristic of the village is its rock-cut architecture. While the origins of the village are still debatable, records show Kandovan has been inhabited continuously for the past eight to nine hundred years [31]. The morphology of the village followed the natural formation of the cone-shaped rock structures carved to create the settlements. Meanwhile, additional extensions and newly built settlements threaten the original layout of the village.

Amid the cold and snowy winters, the stone structures protect the residents from harsh winds in these seasons. The village's pathways follow the natural topography of the hills (Fig. 7). There is an apparent blurring between the private and public spaces

Fig. 7 Panoramic view of Kandovan (top left) by the author, The layout of Kandovan and its surroundings (bottom left) from Google Earth Pro, views of the alleys and the rock-cut Cone-shaped settlements. Photographs by the author (right)

in the village, and by entering narrow alleys, the boundaries disappear entirely in the village. Also, humans and non-human residents of the village live in very close proximity, keeping the livestock attached to the houses. Studies on the geological properties of the houses are threatened by damage and destruction due to weak internal structures and atmospheric factors and need immediate attention for their preservation [32].

4.2 Vinaq

By travelling about 150 km to the north-east of Kandovan, we reach Vinaq. Here, the altitudes drop significantly to the range of 1200 m. This distinct shift in the altitudes creates meso-climates in the region and represents different characteristics in the climatic conditions. Vinaq is located at the Southern side of the Aras River and the junction of three countries, Iran, Azerbaijan and Armenia. While the village is majorly absent in scholarly works, some of its significant buildings reveal its multi-layered forgotten history *vis a vis* different construction techniques and materials. Perhaps the missionaries who reached these lands to spread their religions and made chapels also carried their construction techniques, which gave rise to some of these unique structures. Unlike the ordinary mud-brick village homes, most buildings are out of stone in this particular village.

Fig. 8 The layout of Vinaq (left) from Google Earth Pro. Four views and stone-masonry buildings in Vinaq (right). Photographs by the author

The general morphology of the village also does not follow the regular pattern in other villages that try to increase the heat mass by building houses attached and closer to each other to increase the effect (Fig. 8). The chapel, arches and stone-masonry buildings could be traced to the Christian missionaries who came to the region to establish their places of worship. These parts of the geography are usually warmer than other parts of Azerbaijan in some seasons, up to 10 °C. All these differences are reflected in the dominant vernacular typologies in this part of Azerbaijan.

4.3 Zonuzaq

With a distance of around a hundred kilometres towards the southwest of Vinaq, the altitudes reach around 1900 m high after passing mountain ranges with peaks as high as 3000 m. Among these valleys resides the stepped village of Zonuzaq. The village is strategically located using the steep slope of this hilly area to benefit from the existing topography. Surrounded by lush green gardens, the village faces south and stretches on an east–west axis. The village morphed itself into the natural topography and created a step-like structure. This strategic move benefits the village on two fronts: Firstly, the houses are covered by eath and balance the temperature with the ground in all seasons of the year; secondly, the step-like structure of the houses in the village maximises the thermal masses of the houses and avoid loss of heat and energy during the colder periods of the year (Fig. 9).

Fig. 9 The layout of Zonuzaq (left) from Google Earth Pro. Three views of Zonuzaq, showing its characteristics and occuring transformation (right). Photographs by the author

5 Concluding Marks

This chapter outlined alternative narratives about the diverse geography of Iran by focusing on one of its underrepresented regions in the literature, Azerbaijan, despite its rich history, geography and diverse vernacular typologies. This study has provided the thread of continuous descriptions of different climates in this region dating back to the tenth century. The chapter also highlighted the climatic diversity of Azerbaijan, and, in a broader sense Iran, which could be traced forward to modern-day climate classifications.

The chapter contains three detailed case studies of Kandovan, Vinaq and Zonuzaq in contemporary times and two historical narratives of scholars from the Balkhi School of Geography. By overlaying those regional descriptions with the Köppen-Geiger climate classification, the chapter shows how this region differs from the mainstream understanding of the climate classification and typologies of the vernacular architecture related to Iran. All three case studies provide new insights not only into the colder climatic conditions of Azerbaijan but also into the impacts of meso-climates on traditional building design in the region, highlighting the juxtaposed roles of material choice, building techniques and elements, both indigenous to the area and imported from neighbouring regions. Within an area of almost 100 km radius, diversity and variety are showcased with all three villages that had been documented.

It is worth noting that personal interests, along with the support of individuals like Oleg Grabar at the Aga Khan Program for Islamic Architecture at Harvard, initiated scholarly work by figures such as Klaus Herdeg on Iranian and Central Asian cities since the 1970s. However, this fascination with certain cities and regions, like Isfahan, Shiraz, and Yazd, began to create a biased understanding of Iran's regional and climatic conditions. This chapter offers an alternative narrative, not opposing previous scholarship but adding to the dialogue around its diversity and inclusivity. This is crucial to avoid stereotypical approaches, oversimplification, and reductionism regarding vernacular architecture and its heterogeneity in Iran.

References

1. Oliver P (1997) Encyclopedia of vernacular architecture of the world. Cambridge University Press
2. Vellinga M, Oliver P, Bridge A (2007) Atlas of vernacular architecture of the world. Routledge
3. Rudofsky B (1964) Architecture without architects: a short introduction to non-pedigreed architecture. Museum of Modern Art; distributed by Doubled. 1
4. Upton D (1993) The tradition of change. Traditional dwellings and settlements review, 5(1):9–15
5. Oliver P (2006) Built to meet needs: cultural issues in vernacular architecture: cultural issues in vernacular architecture. Architectural
6. Asquith L, Vellinga M (2006) Vernacular architecture in the twenty-first century: theory, education and practice. Taylor & Francis
7. Frey P (2010) Learning from vernacular: towards a new vernacular architecture. Actes Sud
8. Sheehan S (2018) A guide to reading Herodotus' histories (1st ed.). Bloomsbury publishing Plc. https://doi.org/10.5040/9781474292702
9. Milani A (2004) Lost wisdom: rethinking Persian modernity. Mage Publishers
10. Provencal V (2015) Sophist Kings: Persians as other in Herodotus (1st ed.). Bloomsbury Publishing Plc. https://doi.org/10.5040/9781472593092
11. Colburn HP (2013) Connectivity and communication in the Achaemenid empire. J Econ Soc Hist Orient 56(1):29–52. https://doi.org/10.1163/15685209-12341278
12. Anonymous (982) Hudud al-Alam "The Regions of the World". (Minosrky V, Trans. & Exp. 1937). London, Lucaz. https://archive.org/details/in.ernet.dli.2015.281514
13. Kashani-Sabet F (2014) Frontier fictions: shaping the Iranian nation, 1804–1946. 1st edn, Princeton University Press. https://doi.org/10.1515/9781400865079
14. Hopkins JFP (1990) Geographical and navigational literature. In Young MJL, Latham JD, Serjeant RB (Eds.), Religion, learning and science in the 'Abbasid Period (pp. 301–327). Chapter, Cambridge: Cambridge University Press
15. William Ouseley (1800) The oriental geography of Ebn Haukal, an Arabian traveller of the tenth century. Translated from a manuscript in his own possession. London: Oriental Press
16. Ibid.
17. Tibbetts GR (2016) Balkhī school of Arab Geographers. In: Selin H (ed) Encyclopaedia of the history of science, technology, and medicine in non-western cultures. Springer, Dordrecht. https://doi.org/10.1007/978-94-007-7747-7_8993
18. Barthold V (1984) An historical geography of Iran. Princeton University Press, Princeton. https://doi.org/10.1515/9781400853229
19. Raziei T (2022) Climate of Iran according to Köppen-Geiger, Feddema, and UNEP climate classifications. Theoret Appl Climatol 148(3–4):1395–1416. https://doi.org/10.1007/s00704-022-03992-y

20. Rahimi J, Laux P, Khalili A (2020) Assessment of climate change over Iran: CMIP5 results and their presentation in terms of Köppen-Geiger climate zones. Theoret Appl Climatol 141(1–2):183–199. https://doi.org/10.1007/s00704-020-03190-8
21. Raziei T (2022) Climate of Iran according to Köppen-Geiger, Feddema, and UNEP climate classifications. Theoret Appl Climatol 148(3–4):1395–1416. https://doi.org/10.1007/s00704-022-03992-y
22. United Nations Human Settlements Programme (UN-Habitat) (2022) National urban policy and smart city strategy summary report. UN-Habitat. https://iran.unhabitat.org/wp-content/uploads/2022/08/NUP-Summary-Report-English.pdf
23. Abrahamian E (1982) Iran between two revolutions. Princeton University Press
24. Scott JC (1998) Seeing like a state: how certain schemes to improve the human condition have failed. Yale University Press
25. Floor W (2009)0 Guilds, Merchants, & Ulama in Nineteenth-Century Iran, Mage Publishers
26. AlSayyad N (2006) Cinematic urbanism: a history of the modern from reel to real. Routledge, London, New York. [Web] Retrieved from the Library of Congress. https://lccn.loc.gov/2006007637
27. Mehrjui D (Director) (1969) The cow [Film]. Dariush Mehrjui Production
28. DW Farsi (October 2015). Ali Nasirian's narrative on acting and Iranian cinema before and after the revolution. [Video] YouTube. https://www.youtube.com/watch?v=jktgxeRDLog
29. Panahi J (2018) Three Faces [Film]. Jafar Panahi Film Production
30. Panahi J (Director) (2022) No bears [Film]. Jafar Panahi film Productions
31. Hajirasouli A (2018) The cone-shaped settlement of Kandovan: a continuation of building traditions. Doctoral dissertation, Queensland University of Technology. QUT ePrints. https://eprints.qut.edu.au/121354/
32. Asghari Kaljahi E, Amini Birami F (2015) Engineering geological properties of the pyroclastic cone-shaped rocky houses of Kandovan, Iran. Bull Eng Geol Environ 74:959–969. https://doi.org/10.1007/s10064-014-0679-4

Persian Vernacular Architecture for Climate Resilient Design Thinking: Looking Back, Looking Forward

Ali Cheshmehzangi and Sue Roaf

Abstract In this closing chapter, we summarise how the contributions delved into the ways in which Persian or Iranian vernacular architecture was established over many years to adapt to extreme temperatures and harsh climates. It certainly offers crucial insights for modern sustainable design, which is in need of serious revisions and reconsiderations. Thus, the book examines how Persian architectural principles can influence contemporary practices from multiple perspectives. Key topics include the adaptation of effective systems and design strategies for modern use, the role of traditional materials in energy-efficient design, and the importance of culturally sensitive approaches to architecture. By linking traditional and indigenous knowledge in Iran with present-day requirements, this book aims to promote a deeper comprehension of how past architectural practices can shape the future of sustainable architecture, design, and urbanism. The ultimate goal of the book is to value, recognise, and celebrate the contributions of Persian vernacular architecture to the world; and more importantly, how we can continue to learn from its best practices from climate resilient design thinking.

Keywords Persian architecture · Climate-resilience · Sustainable design · Cooling · Vernacular architecture · Sustainability

1 To Rethink the Benefits of Vernacular Design

The architectural knowledge found in Persian/Iranian vernacular presents valuable insights that are highly relevant to our current climate challenges. This overview serves as a starting point for exploring Persian or Iranian vernacular architecture, emphasising its exceptional ability to withstand climate conditions and its focus on

A. Cheshmehzangi (✉) · S. Roaf
School of Architecture, Design and Planning, The University of Queensland, Brisbane, Australia
e-mail: A.Chesh@uq.edu.au

S. Roaf
Heriot Watt University, Edinburgh, Scotland, UK

A. Cheshmehzangi and S. Roaf (eds.), *Persian Vernacular Architecture*, Urban Sustainability, https://doi.org/10.1007/978-981-96-1116-4_17

design principles that prioritise people. Persian architectural practices offer a rich source of expertise, especially in their advanced passive cooling techniques and their ability to maintain comfortable and healthy living conditions on a large scale, techniques that have been developed over many centuries.

Persian vernacular architecture demonstrates a deep awareness of climate adaptation through innovative passive design methods. This architectural tradition is characterised by its capacity to create comfortable temperatures in extremely hot climates without needing modern mechanical systems. The incorporation of features such as wind catchers (badgirs), qanats (underground aqueducts), and courtyards showcases an expert integration of architectural design with environmental factors. These elements not only alleviate the adverse effects of extreme temperatures but also enhance living standards by creating enjoyable, climate-responsive surroundings.

As seen through the chapter contributions, the focus on people-centered design is crucial to understanding Persian or Iranian architectural practices. These traditional designs prioritise human-scale spaces and communal living, fostering a strong sense of community and connection. The careful consideration of privacy, social interaction, and spatial organisation reflects a dedication to creating spaces that meet the needs and well-being of occupants. This emphasis on the human scale and social functionality aligns with modern values of inclusivity and environmental awareness, presenting a blueprint for contemporary architecture that prioritises user experience and ecological balance.

In recognising the enduring relevance of Persian vernacular architecture, this reflective summary aims to encourage a re-evaluation of current architectural paradigms—some that are not only not sustainable but entirely irrelevant to human well-being and life. The insights gleaned from this rich tradition can inform new approaches to design, emphasising the importance of passive strategies, human-centered spaces, and climate resilience. By drawing inspiration from the past, architects and designers can develop solutions that address today's challenges while valuing the legacy of architectural ingenuity that has shaped our built environment for centuries. This book invites readers to delve into the intricate world of Persian vernacular architecture, exploring how its principles can guide the development of more sustainable and people-oriented designs for the future.

1.1 The Timeless Wisdom of Persian Vernacular Architecture

The global architecture community must know that advancements in architecture did not simply start from Greek or Roman architecture. In fact, they were heavily shaped and influenced by their predecessors, including Persian or Iranian architecture. There were many architectural advancements and contributions before their time. We note that Persian vernacular architecture is more than a historical relic; it is living evidence of how buildings can be designed to thrive in extreme climates. Beyond that, we see traces of intelligence in indigenous design thinking and climate-responsiveness considerations in architectural innovation that cannot be left untold in the world of architecture today. The creativity of Persian architects and master builders, known as Ostads, who mastered the art of creating thermal paradises amidst harsh climates, offers invaluable lessons for contemporary design. This architecture, with its intricate wind catchers, subterranean cool storage systems, ingenious use of materials, etc., provides a blueprint for creating comfortable living spaces even in the most extreme temperatures.

Undoubtedly, vernacular design is the essence of 'good architecture', 'climate-sensitive design', and 'people-centric built environments', which requires the revitalisation of our autochthonous knowledge; something that could help us create—with adequate awareness and a progressive mind—contemporary interventions from Indigenous design. Hence, for us, it is never a better time now to get to grips with what Persian vernacular architecture actually encompasses and how it is experienced in the real world. We need all the help and knowledge out there, in the past and present, to combat and mitigate the challenges of global warming, extreme heat, and unsustainable development and design paradigms. The Persian approach to (vernacular) architecture was deeply intertwined with the environment and country, reflecting a harmonious balance between human needs and the natural world. Thus, we believe there are lessons to be learnt while there is an urgent need for developing new ideas and directions, and most certainly, new business-as-usuals than the ones we currently have.

2 A Request from the Authors, a Simple Reflection for the Readers

Western architectural models, often grounded in different climatic and cultural contexts, may not fully capture the depth of sophistication present in Persian vernacular solutions. The question arises: Do Western models address the complex interplay of airflow, material performance, and spatial utilisation as effectively as Persian systems? Western engineering and architectural practices frequently focus on the efficiency of mechanical systems and the sustainability of materials. While these approaches have their merits, they may fall short in addressing the nuanced needs of extreme heat. For example, concepts such as low embodied energy—prominent in Western discussions—might not directly translate into solutions for extreme heat, where material performance and traditional passive cooling methods often prove more effective.

Moreover, the Western focus on privacy and spatial organisation might not align seamlessly with the traditional Persian approach, which emphasises the contextual use of spaces at different times of day and year. Persian/Iranian architecture's focus on creating adaptable spaces that respond dynamically to environmental conditions could offer new perspectives on designing modern buildings that are both climate-resilient and sensitive to the rhythms of daily life.

Our request to all authors and readers is to be bold. Many contemporary engineers and architects have no idea about how to design buildings that will stand up to the temperatures that the historic buildings of Iran (and many places alike) cannot only cope with but can create a sort of thermal and emotional paradise within—even during extremely hot summers. We know, because we have lived through them there and we have seen and experienced many excellent architectural masterpieces. Through continuous dialogue with our enthusiastic fellows and experts who contributed diligently and intellectually to this book, Sue and I questioned them with supplementary push and encouragement:

- Do Western models even touch on the sophistication of systems like airflow in wind catchers/towers?
- Should we even be trying to understand how traditional buildings work using models developed for very different buildings in very different climates and societies?
- Hugely important is the way spaces in buildings are used at different times of day and year. But, how do traditional design principles help us design modern buildings?
- How does fashionable Western thinking, e.g., low embodied energy, green buildings, Passivhaus, etc., relate to designing for the extreme heat? Is it even relevant? Should there be more focus on how materials perform?
- Are we understanding the importance of Iranian lessons for the future of other regions in a heating world—e.g., water cisterns in a water-challenged future—should every city have them and use them for a wide range of twenty-first century purposes?

As we continued to encourage the authors for months, Sue reminded them that we would love it if they could all go and put their heads in their hands and think hard before submitting their final version of their revised chapters. Sue receptively reminded them that many around the globe have to learn from their expertise, knowledge, and experience in Iran. She excitedly asked, "Do Western Design Models and Tropes simply relate to all climates—or even to their own buildings in a heating world?". Here, in Iran, is how we built heat-resilient buildings—building on the traditional brilliance of our Ostads (evolved over time) can help. We hope the rest of the world can benefit from their sharing of the gift of interpreting the design lessons of the best traditional master builders—with their invaluable techniques and living technologies—in the world of passive design for hot climates. We hope that our readers also spend a little time with their heads in their hands, thinking about how they can add extra value or even contribute to the knowledge that needs to be alive, active, and informative.

This book intends to flag how Iran can help people adapt to our rapidly heating climate. Hence, we want the world to learn from the vernacular; we want architects and designers to learn from good architecture, and learn from master builders of Iran on climate resilient design. As Sue puts it well, we have to respect where it is due, i.e., those masters were trained over centuries and decades to get it right! and they surely did. There is a lot to learn, there is a lot to do. But we must stop reasoning with fast architecture, poor quality design, and quick fixes that do not create continuity nor they can provide us with enough knowledge in the world of good architecture.

Here is a reflection on our preface where we summarised our intention and expectations. We continue to assert that science, art, and philosophy assume from derelict ideas and topographical ready-made compositions; and yet, some believe academia is failing in its responsibilities to create new knowledge, new thinking, and novel methods. But there is still optimism with impetus if we move forward looking back; in a way, learning from the past, expanding our knowledge of vernacularity, and developing new paradigms that are not just beautiful and functional but are purposeful, enlightening, and inspirational. The impacts would be amazing if we could share a common understanding of how the future can be; perhaps, even better than what the neoliberal ideas of politically motivated algorithm complexities dictate to us. We believe with the mindset of learning from the vernacular, we could solve many of our complexities rather than developing endlessly, promoting sustainability politically, and designing carelessly.

The idea of this book started by arguing how we both, as the co-editors in this fascinating journey, very much disliked modelled studies unless they are consciously written with a way of mitigating their limitations. Sue nicely argued that modelled studies are unvalidated in real buildings, they are almost worthless as they just superimpose Western assumptions in crude spreadsheets and pretend they can extract something from that knowledge. We debated passionately that (vernacular) buildings in Iran work because of the symbiotic relationship between the people in them, the climate outside them, and the building itself. Then we collectively decided to put a proviso on the abstracts and later submissions. We debated that purely modelled studies are not by themselves acceptable unless they provide some validation from real buildings of the assumptions they make. We had back-and-forth discussions suggesting there is a better way to think and create through vernacular knowledge. Thus, we don't think there is much value in just harvesting mind games from people who think models reflect reality. In a way, to learn from the vernacular we have to go back to the vernacular; to teach us those lessons that are simply essential, i.e., not just to provide us with resaid story-telling of those old buildings and how they work—because in fact—the models both give the wrong answers and fail to probe the depths of how these amazing buildings work at all.

Since its inception, we highlighted on overarching fact that "our challenge is to learn from the vernacular". We must extract knowledge about what matters in design in the process where we can validate ideas step by step with a range of tools and techniques. We think that for now, most vernacular styles are about 'extreme' design because they must support life without mechanical heating and cooling. Simple things like door angle to the wind or orientation of the building become critical. Then, the design process becomes more meaningful and the design itself becomes more impactful.

Last, but not least, this book is a call out for genuine vernacular design thinking to push architects, designers, and planners to think beyond their comfort zones and those trendy ideas they can see or explore in magazines and catalogues on their coffee tables or in their confined office spaces. In case it is forgotten, we would like to remind everyone that architecture is beyond aesthetics and is borderless in many ways. We want to call for real action, a genuine paradigm shift(s) beyond the politically-diluted pathways or ideals. We want educators to be aware, become knowledgeable, and grow bolder in teaching design thinking, architectural engineering, and ultimately, 'good architecture'.

3 Embracing the Past to Shape the Future

The Persian/Iranian approach to architecture is not merely about historical preservation; it is a call to action for adapting these lessons to a rapidly warming world. As we look to the future, it is imperative to recognise the value of historical architectural wisdom in addressing contemporary challenges. Persian vernacular architecture, with its emphasis on climate resilience, adaptive design, and sustainable practices, offers a rich source of knowledge that can inform and inspire modern building practices. This book aims to bridge the gap between ancient wisdom and modern needs, providing a comprehensive exploration of how Persian architectural principles can be adapted to create better architecture. By learning from the past, we can navigate the complexities of the present and build a future that responds to both our heritage and our environment.

In essence, the study of Persian or Iranian vernacular architecture is not just an academic exercise but a practical endeavour to harness historical insights for contemporary and future challenges. By integrating these lessons into our design practices, we can contribute to a more sustainable and resilient world, where the wisdom of the past guides the innovations of the future.

This work specifically honours and acknowledges the country, and particularly, the significant role of "Persian Vernacular Architecture" in the field of architecture, both historically and in modern contexts. It aims to emphasise the importance of this architectural tradition—through deep thinking (or knowledge) and best practices—in shaping past architectural advancements and ensuring its continuous relevance in future architectural innovations. Educators in architecture and sustainability fields must communicate to students and educate them with correct knowledge that despite many political complexities, 'Persian Vernacular Architecture' has deeply influenced architectural history, including its impact on Greek and Roman architecture, Western architecture, Islamic and Arab architecture, etc. It hurts the most if our teachings of architecture theory and history do not acknowledge this conspicuous fact. This tradition not only provides valuable insights into historical design practices but also offers essential lessons for developing more considerate, people-centered, and climate-resilient architectural thinking and paradigms. As we face the challenges brought by climate change, the enduring wisdom integrated into Persian architectural practices can motivate and direct modern design approaches toward greater thinking. Lastly, we hope the book and its well-articulated and educational chapters can be used in classrooms, research debates, and help promote vernacular architecture for climate-resilient design.

The manufacturer's authorised representative in the EU is Springer Nature Customer Service Centre GmbH, Europaplatz 3, 69115 Heidelberg, Germany. If you have any concerns regarding our products, please contact ProductSafety@springernature.com

Printed and bound by CPI Group (UK) Ltd, Croydon, CR0 4YY
01/10/2026
02225926-0013